Where Have All the Plastics Gone?

Ménage à Trois in the Sea Surface Microlayer

Nanoparticles as Vectors of Environmental Chemicals

H. G. Brack

Davistown Museum

Department of Environmental History

Phenomenology of Biocatastrophe
Publication Series Volume 5

ISBN 10: 0-9892678-4-9
ISBN 13: 978-0-9892678-4-7

Third edition

Cover photo credits
Front: Lower Manhattan, Brooklyn, Governor's Island, New York City. (1999). Image by US NOAA.

Front, lower left: Reisser, J., Proietti, M., Shaw, J. and Pattiaratchi, C. (2014). Ingestion of plastics at sea: Does debris size really matter? *Frontiers in Marine Science*. 1(70).

Front, upper right: Cole, M., Lindeque, P., Fileman, E., et al. (2013). Microplastic ingestion by zooplankton. *Environmental Science and Technology*. 47. pg. 6646-55.

Front, lower right: Wikimedia Commons. (2010). *Plastic waste at Coco Beach, outfall of Mandovi River into Indian Ocean*. Created by Hajj0 ms.

Back: Wikimedia Commons. (2014). *Plastic pollution caused by May flooding*. Created by Marckos.
Back: UC Davis ChemWiki. Creative Commons Attribution-Noncommercial-Share Alike 3.0 United States License.

This publication is sponsored by

Davistown Museum

Department of Environmental History

www.davistownmuseum.org

and

Engine Company No. 9

Radscan-Chemfall

Est. 1970

Disclaimer

Engine Company No. 9 relocated to Maine in 1970. The staff members of Engine Company No. 9 are not members of, affiliated with, or in contact with, any municipal or community fire department in the State of Maine.

Comments, criticisms, and suggestions are welcomed and may be directed to:
curator@davistownmuseum.org

Pennywheel Press
P.O. Box 144
Hulls Cove, ME 04644

1985-2000 Publications Sponsored by Engine Company No. 9
Station 4, Hulls Cove, ME
and
The Center for Biological Monitoring

Radscan: Information Sampler on Long-Lived Radionuclides: 1990-1999

A Review of Radiological Surveillance Reports of Waste Effluents in Marine Pathways at the Maine Yankee Atomic Power Company at Wiscasset, Maine--- 1970-1984: An Annotated Bibliography

Legacy for Our Children: The Unfunded Costs of Decommissioning the Maine Yankee Atomic Power Station: The Failure to Fund Nuclear Waste Storage and Disposal at the Maine Yankee Atomic Power Station: A Commentary on Violations of the 1982 Nuclear Waste Policy Act and the General Requirements of the Nuclear Regulatory Commission for Decommissioning Nuclear Facilities

Patterns of Noncompliance: The Nuclear Regulatory Commission and the Maine Yankee Atomic Power Company: Generic and Site-Specific Deficiencies in Radiological Surveillance Programs

RADNET: Nuclear Information on the Internet: General Introduction; Definitions and Conversion Factors; Biologically Significant Radionuclides; Radiation Protection Guidelines

RADNET: Anthropogenic Radioactivity: Plume Pulse Pathways, Baseline Data and Dietary Intake

RADNET: Anthropogenic Radioactivity: Chernobyl Fallout Data: 1986 – 2001

RADNET: Anthropogenic Radioactivity: Major Plume Source Points

Integrated Data Base for 1992: U.S. Spent Fuel and Radioactive Waste Inventories, Projections, and Characteristics: Reprinted from October 1992 Oak Ridge National Laboratory Report DOE/RW-0006, Rev 8

2000-2015 Publications Sponsored by
Davistown Museum Department of Environmental History
and Engine Company No. 9, Station 4, Hulls Cove, ME

Essays on Biocatastrophe and the Collapse of Global Consumer Society. Vol. 1. 2010.

Biocatastrophe Lexicon: An Epigrammatic Journey Through the Tragedy of our Round-World Commons. Vol. 2. 2010.

Biocatastrophe: The Legacy of Human Ecology: Toxins, Health Effects, Links, Appendices, and Bibliographies. Vol. 3. 2010.

Antibiotic Resistant Bacteria (ARB): A Republication of CDC's Antibiotic Resistance Threats in the United States, 2013 and Abstracts from Other Studies of the Health physics Impact of ARB. Vol. 4. Work in Progress.

Where Have All the Plastics Gone? Ménage à Trois in the Sea Surface Microlayer: Nanoparticles as Vectors of Environmental Chemicals. Vol. 5. 2015.

Table of Contents

Preface

Where Have All the Plastics Gone? chronicles the research and commentary on two environmental issues obscured by the contentious debate on global warming and cataclysmic climate change:

- The breakdown of microplastics into plastic nanoparticles (PNPs) characterized by sorbed ecotoxins, including the wide variety of environmental chemicals produced by pyrotechnic-petrochemical industrial society

- The nanoparticles (NPs), including nanoplastics, intentionally or accidentally produced by rapidly expanding nanotechnologies, their inherent toxicity, and their tendency to sorb and transport ecotoxins at the end of their life cycle

The annotated bibliography that follows the text provides a topic-specific journey through the world of nanoparticles and nanotechnology and their health physics significance. The special focus of this text is on the hidden role of invisible plastic nanoparticles as vectors of environmental chemicals, and their movement through our hemispheric water cycle.

This volume includes information on the history of plastics production and the physics of nanoparticles and their classification. As nanoparticles decrease in size, their surface to mass ratio increases, causing increasing sorbtion of ecotoxins and facilitating their transport to all microbiomes. Numerous appendices detailing the vast inventory of environmental chemicals in the hemispheric water cycle follow the bibliography. Also included is commentary on the social, political, and ecological context of our rapidly increasing production of plastics and environmental chemicals. The overview of news bites and op-eds includes some of the most interesting quotations from the annotated bibliography. A series of biosphere models that sketch the components and dynamics of biocatastrophe in our round world commons follow the introductory essays. This volume also includes a keyword index, a guide to reporting units and abbreviations, and numerous appendices, charts, and graphs.

Note: The 3rd edition of *Where Have all the Plastics Gone?* includes a few corrections, additions, and updates about the total of world plastics production as of March 2018.

Introduction

The research cited in this annotated bibliography provides commentary on an overlooked component of the impact of our plastic and nanotechnological lifestyles on the vulnerable finite ecosystems of a biosphere in crisis. We now see clearly the highly visible impact of the use of fossil fuels to power our ever-growing global consumer society, manifested in rapid global warming and cataclysmic climate change. The transport of environmental chemicals, including persistent organic pollutants (POPs) throughout the biosphere by evaporation, long-range atmospheric transport, runoff, and other pathways is a well documented result of the impact of our increasing use of petrochemistry as the prime mover of the growth of global industrial consumer society. Much less obvious than greenhouse gas emissions and chemical fallout is the invisible role of nanoparticles, including plastic nanoparticles, in transporting the ubiquitous environmental chemicals produced by global consumer society to all trophic levels of the biosphere.

All nanoparticles are characterized by an increasing surface to mass ratio as their size decreases, as typified by the transformation of microplastics by abrasion, photodegradation, and embrittlement into plastic nanoparticles (PNP). The tendency of nanoparticles, including plastic nanoparticles, to sorb (adsorb and/or absorb) environmental chemicals and then be ingested by microbiota in all microbiomes constitutes the ménage à trois that is an overlooked pathway of human exposure to anthropogenic (manmade) ecotoxins.

The essays in this volume discuss anthropogenic nanotoxins as components of the unfolding Age of Biocatastrophe, an ongoing historic event that is the subject of this publication series. The phenomenology of biocatastrophe is characterized by numerous human activities which alter, degrade, or destroy natural ecosystems. Among the prime movers of our assault on nature is the ever-increasing intensity of the chemical warfare human society is waging on the biosphere we inhabit. Plastic nanoparticles are an important, if invisible, pathway of human exposure to these environmental chemicals.

The omnipresence of environmental chemicals in the global water cycle can be expressed by the metaphor of the biosphere as a hemisphere of organic chemicals—an appropriate description of the current status of the global water cycle in view of the importance and ubiquitous presence of persistent organic pollutants (POPs) such as polychlorinated biphenyls (PCBs), for example, which are among the most biologically significant components of our anthropogenic soup. Associated ingredients include other organic chemicals, such as organophosphates, volatile organic compounds (VOCs), pharmaceuticals, methylmercury, and endocrine disrupting chemicals including phthalates, Bisphenol A (BPA), and obesogens. Other constituents of our hemispheric

4

water cycle include remobilized inorganic toxins such as arsenic, cadmium, lead, and anthropogenic radionuclides. Most of these soup ingredients were not constituents of the global water cycle before 1940.

The health physics impact of environmental chemicals associated with the Age of Plastics and the Age of Nanotechnology raise public safety issues of compelling contemporary interest to all concerned citizens. Mitigation of the impact of our behavior on natural ecosystems must include informed debate about the consequences of our rapidly increasing production of thousands of anthropogenic chemicals and their remobilization by nanoparticles that are difficult to measure or monitor. Awareness of what we are doing to the vulnerable, finite biosphere we inhabit is contingent upon the documentation of the invisible presence of environmental chemicals in a hemispheric water cycle that is the basis for all life on Earth.

This volume also includes citations pertaining to aquifer depletion, our last source of potable water not contaminated by chemical fallout. An increase in the frequency, severity, and ecotoxic content of washover-washout events (flooding), especially in industrialized and urban environments, will result in the replenishment of many aquifers and water supplies with contaminated surface runoff.

The history of the Anthropocene (the epoch characterized by the imposition of human ecosystems on natural ecosystems) can be summarized as the ever increasing growth of pyrotechnic petrochemical nuclear society (PPNS), more commonly denoted as global consumer society, in the context of a vulnerable biosphere in crisis. We can Google the impact of our urban, military, and industrial complexes on our climate, natural ecosystems, and hemispheric water cycle. Global warming, ocean acidification, climate change, smog, and chemical fallout as well as deforestation, desertification, and species depletion are among the highly visible environmental consequences of the growth of global consumer society. The Age of Plastics is characterized by invisible plastic nanoparticles (PNPs), which sorb and transfer environmental chemicals throughout the biosphere, including in all microbiomes. These microbiomes are now characterized by ever-growing debris fields of anthropogenic ecotoxins. Plastic nanoparticles are an unfortunate, but invisible legacy of our exploitation of the natural resources of a finite biosphere in crisis – the downside of the many wondrous products of global consumer society and the Age of Plastics.

The Age of Plastics

Global consumer society and its industrial infrastructure have produced ± 10 billion tons of plastic since the Age of Plastics began after World War II. Worldwide production was estimated at 288 million tons in 2014 and over 300 million tons in 2015 (Rossi 2014). At least 300 million tons of plastic have been accidentally or deliberately disposed of in marine environments since 1950. Billions of tons of plastics remain in landfills in urban, suburban, and industrial landscapes as well as in agricultural areas throughout the world. The legacy of the Age of Plastics is, in turn, a component of the ongoing Age of Petrochemicals, which, as of 2015, is producing billions of tons of petrochemicals per year (see appendices 6-10). The Age of Nanotechnology is the latest era in our ongoing cascade of Industrial Revolutions, each of which adopt and expand the innovations and inventions of the past. What these technological revolutions, including the Age of Digital Technology, have in common is the ubiquitous production of massive quantities of useful and inexpensive plastics, which are a key component of the rapid growth of global consumer society. Derived from the combustion of fossil fuels, the production of plastics is in part responsible for the rapid increase in carbon dioxide (CO_2) levels. Well known results of elevated levels of CO_2 production include global warming, climate change, drought, and flooding, which is now spreading environmental chemicals and their nanoparticulate vectors over many highly populated landscapes. Ocean acidification caused by CO_2 derived carbonic acid looms as an imminent environmental crisis of historic dimensions. Rising sea levels and increasing storm intensity are frequent topics of media coverage of our changing climate.

Only a small percentage of plastic debris destined for the marine environment, possibly ±10%, finds a permanent residence in marine sediments. Fish netting and other heavy macroplastics are often documented or recovered from the sea floor. Most plastic debris is, however, lighter in weight than water, explaining the presence of large quantities of floating debris in our now famous garbage patches. Much of the highly visible plastic clutter in the marine environment has been transformed from macrolitter to microplastics (1 um – 1 mm), and then into invisible plastic nanoparticles (PNP) measured in nanometers, i.e. billionths of a meter. Vast ocean gyres of plastic litter become invisible as particle size diminishes. This litter is now the subject of concern due to its ability to sorb and transport environmental chemicals throughout the biosphere. Plastic nanoparticles, many with a size range of 20-200 nm, are an important vector transporting ecotoxins to all trophic levels of the biosphere, including pathways to human consumption.

The ménage à trois of nanoparticles, environmental chemicals, and microbiota in the sea surface microlayer is recapitulated in all aquatic ecosystems, marine or terrestrial, where microbial communities (microbiomes) constitute the foundation of nutrient transfer throughout the biosphere. Nanoparticles as vectors of environmental chemicals also characterize rivers, lakes, surface runoff, seepage, and wastewater discharges, as well as terrestrial microbial communities in biomes ranging from tropical, boreal, and coniferous forests to arctic tundra and the grasslands that produce so much of the world's food.

A huge residual inventory of nanosized anthropogenic ecotoxins also litter our military, industrial, and urban landscapes, the legacy of pyrotechnic petrochemical nuclear society (PPNS) and the Age of Plastics.

Plastic microbeads, often used as abrasives, are a well known constituent of many consumer products. Microbeads typically range in size from 10 μm to 1000 μm [1 mm]. The U.S. Congress banned production of all microbeads on December 22, 2015 in a unanimous bipartisan vote, including those as large as 5 mm (5,000 um). This legislation provides an opportunity to compare the size of plastic microbeads now in our medicine cabinets, waste streams, and oceans with much smaller plastic nanoparticles, which should not be called microplastics.

Literature on microbeads as an abrasive additive to soaps, cosmetics, and toothpastes notes an 8 ounce bottle of skin cream may contain in excess of 4,000 microbeads. The typical diameter of such a microbead might be 100 μm, or 100,000 nm. If we idealize a plastic nanoparticle as having an exact diameter of 100 nm (such exact dimensions of plastic nanoparticles are very unlikely), 1,000 such nanoparticles can fit on the diameter of our soon to be illegal microbead. How many such nanoparticles will fit into one such circular microbead? 1,000,000,045 (1 billion+).

If our 8 ounce jar of ointment contains 4,000 100 μm-sized microbeads, how many 100 nm plastic nanoparticles can be theoretically derived from this single ±$10 investment in good looks at the beach? 4,000,000,180,000 (4 trillion+).

The very smallest sized plastic nanoparticles (±20 nm in diameter) are composed of fragmented polymers with a wide variety of shapes. These include filaments, chains, cones, ribbons, and may also be random mixtures of amorphous or crystalline structures. Such a plastic nanoparticle fragment may take the form of a polymer spaghetti ball (illustrated below). All are characterized by large surface to mass ratios and continue to be efficient vectors of environmental chemicals, such as polychlorinated biphenyls (PCBs), a molecule of which is ±1 nm in size. If we analyze a plastic nanoparticle fragment of a 20 nm diameter size, how many plastic nanoparticle spaghetti loops can we set on our microbead diameter? The answer would be 100,000/20 = 5,000 spaghetti loops. How many can theoretically fit into our 100 um microbead? Answer: 125,014,206,159 (125 billion+).

Theoretical polymer spaghetti loop form. From the UC Davis ChemWiki. Creative Commons Attribution-Noncommercial-Share Alike 3.0 United States License.

These huge numbers illustrate the challenges of locating and measuring the presence of plastic nanoparticles or any other nano-ecotoxins in any trophic level of the biosphere. This data provides a hint at the large number of plastic nanoparticles that are ever-increasing components of the debris fields of all microbiomes. This is the beginning of their voyage throughout the biosphere. Plastic nanoparticles of all sizes are ever increasing components of the water we drink, the air we breath, and the food we eat.

Other possible shapes of plastic polymer fragments are illustrated below.

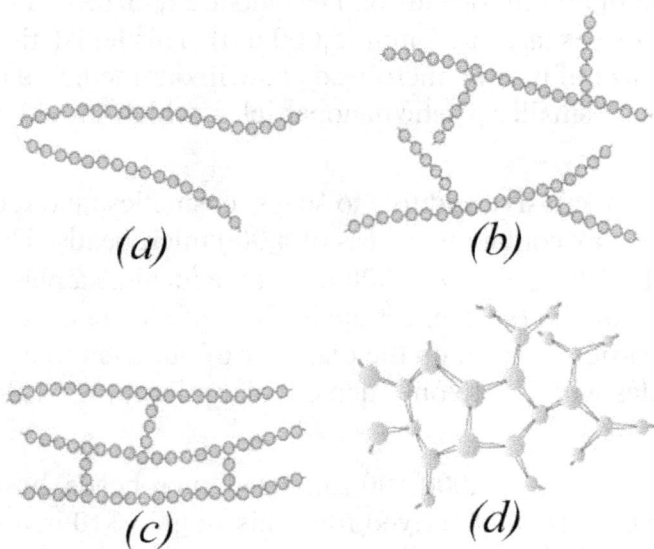

(a)   *(b)*

(c)   *(d)*

Linear (a), branched (b), cross-linked (c), and networked (d) polymers

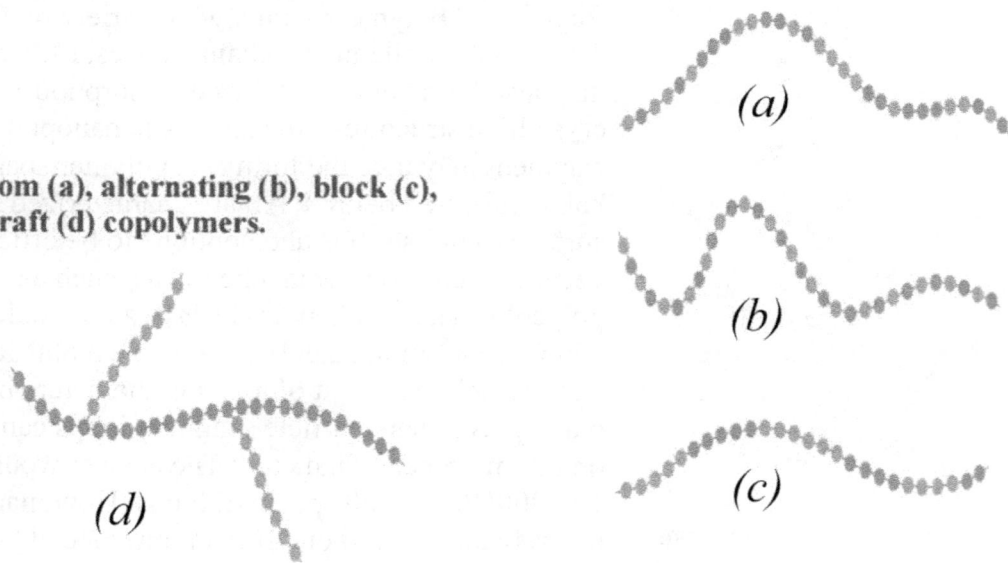

(a)

Random (a), alternating (b), block (c), and graft (d) copolymers.

(b)

(c)

(d)

P.S.: How many 1 nm polychlorinated biphenyl molecules can you fit in our 20 nm plastic spaghetti ball? Any corrections to our math or comments welcomed: curator@davistownmuseum.org

The election of Donald K. Trump as president of the United States and his inauguration (1/20/17) has resulted in a startling and unprecedented historic event: the establishment of a scientific information police state. Taxpayer-funded scientific studies by the EPA, NASA, NOAA, and other governmental agencies on a wide variety of environmental issues, many of them referenced in this publication series, have been terminated, defunded, classified, discontinued, or are otherwise not available for informed debate. All media now have an ethical obligation to report on the growing, if invisible, presence of plastic nanoparticles (PNP) and other nanotoxins in all trophic levels of our biosphere in crisis. The Trump administration's anti-environmentalism, its broad support of polluting industries, and its unwillingness to discuss or document the vast chemical excreta of US and global consumer society as a key component of our changing climate, constitutes a threat to the public safety and future viability of world society.

The Physics of Nanoparticles

The dual focus of this fifth volume in the *Phenomenology of Biocatastrophe* series is the transformation of microplastic debris into plastic nanoparticles and the rapid increase in the intentional production of nanoparticles for industrial, medical, and consumer product use.

Evaluation of the biogeochemical behavior, and thus the potential health physics impact of plastic nanoparticles derived from macro- and microplastic debris, is made more difficult by two factors:

- Plastic nanoparticles in size ranges above 100 nm but below 1,000 nm are frequently referred to as "microplastics" in most scientific studies and almost all news media reports, thus diverting attention from the relationship between decreasing size and increasing toxicity.

- The wide diversity of plastic types, the complexity of their polymer chain positions, and the random mixtures of plastic nanoparticles as environmental contaminants makes scientific analysis of polymer chemistry and behavior extremely difficult, including evaluation of their ability to sorb and then transfer environmental chemicals to air, water, particulate matter and biological media.

As nanoparticles decrease in size, the proportionate sorbtion of environmental chemicals increases as the surface to mass ratio increases, even as their physical presence becomes increasingly difficult to document during translocation to higher trophic levels. Fragments of plastic polymer structures, which might be hundreds of nanometers in length, are particularly hard to monitor due to their small width, often ±20 nm. Oberdörster et al. summarizes the long-known health physics significance of increased surface to mass ratio with respect to nanotoxicology: "Surface molecules increase exponentially when particle size decreases < 100 nm, reflecting the importance of surface area for increased chemical and biologic activity of NSPs" [nanosized particles]. (Oberdörster 2005b, 825). Hett (2004) notes that nanoparticles in the environment, including intentionally manufactured nanoparticles, "could constitute a completely new class of non-biodegradable pollutants…" which are highly reactive and "because of their small size, could have a total surface area measuring up to 1000 square meters per gram." Welcome to the world of quantum physics.

The physical characteristics of nanoparticles noted by Oberdörster and Hett applies to both plastic nanoparticles (PNP) and intentionally manufactured nanoparticles (IMNP) and illustrates why the transformation of plastic debris into plastic nanoparticles is one of the most unfortunate achievements of pyrotechnic petrochemical nuclear society (PPNS). The invisible spread of nanotoxins, including plastic nanoparticles, throughout the biosphere cannot be stopped or mitigated by legislation, environmental awareness and activism, or UN climate change conferences.

The end of life cycle toxicity of the many manufactured nanoparticles (MNP) which are or will be the legacy of the age of nanotechnology replicates the biological significance of plastic nanoparticles which sorb environmental ecotoxins. Of particular note is the potential toxicity of non-plastic carbon nanotubes, fullerenes, nanocrystal clusters, nanobiorods, nanowires, quantum dots and other forms of manufactured nanoparticles as they undergo dispersion and mass wasting at the end of their life cycles. Fullerenes, for example, also sorb ecotoxins and may move through the biotic environment in the same pathways as plastic nanoparticles.

Fine and ultrafine airborne particles in the size range from 50 nm to 2.5 μm, often anthropogenic combustion products (e.g. soot), are also efficient carriers of chemical fallout ranging from volatile organic chemicals (VOCs) to anthropogenic radioactivity and methylmercury (see Appendix 21). Smog is a highly visible form of particulate air pollution in most of the world's cities under certain weather conditions. Diesel exhaust fumes, especially nitrous oxide (NO_2), are incorporated in urban smog and create a particularly noxious form of air pollution. Smog may also incorporate invisible airborne plastic nanoparticles, which have many sources, including landfills, flooding washout events, industrial agricultural activities, and, in fact, from human ecosystems of every description. These airborne particles are also a component of the deterioration of global air and water quality and provide a pathway for the cycling of ecotoxins through the biosphere. Ecotoxins sorbed by airborne particulates are also derived from petrochemicals, industrial pollution, and urban wastelands and are circulated by long range atmospheric transport. Marine snow may provide a re-entry route into the marine environment for airborne plastic nanoparticles; marine aerosols are a tertiary exit route for plastic nanoparticles from marine environments.

It is an unfortunate historical fact that pyrotechnic petrochemical nuclear society (PPNS) is now waging unrelenting chemical warfare on the biosphere. The movement of anthropogenic ecotoxins, especially organic chemicals, within the microbiomes of all trophic levels of a biosphere in crisis is, in part, associated with invisible plastic nanoparticles that are now constituents of all ecosystems.

The Environmental Protection Agency (EPA) provides a generic description of six categories of water contaminants that are often in our drinking water as nanoparticles (see below and Appendix 2). This overview provides a useful starting point for documenting the large number of organic (carbon-containing) chemicals that are moving through the hemispheric water cycle. A more detailed listing of environmental chemicals documented by the U.S. Center for Disease Control (CDC) in the human biome is listed in Appendix 3, *Chemicals in the CDC Fourth Annual Report on Human Exposure to Environmental Chemicals Updated Tables*. The EPA listings are water contaminants of worldwide concern. Organic chemicals are the most ubiquitous biologically-significant anthropogenic ecotoxin in the EPA's listings. They now permeate the hemispheric water cycle, which can now be characterized as a bowl of organic chemical soup, with a wide variety of other ingredients.

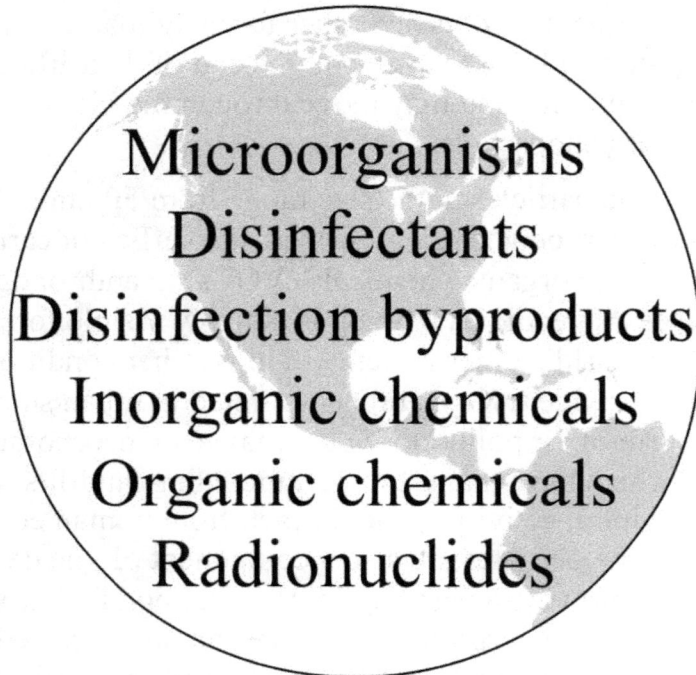

Microorganisms
Disinfectants
Disinfection byproducts
Inorganic chemicals
Organic chemicals
Radionuclides

(adapted from http://water.epa.gov/drink/contaminants/#List. See Appendix 2.)

The vast body of research cited in this volume documents the propensity of nanoparticles, including plastic nanoparticles, to serve as vectors of the environmental chemicals that are the legacy of the Age of Petrochemicals. These studies help define the context of the historical evolution of the Age of Biocatastrophe characterized by ever intensifying "chemical warfare" on the biosphere (McCauley 2015). The key to the future of human society in a biosphere with finite resources lies in the viability of the earth's water cycle. The single greatest challenge to the future of human society is the ongoing but invisible contamination of all aquatic and terrestrial ecosystems with environmental chemicals. All microbiomes are now characterized by growing debris fields of anthropogenic ecotoxins and the many nanoparticles, including plastic nanoparticles, which are their vectors.

The growing presence of environmental chemicals in the hemispheric water cycle is, in turn, only one component of a worldwide water crisis with vast environmental, social, and political ramifications. Ocean acidification, deforestation, desertification, aquifer contamination and depletion, urbanization, warfare, and the massive industrial impact of global consumer society all reflect the influence of human ecology on the viability of our natural ecosystems, including our finite global water cycle. Biocatastrophe is the result of the synergistic impact of many human activities, including pyrotechnology and petrochemistry, on natural ecosystems. A principle legacy of the Age of Plastics is the vast landscape of discarded plastic wastes that gradually evolves into microplastic litter and then into invisible plastic nanoparticles, many with a size range of 20 to 200 nanometers (nm). These plastic nanoparticles are now the invisible vectors for the transport of ecotoxins throughout the biosphere.

Classes of Nanoparticles

The following classification of nanoparticles provides a guide to the research topics cited in this text. While the focus of this annotated bibliography is on the transformation of visible microplastics into invisible nanoplastics, all world citizens have a constant daily interreaction with nanoparticles of every description, ranging from environmental chemicals to the microbial populations in all biomes.

Nanoparticles are defined as having a size range of 1 nanometer to 1,000 nanometers (a nanometer being a billionth of a meter), abbreviated as nm. Nanotechnology utilizes a wide range of nanoparticles ranging in size from 1 nm to 100 nm. Wikipedia contains an extensive description of the nanoparticles used in nanotechnology, noting their size range is 100 nm or less. Immense quantities of plastic nanoparticles derived from the breakdown of microplastics are also present in all environments, many of which are larger than the 100 nm nanoparticles utilized in the world of nanotechnology.

Microparticles are defined as having a size range of 1 micrometer to 1,000 micrometers, expressed as microns, and abbreviated as um (i.e. millionths of a meter). 1,000 μm is the equivalent of 1 millimeter. Particulate matter with a size range above 250 μm to 500 μm in size can be seen by the human eye. Airborne particulates can take the form of naturally occurring smoke or volcanic ash but are usually complex agglomerates of anthropogenic combustion-derived smog, which often includes soot, water vapor, and associated environmental chemicals, including diesel exhaust (NO_2). Combustion products in the form of particulate matter (PM) may also be vectors of plastic nanoparticles, especially in size ranges below 200 nm. It is widely acknowledged that airborne particles in size ranges less than 2.5 microns have the most significant health physics impact; the smaller their size, the greater the health hazard associated with their inhalation. The smallest sizes of these contaminants, ultra fine airborne particles (≤ 100 nm), pose the greatest threat to public health. The very small sizes of the airborne toxins associated with ultrafine particles make it extremely difficult to evaluate their presence or their health physics significance. The association of fine and ultrafine particles with airborne chemical fallout is infrequently noted in news media commentary on air pollution, as in reports depicting smog in cities in China, India, and elsewhere. Also often not mentioned in mainstream media is the phenomena that highly visible macro and microplastics in oceanic gyres, for example, gradually break down into invisible microplastics < 250 μm and then into smaller nanoparticles < 1μm that sorb environmental chemicals.

The following definitions and acronyms are used in this text to describe the various categories of nanoparticles.

Naturally occurring nanoparticles (NONP):

- Naturally occurring nanoparticles from forest fires and volcanic ash (± 1 μm) and colloidal particles in water, among many NONP. Ultrafine carbon NONPs have the same size range and solution components as anthropogenic (manmade) carbon ultrafine nanoparticles (ANPs) in air pollution events.

Anthropogenic nanoparticles (ANP) and plastic nanoparticles (PNP):

- **Man-made nanoparticles (NP):** nanoparticles such as ultrafine particles (UFP) inadvertently produced by fossil fuel combustion and other anthropogenic activities. The US EPA and world environmental agencies classify air pollution in three size ranges:
  - PM_{10}: Particulate matter having a diameter smaller than 10 microns (10 μm).
  - $PM_{2.5}$: Particulate matter having a diameter smaller than 2.5 microns (2.5 μm). The EPA air quality standard for $PM_{2.5}$ is 15μg/m^3.
  - $PM_{0.1}$: Particulate matter having a diameter smaller than 0.1 microns (100 nm).
- **Intentionally manufactured nanoparticles (IMNP):** carbon nanotubes, fullerenes, nanocrystal clusters, nanobiorods, nanowires, quantum dots, and many other nanoparticles produced for consumer products and medical and manufacturing uses. Manufactured nanoparticles, including manufactured nanoplastics, are produced using specialized molecular structures that may or may not incorporate non-plastic materials such as carbon, kaolinite clay, and metals.
- **Plastic nanoparticles (PNP):** derived from the breakdown of macroplastics to microplastics in oceanic and terrestrial environments resulting from the production of ± 10 billion tons of plastic since 1950.

Where Have All the Plastics Gone? is a question yet to be asked by a self-indulgent global consumer society where the mantras remain "growth" and "profit" rather than sustainability and survival. The invisible presence of environmental chemicals as nanoecotoxins, their association with PNP, and their movement throughout the biosphere, unlike highly visible hurricanes, droughts, and tropical rain events, is not often a subject of informed debate. The rapid spread of anthropogenic ecotoxins now permeating the hemispheric water cycle is an unfortunate historical event. This most important achievement of humanity in the Anthropocene is an invisible, but important, component of a growing world water crisis now threatening the future of humanity.

The Legacy of Environmental Awareness

As a preface to the following commentary about nanotoxins and the unfolding Age of Biocatastrophe, a number of observations need to be made about the growing awareness of a biosphere in crisis that can be dated to the mid-1960s. This increasing awareness of the environmental issues pertaining to a biosphere under attack by human activity may be called the good news. The bad news is that the fundamental activity of humanity in the Anthropocene constitutes a massive attack on the diversity and viability of most natural ecosystems.

Most environmental issues of the 1960s are now lost in the fog of anti-Vietnam War activism. However, there was a growing awareness of the dangers of a global network of nuclear power plants, which followed the widespread opposition to nuclear weapons testing that was ended by the US in 1963. By the late 1960s, the persistent organic pollutant (POP) polychlorinated biphenyl (PCB) became the subject of extensive health physics research, and was eventually banned from production in the US by 1972 and in Europe by 2001. Other POPs eventually became the focus of the Stockholm Convention in 2001 and were also banned (Appendix 1). Huge quantities of PCBs, as well as pesticides and other persistent toxic chemicals, had been in production for several decades. Large quantities of PCBs were widely used as coolants in electrical transformers, machinists' cutting oils, and floor varnishes, and for other military, industrial, and consumer product uses. Many residual point sources continue to release this and other highly toxic chemicals to the environment.

In the late 1960s ozone depletion caused by the high flying supersonic transport (SST aircraft) became a subject of concern. The first Earth Day was held in Boston in April, 1970, and included the first die-in held at Logan Airport to protest the upcoming use of the SST in America. In the fall of 1970, Congress passed legislation barring its use. In the early 1970s, the threat of chlorofluorocarbons as another ozone depleting substance became the subject of widespread environmental concern. Chlorofluorocarbons were frequently used as coolants, to make Teflon, and in a wide variety of other applications. Their use and production were gradually restricted during the next 15 years. Their continued presence in many water supplies illustrates their status as a persistent organic pollutant (POP). President Donald Trump, please note "Persistent Organic Omnipresent Pollutants" = POOP.

Between 1970 and 2000, a robust environmental movement, including the Natural Resources Defense Council, the Environmental Working Group, Greenpeace, Union of Concerned Scientists, and many other non-government organizations (NGOs) helped foster an era of governmental and private research, documentation, and regulation of numerous environmental chemicals, many of which are referenced in this publication

16

series. The Toxic Release Inventory was established in 1986 under the Emergency Planning and Community Right-to-Know Act (US Environmental Protection Agency 1997a). Of most importance was the establishment of numerous superfund sites in 1988 as a result of the Comprehensive Responsibility and Liability Act, also known as the Superfund Act. The Food Quality Protection Act and the Safe Drinking Water Act Amendments were passed in 1996.

The Centers for Disease Control (CDC), established in 1946, was renamed the CDC in 1970, and began its extensive documentation of environmental chemicals including dioxins, the 209 cogeners of PCBs, and numerous other POPs and persistent bioaccumulative and toxic chemicals in human urine, serum, and blood. During and after the heyday of the environmental movement, numerous important legislative mandates have greatly curtailed sulfur dioxide (SO_2), diesel and automotive exhaust emissions, leaded gasoline, the use of toxic organotins in marine paints and varnishes, and cigarette smoking in public. Most forms of PBDE fire retardants are no longer manufactured; phthalates, bisphenol A, and many hormone-disrupting chemicals (HDCs) and pharmaceuticals are the subject of continuing efforts towards restrictions and/or elimination.

We now live in the Age of Income Inequality characterized by the simmering resentments of a large underclass of marginalized citizens. Donald Trump and Ted Cruz now (2016) give voice to their resentment. None the less, we may celebrate the many achievements of that other 1%, those ±3,000,000 concerned citizens, activists, scientists, and students who were and are at the core of a growing awareness of the many environmental crises we now face.

Unfortunately, much more powerful economic and political forces are undercutting the effectiveness of the environmental movement in the US and elsewhere. Entrenched anti-environmental interests now range from conservative rural communities, including the Tea Party movement, to entrenched fossil fuel industries and diesel exhaust producing automobile manufacturers, such as Volkswagen. There is now widespread support for continuing and even increasing the use of oil, gas, and coal to fuel our global consumer society. Our finite and vulnerable biosphere in crisis is now the subject of a rising tide of political anti-environmentalism in the Obama era. A combination of social stress, political paralysis, indebtedness, lack of public resources, and widespread ignorance of industrial history, chemistry, biology, and the impact of human ecology on the natural world are all components of a now inevitable historical event, the genesis of biocatastrophe, where human ecosystems envelope and overwhelm natural ecosystems. Nanoparticles as vectors of environmental chemicals are an invisible component of the Anthropocene's war on the biosphere.

We now live in an Age of Irony. Much of the terrestrial environment looks cleaner, especially in the US and Europe, if one ignores the smog in Paris. New restrictions are being implemented to restrict greenhouse gas emissions from US coal-fired power plants. If they survive legal challenges in the US court system, they will constitute an important step in implementing the Paris climate change agreement. Automotive energy efficiency is rapidly improving. The world of nanotechnology is introducing a vast array of energy saving innovations and inventions. All nations also share a growing public awareness of the reality of global warming, cataclysmic climate change, oceanic acidification and eutrophication, coral reef die-off, and fisheries depletion. This environmental awareness includes an often inarticulated intuitive knowledge of the vulnerability of our fragile world commons now under constant assault in the late Anthropocene. Invisible nano-molecular ecotoxins now dance and mate with plastic nanoparticles in all ecosystems.

These ecotoxins, egested by our pyrotechnic petrochemical industries, constitute a chemical attack on the biosphere in their proliferation and long-range atmospheric transport. They are now a constituent of all biotic media ranging from microbial communities to human cord blood. Plastic nanoparticles are an invisible vector for the movement of these environmental chemicals in all trophic levels of the biosphere. Their growing presence is, unfortunately, also a phenomenon that can't be mitigated by the research and activism of the one percent of environmental activists.

Given the rapid increase in the world's population since 1970, the accompanying expansion of urban and suburban landscapes, including rapidly growing urban wastelands, and the reality of a finite round world biosphere in crisis, the good news about environmental activism and governmental research and legislation during the last 5 decades is not enough to counteract the consequences of the accelerating destruction of the natural ecosystems that are essential to the viability of human society in the coming decades and centuries. The recent conference in Paris (The 2015 United Nations Climate Change Conference from November 30 to December 12) clearly articulated the hope of all nations that radical steps to change the lifestyle of global consumer society will result in a last-minute avoidance of climate disaster due to increasing CO_2 emissions.

The invisible but very real threat of chemical fallout, including ecotoxins carried by unseen nanoparticles of all kinds, must now be included as a key component of a "changing climate." This is unlikely despite our hope to avoid biocatastrophe. The "B-word", as well as the term "plastic nanoparticles", have not yet appeared in The New York Times, Huffington Post, or on Wikipedia (April 1st, 2016). Perhaps plastic nanoparticle denial and biocatastrophe denial will join climate change denial as phenomena we are just imagining. We will then live happily forever. Tilapia anyone?

18

Biocatastrophe: What is it?

Historically speaking, many events have been the subject of intense rituals of evasion while they were taking place. No new stories, op-eds, or concerned citizens meetings occurred while the concentration camps in Germany were constructed and utilized to incinerate millions of Jewish citizens. So to the present age we have an elaborate ritual of the evasion pertaining to the fate of pyrotechnical man. It was the use of fire and fuel (wood, coal, oil, and gas) that created industrial society. We have documented the romance of the Wooden Age with its bronze, iron, and steel tools that created a maritime society that explored and conquered the new world. The petrochemical industrial state, which evolved in the 20th century utilizing coal, oil, and now, natural gas as its prime movers, has evolved into a global consumer society. And global is the key word in the ongoing rituals of evasion as to what is presently occurring in a round world biosphere with limited resources now under assault by the many anthropogenic ecotoxins produced by pyrotechnic-petrochemical-nuclear society (PPNS). The hemispheric water cycle provides the context for the movement of environmental chemicals in all trophic levels of the biosphere.

 Collectively, 7,350 million of us live in a finite biosphere with a vulnerable global water cycle and a terrestrial landscape undergoing assault from every direction. If we traverse to a high tech National Geospatial Intelligence Agency (NGA) satellite and look down on the earth's surface, we can Google 29 rapidly expanding cities with over 10 million inhabitants and another 425 urban areas with more than 1 million inhabitants, soon to be joined by many more. If we look a little closer, we see the vast landscape of industrial activities based primarily on pyrotechnic petrochemistry, with a little help from nuclear reactors, which boil water to create electricity. If we Google our industrial landscape, including the growing multitude of huge waste landfills, we see that petrochemistry is the prime mover of a now globalized consumer society. We collectively produce at least 750 million tons of gasoline, chemicals, and plastics every year to help at least a few of us to live comfortably within our gated communities or fifty story penthouses in New York City. 750 million tons of petrochemicals are only a small fraction of the vast production of machinery, transportation equipment, building products, and consumer goods that are the constituents of our ever-growing consumer society. The end of use fate of the many products of military-industrial-consumer society involve the dispersion of tens of thousands of environmental chemicals in our vulnerable finite global water cycle, i.e. chemical warfare against the biosphere. Within this gestalt of the vast productivity of our pyrotechnic petrochemical industrial society, there is a particular topic of special interest, the production of plastics, that wondrous facilitator of the many conveniences of global consumer society. The world commons now produces in excess of 300 million tons of plastics a year (2015), soon to reach a billion tons every three years. As these plastics are discarded and fragment into smaller nonbiodegradable particles, they become a prime mover of ecotoxin transport throughout the hemispheric water cycle, beginning in the microbiomes that are the basis of life on Earth.

If we Google our aquatic, and especially our marine, ecosystems, we see that ±10% of our annual global production of plastics has been translocated to the marine environment, now manifested in highly visible oceanic gyres of metaplastics. These in turn degrade into mesoplastics (± 10 mm) and then to microplastics (≤ 1mm). Microplastics also include the ubiquitous microbeads derived from cosmetics, toothpaste and other consumer products. These gyres, the most famous of which is the great Pacific garbage patch, along with extensive shoreline deposits of plastic debris, have been the subject of intensive scrutiny by concerned environmental and scientific organizations during the last two decades (see bibliography). Unfortunately, systematic measurements of oceanic plastic inventories, including microplastics utilizing nets that capture debris as small as 380 um, have only accounted for a small percentage of the plastics estimated to be in the marine ecosystem. As of 2015, the total of plastics accidentally deposited to marine ecosystems since 1950 is at least 300 million tons. The question that has arisen is "where are all the plastics?" (Thompson 2004).

Look again from our GSA satellite perch, and Google what happens to those non-degradable synthetic consumer products whose useful lifetime varies from a few minutes to possibly a decade. Less than ten percent of the total global production of plus or minus ten billion tons of plastics have been recycled. Possibly another twenty percent have been disposed of in high-tech municipal waste incinerators, including waste to energy facilities, some of which, with the assistance of high temperature combustion, can eliminate most of the hazardous emissions from burning plastics and other semi-toxic wastes. Congratulations to a half-dozen European nations who lead the world in recycling and the safe disposal of plastics (e.g., Germany 80%??).

Most mega cities including New York City are not doing so well…so where do all the other plastics go? Massive landfills such as those in Pennsylvania and elsewhere are now the destination of New York City wastes. Landfills of all sizes are the repository of most urban and suburban waste streams, which are capped temporary repositories for plastics that will gradually disintegrate into small particles if not incinerated in inefficient dioxin producing municipal waste incinerators. Smaller landfills in every community join backyard dioxin emitting incinerators as another destination of plastic wastes. General litter is a worldwide problem, especially in third world locations, which often lack municipal landfills. A significant percentage of our plastic wastes (±10%) are translocated, especially by rivers, from waste laden terrestrial environments to our marine environments.

Take another look at the rapidly growing cities that are in our pyrotechnic landscape. So where have all the plastics produced by global consumer and industrial society gone? The answer comes in the form of an epiphany: using trolling nets with a smaller opening (± 63 um vs. 380 um), immense numbers of smaller plastic particles have been recovered in proportion to the smaller number of microplastics recovered utilizing the larger net size. Could much of our missing plastic – our beloved cosmetic microbeads and Saran Wrap – have abraded into microplastics smaller than 63 um?

In fact, in both aquatic and terrestrial ecosystems microplastics continue to abrade and fracture into particles smaller than 1 um, i.e. plastic nanoparticles. Ultraviolet light (sunlight) plays a major role in embrittling plastics and allowing them to evolve into plastic nanoparticles (PNP). Not yet defined in Wikipedia (4/1/2016), plastic nanoplastics are invisible, difficult to measure, and continually evolve into smaller nanoparticles, eventually reaching a size range from 25 nm to 100 nm or smaller.

Here we return to the phenomenon of our sacred rituals of evasion. We now have widening ongoing discussions about global warming, climate change, oceanic acidification, and marine species depletion. The recent conference in Paris expressed the worldwide concern about increasing CO_2 emissions and their impact on climate. One of the inadvertent achievements of industrial society is the production of greenhouse gases in the form of CO_2 and many other chemicals that are not so frequently the subject of media reporting. Global warming due to these emissions is no longer the subject of ritual of evasion, at least outside the world of the American Tea Party Taliban. But if we do a data search on components of chemical fallout, including POPs, the inevitable result of our pyrotechnic rituals of evasion again become apparent. Ocean acidification is the specific focus of ongoing research, so too glacial and sea ice melting, increasing sea level change and rising global and oceanic temperatures. We are also experiencing increasing frequencies of droughts, permafrost melting, and methane release. Hurricanes (Katrina, Sandy, Maria, Harvey, Irma, etc.), polar vortexes, and increasingly frequent blizzards, are on the news and the weather channels. Tropical forest destruction is now estimated at 1 billion acres in the last 40 years. So why not a discussion about the synergism of the components of our evolving environmental debacle and a more informed debate on the proliferation of environmental chemicals and their transport by invisible nanoparticles? That climate disaster is only one component of the historical evolution of biocatastrophe is an ecological reality yet to be reported by our mainstream media. Our sacred rituals of evasion continue.

The presence and the ecological significance of plastic nanoparticles are also the subject of an ongoing ritual of evasion. What are plastic nanoparticles, sometimes called nanoplastics? If they are not in Wikipedia or written about in the New York Times, how can they exist? Herein lies the answer to two questions. One, where have all the plastics gone? Two, what is their ecological significance in our finite world commons? Microplastic debris continues to abrade, undergo photodecomposition (embrittlement), and disintegrate, evolving into invisible particles measured in nanometers (nm) or billionths of a meter – of lesser size than those captured in that 63 um net. As they evolve into particles less than 1 um (1 micrometer in diameter), they become plastic nanoparticles (less than 1000 nm), with continued transformation into sizes in the 25 - 100 nm range. Whether macro, micro, or nanoplastics, these particles have the same biopersistence; they don't suddenly decay, and their semipermanent longevity makes them a long lived component of the biogeochemical cycles of marine and terrestrial environments. All microbiomes are now characterized by ever-growing debris fields of manmade ecotoxins, including plastic nanoparticles. In this size range they join the world of ultrafine particulate matter (≥ 100 nm), typified by smog, usually generated by combustion of coal, gasoline, and petrochemicals. Many other nanoparticles are released as deliberately manufactured nanoproducts, including nanoplastics, all of which reach their end-of-life-cycle use and disintegrate by abrasion and mass wasting in a similar manner as plastic wastes. Both deliberately and accidentally produced nanoparticles are characteristic of our brave new world of nanotechnology, the cutting edge of the ongoing inventiveness of pyrotechnic petrochemical man. In this context, the Age of Plastics (1950) is a precursor as well as a participant in the Age of Nanotechnology (2015).

Manufactured nanoparticles (MNPs) include quantum dots, fullerenes (buckyballs or CO_{60} particles), carbon nanotubes (single walled or multi-walled), nanoboxes, nanoclays, nanocrystals, nanocubes, nanosquares, and nanowires, many of which incorporate intentionally manufactured nanoplastics (IMNP). Typical products of nanotechnology also include nanoarrays, nanoclusters, nanodevices, nanofillers, nanomachines, nanometal, nanopharmaceuticals, nanoporous membrane, nanoprobes, and nanosensors. These are the tools of nanotechnological industries, the use of which is now revolutionizing fields such as medicine (nanomedical products), personal care products (sunscreens, etc.), packaging, and consumer products of every description.

Plastic nanoparticles (PNP) (± 200 nm) have an important characteristic they share with the intentionally manufactured nanoparticles of nanotechnological industries. As noted, as their size becomes smaller, the ratio of their surface area to their mass increases. We now begin to enter the world of Quantum Mechanics. The many variations of manufactured nanoparticles have the capacity for what is called magical feats of conductivity and reactivity. Plastic nanoparticles share some of these characteristics, the most notable of which is their ability to sorb environmental chemicals of every description; the smaller the diameter of the PNP the greater the quantity of environmental chemicals sorbed in proportion to weight. Invisible PNP and their ecotoxic contents are then incorporated in the aquatic food chain by grazing pico and nano zooplankton, generically referred to as marine microbes, and then are translocated to higher trophic levels within pelagic and benthic environments (see biosphere model 8). The highly bioactive sea surface micro layer ($\leq$ 1 millimeter in thickness) is an especially important location of the nanoplasticizing of microbial ecosystems. The translocation and bioaccumulation of PNP transported ecotoxins begins at this bottom layer of the aquatic food chain. The United Nations-associated Joint Group of Experts on the Scientific Aspects of Marine Environmental Protection provides this description of the sea surface microlayer:

> The sea surface of the ocean comprises a series of sublayers. These include a thin surface nanolayer ($< \sim 1\mu m$) containing high densities of particles and microorganisms; and the surface millilayer ($< \sim 10mm$) inhabited by small animals and the eggs and larvae of fish and invertebrates. The sea-surface microlayer is operationally defined in this report as the uppermost $\sim 1000\mu m$ (1mm) of the ocean surface. It, together with an overlying atmospheric layer of thickness 50-500μm, constitutes the boundary layer between the ocean and atmosphere...Material accumulated in the sea-surface microlayer is ejected into the atmosphere in an enriched form as part of the sea-salt aerosol produced by bursting bubbles. This provides a mechanism for the selective transfer of materials to terrestrial environments. Documented examples of such aerosol transport from sea-surface microlayers include bacteria, viruses, 'red tide' dinoflagellates and artificial radionuclides... High concentrations of toxic chemicals are also often found in the surface microlayer compared to the subsurface bulkwater in coastal environments. (GESAMP 1995)

We might describe this biochemical event as a dance of quadrillions of quadrillions of microbes with xenobiotic substances (i.e. ecotoxins infused in or sorbed by nanoparticles, including plastic nanoparticles [PNPs]). The word microbes is insufficient to describe the trophodynamic behavior of the lowest levels of the marine food chain, the most dynamic component of which is ironically in the upper millimeter of the ocean's surface. Autotrophic photosynthetic cynobacterium may dance with our xenobiotic visitors, but marriage is unlikely. In contrast, the vast array of heterotrophic organisms in our biogenic dance will graze on or otherwise ingest tiny PNP (50-200 nm) and other round particles. These PNP have already married (sorbed) thousands of biologically significant organic chemicals that are an invisible component of marine snow, that never ending stream of carbon from atmospheric deposition, which combines with the egested particulates in oceanic microbial ecosystems. The chemistry of carbon recycling by atmospheric depositions includes a huge variety of airborne anthropogenic chemicals, including persistent organic pollutants (POPs), such as PCBs, which were as noted, the subject of restriction by the Stockholm Convention. This text uses the metaphor of a hemisphere of organic chemicals to describe the ongoing deterioration of the earth's water cycle as these and other chemicals move throughout the biosphere.

What we have is in fact the phenomenon of PNP-Chemfall couples dancing with and being ingested by heterotrophic nanoflagelates, dinoflagellates, and picoplankton. Herein we have our unfortunate ménage à trois, an invisible historical event with vast repercussions for human society in the Anthropocene. As saprophytic bacterioplankton, grazing protozoans, ciliates, and many types of nanoplankton join our dance, they give rise to asexual threesomes that then translocate up the food chain. These nanobiotic marriages recycle anthropogenic petrochemical ecotoxins, which include HDCs and obesogens, to higher levels of the marine food chain. Where did all those polychlorinated dibenzofurans (PCDFs) in that tasty black Grouper or shipwrecked seals come from? The answer is the eco-technological dances and marriages of the nearly invisible inhabitants of our globally contaminated sea surface micro-layer, to name the most important of all impacted marine and terrestrial micro ecosystems. All microbiomes are now characterized by ever-growing debris fields of anthropogenic ecotoxins. Welcome to the free-enterprise age of "where have all the plastic nano particles gone?"

The sea surface microlayer is not the only marine ecosystem impacted by chemical fallout laden plastic nanoparticles and other nanoparticles. Benthic communities such as coral reefs are rapidly disappearing. As of 2015 approximately 60% of all coral reefs have been destroyed or are severely damaged. Rising ocean temperatures and greenhouse gas-related acidification, the result of rapidly increasing levels of carbonic acid, are considered major causes of coral reef die off. Any underwater Google search in the form of a toxological profile of the biochemistry of benthic ecosystems would reveal that coral reefs are also a destination of anthropogenic chemicals that now number ±100,000, of which at least 10,000 are biologically significant in their persistent toxicity. Many others are short lived chemicals which have a health physics impact in terrestrial and marine ecosystems before dissipating, often within a few weeks or months. So too with the pelagic and benthic courting ceremonies in biotic communities in coral reefs and their vulnerable microbiomes. Chemicals that we can no longer detect, as well as the congeners, which are their daughter products, may have already had a biologically significant impact. The increasing acidity of the marine environment from greenhouse gas emissions is only one component of coral reef die-off. The ease of documenting increasing acidity levels helps us ignore the growing inventories of anthropogenic ecotoxins in coral reef communities as well as in all microbiomes. These environmental chemicals are then translocated to all trophic levels of our biosphere.

In the coming age of nanotechnology, growing inventories of intentionally manufactured nanotoxins will join plastic nanoparticles in a trophodynamic dance of ecotoxins of every description. We hopefully presume that many of the nanotoxological safety issues referred to in the following annotated bibliography will be addressed as nanotechnology bestows its many benefits on world society. As with the end of the life cycle fate of our beloved plastics, the end of life fate of the nanoparticles derived from our plastic bumpers, tennis rackets, self-cleaning glass and dental and medical implants will also eventually contaminate the vulnerable biogeochemical cycles of a biosphere in crisis.

We are now entering the Age of Biocatastrophe where many unfortunate events occur simultaneously. Our oceanic ménage à trois, as it were, of micro biota, plastic nanoparticles, and environmental chemicals is a tiny slice of the biocatastrophe pie chart.

The National Security Agency (NSA), with the help of the General Services Administration (GSA) Satellite Google masters, have extensive classified inventories of environmental chemicals in pathways to human consumption. These include toxological profiles of CAFO (caged animal farm operation) products and on other foods of every description, including corn, milk, eggs, high fructose corn syrup, cereals, etc. Unclassified detailed surveys of environmental chemicals in human urine, blood, and serum compiled by the CDC are available to any interested citizen. While many governmental surveillance programs that are a legacy of the Age of Environmental Activism (1970-2000) will soon be curtailed or defunded in the chaos of post-Obama environmental politics, the databases compiled by numerous environmental organizations as well as federal and state programs (e.g., Maine on methylmercury or New York on plastic microbeads) are still accessible. Notable among continuing chemical fallout events which are well documented by both NGOs as well as by governmental entities such as the CDC include the following. See *Biocatastrophe: The Legacy of Human Ecology* (Brack 2010c).

- Cord blood contaminants (Environmental Working Group)
- Hormone and endocrine disrupting chemicals (Colburn 1997, 194) (www.ted.com) (Krimsky 2000) (Brack 2010)
- Hydrofracking ecotoxins and impacted ground water (Nalbone 2014, 194)

- Greenhouse gas emissions (Intergovernmental Panel on Climate Change)
- The proliferation and pathways of growth hormones (Torkelson 1988)
- Pharmaceuticals in our wastewater discharges (US Geological Survey)
- Methylmercury and its biomagnifications (Watras 1984)
- Volatile organic contaminants in aqueous environments (Biodiversity Research Institute)
- PBDE in biological media including seals and humans (Shaw 2002; 2003)
- Drinking water pollutants (US Geological Survey, US Environmental Protection Agency)
- Ozone depleting chemicals (Miller 2008)
- Enough POPs and PBT surveys to fill Fenway Park

If we return to our GSA satellite perch and Google the synergistic relationship of the many components of the ongoing geochemical contamination of our biosphere we see the basic dynamics of biocatastrophe emerging as the history of human ecology unfolds. Documenting the history of biocatastrophe necessitates exploring the interrelationship of the multitude of unfortunate events. The synergism – the interrelationship – of these historical events is the subject of a continuing ritual of evasion by a pyrotechnic global consumer society and its mass media digital communication systems. The toxic dance of ecotoxin laden plastic nanoparticles with the many biota in our marine and freshwater ecosystems is not yet a subject of informed public debate. That these accidental byproducts of the Age of Plastics will soon be joined by the end of the life cycle nanotoxic remnants of the ongoing Nanotechnological Industrial Revolution is another small slice of the biocatastrophe pie.

A larger slice of the biocatastrophe pie is the biogeochemical cycles of the tens of thousands of PBT chemicals produced by pyrotechnic-petrochemical industrial society. Their atmospheric transport after evaporation, or riverine transport after use and disposal, are pathways to human consumption that have vast health physics, economic and social impacts. Other slices of our biocatastrophe pie, in addition to cataclysmic climate change and its many components, include the world water crisis and the dwindling resources of potable water for the world's growing urban environments. The looming impact of the world's water crisis on industrial agriculture is mirrored by the growing unavailability of fresh water for non-industrial agriculture, especially in the war torn mid-East (e.g. Syria) and sub-Sahara Africa. Political chaos, social stress, and civil and regional warfare follow in the footsteps of depleted water supplies, denuded landscapes, and the declining quality of soils.

We also live in a world of the growing presence of antibiotic resistant bacterial and viral infections, something not visible from our Google satellite perch. Growing income inequality and a lack of employment opportunities for over 50% of the world's population, and the accompanying social stress and political paralysis are also not visible from our Google perch. Also hidden in the panorama of urban landscapes is a flourishing shadow banking network operating in the context of the world economy with 750 trillion dollars in debt and 250 trillion dollars in world assets. Vast profits are made by a tiny minority selling debt as an asset, while the supposed assets of pension and retirement funds are, in fact, a lucrative source of profit. The hidden costs of our parasitic shadow banking kleptocracy join the unacknowledged hidden costs of the Age of Plastics and the vast debts of unfunded pensions and collapsing infrastructure to constitute a threat to the future viability of global consumer society.

While the vulnerability of a debt ridden world economy and its parasitic hedge and equity funds are hidden within the silicon chips of our highly profitable information technology economy, the soaring skyscraper penthouses in Manhattan and elsewhere are highly visible from our GSA Google perch. Recently the subject of a five part series in *The New York Times* (February 8-12, 2015) the kleptocrats from Malaysia, India, Russia, China, and elsewhere who occupy these penthouses join the ranks of several thousand successful shadow bankers and oil billionaires who now control many of the monetary assets in a debt ridden global economy. Look again at the history of pyrotechnic-petrochemical man as the world population reaches and then exceeds the limits of a finite world biosphere. Disparities in world income equality highlight an essential characteristic of modern society: self-indulgence. In India, ±600 million citizens still practice open defecation (politely called egestion) due to a lack of sanitary facilities. Global consumer society recapitulates this phenomenon by indulging in the magnificent benefits of combustion engines and the world of plastics without accounting for the hidden costs of petrochemistry, i. e. petrochemical egestions in the form of a hemispheric water cycle as an ecotoxic soup.

Herein lies the key to success of our round world pyrotechnic-petrochemical consumer society. The world of plastics has been a prime mover in the successful evolution of a rapidly growing world consumer society based on the imposition of human ecology on natural ecology. Now the clock is ticking. Civil unrest and displaced populations are spreading. Paralysis, resentment, and anger characterize our political system. The growth and spread of antibiotic resistant bacteria and viruses parallel the invisible spread of anthropogenic chemical fallout, including ecotoxins transported by nanoparticles. The looming threat of cataclysmic climate change cannot be evaded. Resource exploitation and depletion and growing urban and suburban wastelands are highly visible. Most importantly, the world's water cycle – a finite resource that cannot be replaced by innovate information technology – is now being irrevocably contaminated by the invisible environmental chemicals produced by highly profitable petrochemical industries. The imposition of human ecology on the natural ecology of a vulnerable finite world biosphere is symbolized by the success, comforts, and profits of global consumer society and its world of plastic. Ask the question "where have all the plastics gone" and open a Pandora's box of dancing microbes, anthropogenic nanoparticles, and environmental chemicals, a ménage à trois that is the bequest of pyrotechnic-petrochemical-nuclear society (PPNS).

This marriage of environmental chemicals, plastic nanoparticles and manufactured nanoparticles, and microbiota is the progeny of what is most accurately labeled predatory petrochemical consumer society. We all indulge in the rituals of celebrating the many temporary benefits of this milieu. Few of the earth's human inhabitants do not have a relationship with combustion engines, electronic equipment, such as cell phones, iPads, and computers, fantastic plastic, and many other accoutrements of a global consumer society. The values of a rapidly growing and highly productive consumer society are characterized by three words: growth, profit, and free enterprise. For our global kleptocracy and their 1%-ers and their 10%-ers, these words are their pledge of allegiance to a predatory system of human ecology. For the other 90% of human populations, who also practice open petrochemical egestion, as well as for our natural ecosystems, another word needs to be substituted for this anthem: survival.

Overview: News Bites and Op-Eds

The following commentary consists of editorial opinions and observations by the editor of this publication, by other governmental and NGO sources, or selected quotations (in alphabetical order) from the annotated citations included in our survey of nanoparticles as vectors of environmental chemicals.

In the century 1915 to 2015 we have witnessed world and regional warfare and destruction, as well as industrialization and the growth of a global consumer society, all of which produced vast quantities of anthropogenic ecotoxins. Biocatastrophe is an historical event.

The most important of many finite resources of the Earth's biosphere is water. Without water, there would be no biosphere.

The second law of thermodynamics expresses the reality that environmental chemicals, including organic chemicals, when released within the hemispheric water cycle, gradually reach equilibrium in biotic environments.

The movement of water through all living matter—whether in microbiota or in cord blood in human embryos—is a part of the global water cycle now being contaminated by human activities, including pyrotechnology, petrochemistry and the ubiquitous use of pharmaceuticals by global consumer society.

The world water crisis has a number of ecologically destructive and thus socially disruptive components:

- Increasing air and water temperatures

- The increasing intensity of rainfall and snowfall events due to global warming and increasing atmospheric moisture content

- The increasing intensity of drought due to global warming

- Declining surface water supplies due to desertification as a result of human activities including farming, cattle grazing, and deforestation

- The increasing intensity and complexity of atmospheric chemical fallout due to anthropogenic activity. Declining worldwide terrestrial water quality is a result of petrochemical production, use and combustion, industrial agriculture, and the vast inventories of environmental chemicals produced by global consumer society

- The future threat of vast inventories of environmental chemicals that are now stored in waste sites, landfills, urban industrial landscapes, and residential households and workshops in most urban and suburban areas

The uptake of environmental chemicals by nanoparticles including plastic nanoparticles is only one of many pathways of ecotoxins to human consumption.

Black Swan Event (BSE): The mother of all Black Swan events is a severe solar storm. Such an event, which occurs every ±150 years, would cause a sudden drop in the economic and industrial activities of global consumer society, impact the viability of the shadow banking network, and incapacitate the everyday conveniences of the Age of Information Technology (computers, GPS, cell phones, etc.) Homeland security has recently released or declassified a series of reports depicting what could happen to our vulnerable network of high voltage transformers in the event of a solar storm as intense as the 1859 Carrington event. Loss of electricity in impacted urban areas in the Eastern United States could last for 2 years or more while damaged transformers were being replaced. See the Solar Storms link on the home page of www.davistownmuseum.org.

The world water crisis is not a black swan event because its many components are well documented, even if the growing threat of chemical fallout is often the subject of denial or inattention.

Organochlorines and other organic chemicals are fat-soluble chemical compounds resistant to degradation, and are stored in the adipose tissue of most animals on the planet, including humans. Accumulation of these compounds in the body is related to fat mass, obese individuals having a higher plasma organochlorine concentration than lean subjects.

"Polychlorinated biphenyls (PCBs) are a family of 209 congeners. Polybrominated diphenyl ethers (PBDEs) are another family of 209 congeners. Any significant quantity of PCBs or PBDEs is by default a blend of multiple molecule types because each molecule forms independently, and chlorine does not strongly select which site(s) it bonds to." (Wikipedia 2015)

"Nanoplastics is just one industry under the much larger umbrella of nanotechnology. Nanotechnology creates biochemical machinery on the nanoscale level. A nanometer (nano for short) is one billionth of a meter, so nanotechnology deals with molecular or atomic processes. At this level of life, proteins and other chemicals interact to form bonds and carry out processes. Nanotechnology harnesses those natural processes through direct manipulation to create unique configurations that can have profound effects on the macro scale. Nanoplastics is nanotechnology applied to plastics." (http://www.wisegeek.com/what-are-nanoplastics.htm)

"U.S. Geological Survey (USGS) and U.S. Environmental Protection Agency (EPA) scientists have shown that wastewater treatment plants are a significant source of pharmaceuticals and other emerging contaminants to rivers." (http://toxics.usgs.gov/highlights/tracing_wastewater.html)

A fundamental threat to the viability of industrial agriculture is the ongoing depletion of fresh water aquifers in the central valley of California, the American southwest and the central plains, as well as in any nation where surface water supplies and rainfall are being supplemented by aquifers.

Off limit Googles: Some, if not most, GSA satellite Googles are classified as a matter of national security interests. Comprehensive aerial radiological surveys were already being executed before the National Imaging and Mapping Agency (NIMA) was established in (1994). NIMA had the capability of locating radiation sources as incidental as radioactive cerium 144 camping lanterns in US hardware stores. To establish baseline information on toxic chemicals in the biosphere for evaluation of a possible terrorist attack, GSA currently provides the NSA with extremely detailed surveys of biologically significant environmental chemicals in all surveyable components of the environment: atmospheric, terrestrial, and aquatic. The NSA should now declassify all environmental chemical databases pertaining to biotic and abiotic media collected by its NSA.net and compiled by its laboratories throughout the US.

Numerous domestic and international repositories and point sources of anthropogenic radioactivity remain classified due to the necessity of maintaining the secrecy of radiological surveillance programs. RADNET: Nuclear Information on the Internet provides a comprehensive survey of known US and English point source and radioactive waste repositories, including those impacting US aquifers. (http://www.davistownmuseum.org/cbm/index.html)

"The size and number of marine dead zones—areas where the deep water is so low in dissolved oxygen that sea creatures can't survive—have grown explosively in the past half-century." (http://earthobservatory.nasa.gov/IOTD/view.php?id=44677)

Cyanobacteria, an important component of the microbial food web and the sea surface microlayer, have been documented in sea water at 100×10^6 cells per liter.

"If you give tributyltin to pregnant mice, their offspring are heavier than those not exposed...Tributyltin is changing the metabolism of exposed animals, predisposing them to make more and bigger fat cells." (Holtcamp 2012)

Protozoans are ubiquitous single celled microscopic animals, which as grazers play a key role in the uptake and translocation of contaminated marine snow.

Copper becomes a catalyst for dioxin formation when flame retardants are incinerated.

"Toxic equivalency factor (TEF) expresses the toxicity of dioxins, furans and PCBs in terms of the most toxic form of dioxin, 2,3,7,8-TCDD. The toxicity of the individual congeners may vary by orders of magnitude." (http://en.wikipedia.org/wiki/Toxic_equivalency_factor)

The ecotoxicity of plastic nanoparticles (PNP) is a new and emerging issue.

8% of oil production is used in the fabrication of plastics.

The failure of microorganisms to catabolize synthetic macromolecules is the basis for trophic level transfer of environmental chemicals to the human biome.

"Older age is a well established risk factor for increased serum organochlorine concentrations, presumably as a consequence of cumulative exposure and temporal trends in exposure." (Choi 2006)

"Widespread occurrence of neuro-active pharmaceuticals and metabolites in 24 Minnesota rivers and wastewaters has been documented by the Minnesota Department of Natural Resources." (Writer 2013)

CO_2 is only one of thousands of chemicals in the gaseous component of the chemical war on the biosphere.

"Zooplankton may play an important role in the biomagnification of pollutants up food webs." (Almeda 2013b)

"In the Great Lakes Basin sport fish consumption has been demonstrated to be an important source of PCB and DDE exposure." (Anderson 2008)

"Due to the hydrophobic nature of the micro plastics, they attract persistent organic pollutants, known as POPs." (Andersson 2014)

"Micro plastics are the perfect size (Andrady et al. 2011) to be mistaken for planktons and could thus enter the very foundation of the food chain, should they be able to migrate from the gastrointestinal area to other bodily tissues." (Andersson 2014)

"Through mechanical attrition, often assisted by photodegradation, plastics in the marine mileu may undergo very slow embrittlement and break down into inconspicuous, very fine detritus, which ultimately disappears from view—but not from the environment." (Andrady 2003, 384)

"There is little doubt that nanoscale particles are produced during weathering of plastics debris." (Andrady 2011)

"Nano- and picoplankton are not only the predominant group of plankton biomass but are also the predominant contributors to primary production. As plastic nanoparticles in the water are of a comparable size scale, understanding their mechanisms of interaction with the nano- or picofauna is particularly important." (Andrady 2011)

"PCBs were produced commercially in the United States from 1929 until 1977. Marketed worldwide under trade names such as Aroclor, Askarel, and Therminol, the annual U.S. production peaked in 1970 with a total production volume of 85 million pounds (39 million kg) of Aroclors." (ATSDR 2015)

"Bisphenol A at low environmentally relevant levels can transfer across the human placenta, mainly in active unconjugated form." (Balakrishnan 2010)

"Detection of synthetic organic compounds…occurred only in wastewater treatment-plant effluents and at downstream sites. Concentrations ranged from nanograms per liter to milligrams per liter." (Barber 2011b)

"In municipal landfill leachates the DOC [dissolved organic carbon] is much more important as a transport vehicle for hydrophobic phthalate esters than the suspended particles present." (Bauer 1998)

"Nanoparticle mobility and toxicity have been shown to be a function of aggregate size, and generally increase as size decreases." (Bennett 2012)

"The use of brominated flame retardants (BFRs) has increased over the last 30 years with present global production about 310,000 tons year…Additive BFRs, including PBDEs, are mixed into plastics and foams but do not form chemical bonds. This makes them much more likely to leach out of goods and products." (Besis 2012)

"The amount of nano- and microplastic in the aquatic environment rises due to the industrial production of plastic and the degradation of plastic into smaller particles. Concerns have been raised about their incorporation into food webs. Little is known about the fate and effects of nanoplastic, especially for the freshwater environment." (Besseling 2014).

"Because the plastic that enters the ocean tends to fragment, it is likely to remain in the environment 'for hundreds, if not thousands, of years.'" (Betts 2008a)

" In 2010, biocidal plastics and textiles are predicted to account for up to 15% of the total silver released into water in the European Union. The majority of silver released into wastewater is incorporated into sewage sludge and may be spread on agricultural fields." (Blaser 2008)

"Plastic particles tend to accumulate persistent, bioaccumulating and toxic contaminants such as PCBs, DDT and PBDEs. Microplastics have larger surface to volume ratios, potentially facilitating contaminant exchange and have been shown to be ingested by a range of organisms." (Bowmer 2010)

"Dioxins are a group of chemicals that are similar in their chemical structure and their toxic effects on biological tissues (EPA, 2010). They are formed by the incineration of products containing polyvinyl chloride (PVC), polychlorinated biphenyls (PCBs) and other chlorinated compounds, by industrial processes that use chlorine, and by the combustion of diesel and gasoline." (Breast Cancer Fund 2014)

"Plastics debris is accumulating in the environment and is fragmenting into smaller pieces; as it does, the potential for ingestion by animals increases." (Browne 2008)

"Enhanced affinity to soot may not be limited to polycyclic aromatic hydrocarbons but may extend as a significant process for a wider range of hydrophobic organic compounds." (Bucheli 2003)

"Microplastics within the size range 2 – 230 µm can be ingested by and cause harm to an array of marine life, including zooplankton, mussels, polychaetes and crabs" (Cassone 2014)

"Up to 10% of all newly produced plastics will eventually find their way to our seas and oceans." (Cauwenberghe 2013a)

"Microplastic pollution has spread throughout the world's seas and oceans, into the remote and largely unknown deep sea." (Cauwenberghe 2013b)

"Urgent efforts are needed to make manufactured nanoparticles less reactive if there is any risk that they enter natural environments." (Cedervall 2012)

"Triclosan (TCS), a high-production-volume chemical used as a bactericide in personal care products, is a priority pollutant of growing concern to human and environmental health." (Cherednichenko 2012)

"To date, more than 1000 organic chemicals have been identified in groundwater contaminated by landfills, most of which fall into the categories: aromatic hydrocarbons; halogenated hydrocarbons; phenols; and pesticides." (Christensen 2000)

"Microplastic particles (MPPs; <5 mm), also known as microbeads, are found in skin cleansing soaps and are released into the environment via the sewage system…MPPs in the environment can sorb persistent organic pollutants (POPs) that can potentially be assimilated by organisms mistaking MPPs for food." (Chua 2014)

"At all the stations, sea-surface microlayer (SML) concentrations of the selected organic compounds were significantly higher than sub-surface water (SSL) values." (Cincinelli 2001)

"Contaminants contribute to the increasing prevalence of attention deficit hyperactivity disorder, autism, and associated neurodevelopmental and behavioral problems in developed countries." (Colborn 2004)

"The outmoded 'Maximum Tolerated Dose' or the U.S. Agency for Toxic Substances and Disease Registry's (ATSDR's) Minimum Risk Levels (MRLs) are based on crude, traditional toxicological protocols and endpoints that have almost completely missed low-dose, endocrine system-mediated effects." (Colborn 2007)

"Perpetual fragmentation of plastic litter, coupled with the increasing popularity of household products containing microscopic plastic exfoliates suggests marine plastic debris is becoming, on average, smaller over time." (Cole 2013)

"Leachate from municipal landfills can create groundwater contaminant plumes that may last for decades to centuries…Of greater concern are the closed landfills in the United States, estimated at more than 90,000 two decades ago. These closed landfills are typically unlined so that exposure of buried waste to precipitation and groundwater seepage is expected to create leachate plumes containing complex mixtures of organic and inorganic contaminants." (Cozzarelli 2011)

"The average size of plastic particles in global environments seems to be decreasing, while abundance of such particles is increasing due to continuous fragmentation." (Doyle 2011)

"Microplastics are shed from all plastic-coated paper products during composting." (Ecocycle 2011)

"Plastic debris also acts as a sink for toxic chemicals. Plastic sorbs persistent, bioaccumulative, and toxic substances (PBTs), such as polychlorinated biphenyls (PCBs) and dioxins, from the water or sediment. These PBTs may desorb when the plastic is ingested by any of a variety of marine species." (Engler 2012)

"Plastic debris appears to act as a vector transferring PBTs from the water to the food web, increasing risk throughout the marine food web, including humans." (Engler 2012)

"BPA is at unsafe levels in one of every 10 servings of canned foods (11%) and one of every 3 cans of infant formula (33%)." (Environmental Working Group 2007)

"Gas exchange seems to be the main input mechanism of organochlorine compounds from the atmosphere to terrestrial and aquatic systems." (Fernández 2003)

Flat world digital technology is derived from round world petrochemistry.

"Chemicals implicated in endocrine disruption include biocides, industrial compounds, surfactants, and plasticizers including bisphenol A (BPA)…Global consumption of BPA in 2011 was predicted to exceed 5.5 million metric tons." (Flint 2012)

"Humanity has recognized that our own climb up the ladder of technological sophistication comes with a heavy price. From climate change to resource depletion, our evolution into a globe-spanning industrial culture is forcing us through the narrow bottleneck of a sustainability crisis." (Frank 2015)

"Engineered nanoparticles (ENPs) may… find their way into the soil environment *via* wastewater, dumpsters and other anthropogenic sources." (Frenc 2013)

"Very high concentrations of BPA and phthalates were confirmed in waste dump water and compost water samples as well as in the liquid manure samples." (Fromme 2002)

"The scientific basis for the cytotoxicity and genotoxicity of most manufactured nanomaterials are not understood." (Fu 2014)

"Inhaled 100nm particles can negotiate the lung barrier within 60 seconds, then show up in the liver and other internal organs within an hour." (Gatti 2008)

"Substantial PBDE burdens may be incurred by insects in contact with current-use and derelict treated polymers within human spaces and solid waste disposal sites (e.g. landfills, automotive dumps, etc.)." (Gaylor 2012)

"Continuous input of pharmaceuticals into rivers, through wastewater treatment systems, may cause adverse effects on the aquatic ecosystems of the receiving waterbodies, due to the intrinsic biological activity of these compounds." (Ginebreda 2009)

"PBDEs, like other POPs, can cross the placenta barrier…The presence of PBDEs in cord blood and placenta samples indicates that there is prenatal exposure of PBDEs, which could continue after birth via breast milk." (Gomara 2007)

"Environmental endocrine disruptors (EEDs) are exogenous chemicals that mimic endogenous hormones such as estrogens." (Gorelick 2014)

"A potential cause of concern…is the influence that microplastic may have on enhancing the transport and bioavailability of persistent, bioaccumulative, and toxic substances (PBT)." (Gouin 2011)

"Susceptibility to nanometal toxicity differed among species, with filter-feeding invertebrates being markedly more susceptible to nanometal exposure compared with larger organisms (i.e., zebrafish)." (Griffit 2008)

"The modern world is plagued with expanding epidemics of diseases related to metabolic dysfunction…[including] obesity, diabetes, cardiovascular disease, hypertension, and dyslipidemias (collectively termed metabolic syndrome)…The environmental obesogen hypothesis proposes that exposure to a toxic chemical burden is superimposed on these conditions to initiate or exacerbate the development of obesity and its associated health consequences." (Grün 2007)

"Tens of thousands of organic chemicals are currently in use, however [sewer]sludge concentration data could only be found for 516 organic chemicals in the peer reviewed literature and official government reports." (Harrison 2006)

"Particles in the nanometer range have two particular properties. First, anything smaller than about 50 nm is no longer subject to the laws of classical physics, but of quantum physics… Second, with decreasing size, the ratio between mass and surface area changes… As size decreases and reactivity increases, harmful effects may be intensified, and normally harmless substances may assume hazardous characteristics." (Hett 2004)

Nanoparticles in the environment "could constitute a completely new class of non-biodegradable pollutants… [They are] highly reactive nanoparticles which, because of their small size, can have a total surface area measuring up to 1,000 square meters per gram." (Hett 2004)

"Absorption [by nanoplastic particles] may be 15-250 times higher than that of microplastic particles." (Hollman 2013)

"Microplastics may concentrate contaminants such as PCBs up to 10^6 fold." (Hollman 2013)

"Obesity is rising steadily around the world. Convincing evidence suggests that diet and activity level are not the only factors in this trend—chemical 'obesogens' may alter human metabolism and predispose some people to gain weight. Fetal and early-life exposures to certain obesogens may alter some individuals' metabolism and fat-cell makeup for life. Other obesogenic effects are linked to adulthood exposures." (Holtcamp 2012)

"Nanotechnology is expected to soon pervade, and often revolutionize, virtually every sector of industrial activity, from electronics to warfare, from medicine to agriculture, from the energy we use to drive our cars and light our homes to the water we drink and the food we eat…There are inherent risks as well. What will happen when nanomaterials and nanoparticles get into our soil, water, and air, as they most assuredly will, whether deliberately or accidentally? …What will happen when they inevitably get into our bodies, whether through environmental exposures or targeted applications?" (Hood 2004)

"By 2020 nanoplastics will be by far the largest section of the nanomaterial market." (Intertech-Pira 2006)

"POPs are lipophilic compounds and accumulate in the fatty tissue of living organisms…Fish oil produced from pelagic fish and their by-products is used as polyunsaturated ω-3 fatty acid source in compounded feeds. High inclusion levels make it the main POP contributor in Atlantic salmon feed." (Jensen 2011)

"POPs are typically 'water-hating' and 'fat-loving' chemicals, i.e. hydrophobic and lipophilic. In aquatic systems and soils they partition strongly to solids, notably organic matter, avoiding the aqueous phase…Volatilisation now exceeds deposition, i.e. the water bodies now act as sources to atmosphere, rather than as sinks." (Jones 1999)

"Toxic substances such as dioxins, mycotoxins, heavy metals, pesticides, veterinary drugs and polycyclic aromatic hydrocarbons are almost ubiquitous in the environment. Thus, they are also present in ingredients for animal feed." (Kan 2007)

"The incidence rates for breast and prostate cancers in the United States have progressively risen since 1975. This trend has been attributed to multiple factors including increased exposure to endocrine disrupting agents." (Keri 2007)

"Ultrafine particles…are unintended particles originating often from combustion processes of diverse sources …ultrafine ambient particles are often composed of a multitude of compounds, which may be structured in a highly complex manner…Particles in the nanometer size range have two particular properties: (a) anything smaller than about 50 nm is no longer subject to laws of classical physics but of quantum physics. (b) Surface to mass ratio increases with decreasing size." (Kreyling 2006)

"The plastic input via the Danube into the Black Sea was estimated to 4.2 t per day." (Lechner 2014)

"Browne et al. (2011) report that in excess of 1900 microplastic fibres from clothing can be released into domestic wastewater by laundering a single garment in a domestic washing machine." (Leslie 2011)

"The abiotic and biotic degradation rates of synthetic polymers are extremely low - the material is expected to persist for hundreds to thousands of years, even longer in deep sea and polar environments." (Leslie 2011)

"Stretch film…may make a significant contribution to contamination of foodstuffs by BPA." (Lopez-Cervantes 2003)

"Initiatives to improve food security through increases in fish consumption in the region may be compromised from bioaccumulation of toxins in coastal and oceanic fishes." (Markic 2014)

"A total of 129 out of 202 CECs [contaminants of emerging concern] were detected [in fresh leachate from landfills in the conterminous United States] during this study, including 62 prescription pharmaceuticals, 23 industrial chemicals, 18 nonprescription pharmaceuticals, 16 household chemicals, 6 steroid hormones, and 4 plant/animal sterols." (Masoner 2014)

"Plastic resin pellets (small granules 0.1–0.5 centimeters in diameter) are widely distributed in the ocean all over the world. They are an industrial raw material for the plastic industry and are unintentionally released to the environment both during manufacturing and transport." (Mato 2001)

"Significant knowledge gaps exist in all areas of nanotech risk assessment… About a third of the hundreds of nanotechnology-related consumer products now on the market are intended to be ingested, or applied to the skin… Are buckyballs (C60) in cosmetics harmful? What are the risks in releasing nano-silver into the environment? Do nanomaterials in foods and food packaging present a risk? Will exposure to engineered nanomaterials lead to ill health? What happens to engineered nanomaterials at the end of a nano-product's life?" (Maynard 2006b)

"PBDE, BPA, and other EDCs exposure have been implicated as contributing to obesity in both humans and model animals, possibly by interfering with estrogen and androgen signaling…Weight gain is associated with several commonly used medications, including psychotropic medications, antidiabetics, antihypertensives, steroid hormones and contraceptives, antihistamines, and protease inhibitors." (McAllister 2009)

"An unimpeded transition toward an era of global chemical warfare on marine eco-systems (e.g., ocean acidification, anoxia) may retard or arrest the intrinsic capacity of marine fauna to bounce back from defaunation" (McCauley 2015)

"Urban rivers are an overlooked and potentially significant component of the global microplastic life cycle." (McCormick 2014)

"The obesogen hypothesis…postulates that pre- and perinatal chemical exposure can contribute to risk of childhood and adolescent obesity…Organochlorine chemicals as well as several classes of chemicals that are PPAR [Peroxisome proliferator-activated receptors] agonists are identified as possible risk factors for obesity." (la Merrill 2011)

"Mercury in fillets of shortnose sturgeon from the Penobscot and Kennebec (mean 0.49 parts per-million, ppm; range: 0.19 to 1.00 ppm wet weight) were elevated compared to freshwater regional and national fish tissue biomonitoring programs…A suggested tissue threshold-effect concentration for mercury in whole-body fish is 0.20 ppm and the Maine Fish Tissue Action Level for consumption is also 0.20 ppm. Both Atlantic sturgeon and one shortnose sturgeon fillets were essentially at the mercury tissue threshold effect concentration and state Action Level. Eight shortnose sturgeon fillets exceeded the whole-body effect threshold concentration and Maine Action Level." (Mierzykowski 2012)

"Manufactured nanoparticles, nano-emulsions and nano-capsules are now found in agricultural chemicals, processed foods, food packaging and food contact materials including food storage containers, cutlery and chopping boards." (Miller 2004)

"The surface properties and very small size of nanoparticles and nanotubes provide surfaces that may bind and transport toxic chemical pollutants, as well as possibly being toxic in their own right by generating reactive radicals." (Moore 2006)

"The list of chemicals measured represents only a small fraction of the approximately 30,000 chemicals widely used in commerce (>1 t/y). The vast majority of existing and new chemical substances in commerce are not monitored in environmental media." (Muir 2006)

"Food contact materials are a major source of food contaminants. Many migrating compounds, possibly with endocrine disruptive properties, remain unidentified." (Munke 2009)

"Comprehensive testing by the Environmental Working Group (EWG) reveals a surprising array of chemical contaminants in every bottled water brand analyzed…Our tests strongly indicate that the purity of bottled water cannot be trusted." (Naidenko 2008)

"In our waters, microbeads persist for decades, acting as sponges for toxic chemical pollutants. Mistaken for food by aquatic organisms, microbeads serve as a pathway for pollutants to enter the food chain and contaminate the fish and wildlife we eat…Plastic debris accumulates pollutants such as PCBs up to 100,000 to 1,000,000 times the levels found in seawater." (Nalbone 2014).

"New nanotechnology consumer products emerge at a rate of three to four per week…New emerging nanotechnology applications will affect nearly every type of manufactured product through the middle of the next decade, becoming incorporated into 15% of global manufacturing output, totaling $2.6 trillion in 2014." (National Institute for Occupational Safety and Health 2009)

"Annual production of plastics topped 265 million tons in 2010 with an expected 40% increase in consumption per capita worldwide by 2015…It has been estimated that 10%

of globally produced plastics in 1997 ended up as plastic oceanic waste. If these estimates are correct and these trends continue, an estimated 38 million tons of debris will enter the marine environment in 2015 alone… Policies to reduce the amount of plastics entering consumer and manufacturing markets are currently unlikely to gain traction." (National Oceanic and Atmospheric Administratin Marine Debris Program 2014)

"~75-80% of the tracked increase in autism since 1988 is due to an actual increase in the disorder rather than to changing diagnostic criteria…Among the suspected toxins surveyed, polybrominated diphenyl ethers, aluminum adjuvants, and the herbicide glyphosate have increasing trends that correlate positively to the rise in autism." (Nevison 2014)

"Obesity is quickly becoming a significant human health crisis because it is reaching epidemic proportions worldwide…The obesity epidemic coincided with the marked increase in use of industrial chemicals in the environment over the past 40 years." (Newbold 2008)

Data "support an association of endocrine disrupting chemicals, such as diethylstilbestrol, bisphenol A, phytoestrogens, phthalates, and organotins, with the development of obesity." (Newbold 2010)

"There is a considerably higher amount of small plastic particles when using an 80μm mesh to concentrate the water samples. Up to 100,000 times higher concentrations of small plastic fibers was retained on a 80μm mesh compared to a 450μm mesh." (Norén 2008)

Obesity levels for adults in the United States was 33.9% to 35.6% in 2013, the highest in the world. (OECD 2014)

"Most plasticizers appear to act by interfering with the functioning of various hormone systems, but some phthalates have wider pathways of disruption." (Oehlmann 2009)

"Biological behaviour of NPs and their effects on living organisms can become totally different when particle size decreases." (Ostiguy 2008)

"Fish and consequently fishmeal and fish oil has been identified as one of the most important contributors to the level of dioxins and DL-PCBs in food and feed products." (Oterhals 2011)

"It has been concluded that humans are exposed to toxic compounds via diet in a much higher degree compared to other exposure routes such as inhalation and dermal exposure…All US government pesticides datasets showed that persistent OCP [Organochlorine pesticide] residues were surprisingly common in certain foods despite being off the market for over 30 years… About one quarter of samples of organically labelled fresh produce contained pesticides residues, compared with about three quarters of conventional samples." (Panseri 2013)

"During body weight loss, lipid mobilization and a decrease in fat mass result in increased concentrations of organochlorines in plasma and adipose tissue." (Pelletier 2003)

"Carbon nanotubes have distinctive characteristics, but their needle-like fibre shape has been compared to asbestos, raising concerns that widespread use of carbon nanotubes may lead to mesothelioma, cancer of the lining of the lungs caused by exposure to asbestos." (Poland 2008)

"Airborne particles are covered with various contaminants, and have been found to penetrate the subcellular environment and induce oxidative stress and mitochondrial damage *in vitro*." (Raz 2014)

"The deep-water Oculina coral reef ecosystem is unique and exists solely off the east coast of central Florida…Submersible and ROV surveys conducted from 2001 to 2006 suggest that much of the Oculina habitat has been reduced to rubble by bottom trawling which unfortunately is a trend for deep-water reefs worldwide." (Reed 2007)

"Feeding on plastic biofilm is not restricted to zooplankton, and possibly occurs with rafting organisms such as amphipods, gastropods, and chitons, which are known to associate with floating debris such as plastics." (Reisser 2014a)

"Terrestrial landmasses are conspicuously empty on maps of global microplastic distribution: they have simply not been studied…Very small particles or fibers could be spread further by becoming air-borne (for example from landfills, or other surface deposits) and then enter terrestrial systems and the soil through atmospheric deposition." (Rillig 2012)

"Plastics are primarily synthetic organic polymers derived from petroleum. When exposed to UV radiation in sunlight, these polymers break into smaller and smaller pieces, but they are still present as plastic, and they are not biodegradable in any practical human scale of time." (Rios 2007)

"E-waste comprises discarded electronic appliances, of which computers and mobile telephones are disproportionately abundant because of their short lifespan…Most E-waste[is] being produced in Europe, the United States and Australia. China, Eastern Europe and Latin America will become major E-waste producers in the next ten years… Burning E-waste may generate dioxins, furans, polycyclic aromatic hydrocarbons (PAHs), polyhalogenated aromatic hydrocarbons (PHAHs), and hydrogen chloride…In 2006, the world's production of E-waste was estimated at 20–50 million tonnes per year." (Robinson 2009)

"Small plastic particles are hazardous and in addition they sorb hazardous chemicals. Thus, aquatic plastic debris is unique to other materials that accumulate priority pollutants, like sediments and algae, because of the combination of plastic with sorbed chemicals. This 'Cocktail of contaminants' may cause effects beyond those caused from each contaminant alone." (Rochman 2013a)

"Largely because of a rapidly growing reliance on fossil fuels and industrialized forms of agriculture, human activities have reached a level that could damage the systems that keep Earth in the desirable Holocene state…We have found nine such processes for which we believe it is necessary to define planetary boundaries: climate change; rate of biodiversity loss (terrestrial and marine); interference with the nitrogen and phosphorus cycles; stratospheric ozone depletion; ocean acidification; global fresh water use; change in land use; chemical pollution; and atmospheric aerosol loading." (Rockström 2009a)

119,557 commercial chemicals, including pesticides, biocides, and pharmaceuticals were evaluated as persistent (P), bioaccumulative (B), or very bioaccumulative (VB), and as persistent organic pollutants (POPs) in this detailed survey. Of these chemicals, the report indicates 64,721 chemicals were "substances for which P- and B-score score could be calculated." (Rorije 2011)

"Bacteria and their consumers are generally important components of energy flow and nutrient cycling in a wide variety of aquatic ecosystems…heterotrophic nanoplanktonic protists (2 to 20 µm microorganisms, primarily flagellated protozoa) have been implicated as the major grazers of bacteria in most pelagic freshwater and marine communities." (Sanders 1992)

"We ranked 2986 different pharmaceutical compounds in 51 classes relative to hazard toward algae, daphnids, and fish…Modifying additives were the most toxic classes. Cardiovascular, gastrointestinal, antiviral, anxiolytic sedatives, hypnotics and antipsychotics, corticosteroid, and thyroid pharmaceuticals were the predicted most hazardous therapeutic classes." (Sanderson 2004)

"The soot when generated [from burning waste plastics] is accompanied with volatile organic compounds (VOCs), semi-VOCs, smoke (particulate matter), particulate bound heavy metals, polycyclic aromatic hydrocarbons (PAHs), polychlorinated dibenzofurans (PCDFs) and dioxins and has the ability to travel thousands of kilometers." (Saskatchewan Ministry of Environment 2012)

"Rorije et al. (2011) identified, in a set of 65,000 industrial chemicals, pharmaceuticals, pesticides and biocides, almost 2,000 substances that may fulfill the persistence and bioaccumulation criteria of the Stockholm Convention…Similarly, Strempel et al. (2012) identified 3,000 to 5,000 potential PBT chemicals in a set of 95,000 industrial chemicals…Our database of 93,144 substances does not contain all relevant types of chemicals." (Scheringer 2012)

"Ingestion of soil particles from environmentally contaminated areas may contribute to elevated dioxin levels in free-range chicken eggs." (Schoeters 2006)

"Antidepressant pharmaceuticals are widely prescribed in the United States; release of municipal wastewater effluent is a primary route introducing them to aquatic environments, where little is known about their distribution and fate." (Schultz 2010)

"Some eight million metric tons of plastic waste makes its way into the world's oceans each year, and the amount of the debris is likely to increase greatly over the next decade." (Schwartz 2015)

"Reproductive disorders of newborn (cryptorchidism, hypospadias) and young adult males (low sperm counts, testicular germ cell cancer) are common and/or increasing in incidence…It has been hypothesized that these disorders may comprise a testicular dysgenesis syndrome (TDS) with a common origin in fetal life." (Sharpe 2008)

"PCBs, dioxins, and mercury (Hg) are prevalent in Maine's marine environment and are of concern because of their documented immune and endocrine-disrupting potential in seals, other marine wildlife, and humans." (Shaw 2002)

"Brominated flame retardants, especially the polybrominated diphenyl ethers (PBDEs), are ubiquitous persistent organic pollutants (POPs) that biomagnify and are associated with endocrine-disrupting and neurodevelopmental effects in rodent studies…Concern has increased about human health risks in tandem with evidence of rising PBDE concentrations in human breast milk, particularly in the United States." (Shaw 2005a)

"Recent studies have shown that concentrations of PCBs, dioxins, and other persistent organic pollutants (POPs) can be significantly higher in farm-raised salmon than in wild salmon." (Shaw 2005b)

"PCBs, DDTs, and CHLs were the major persistent organochlorines in harbor seal blubber…DDT and PCB concentrations have declined from the high levels reported in the early 1970s, but no declines were observed in our samples over the ten-year period 1991–2001." (Shaw 2005c)

"HBCD [Hexabromocyclododecane] was detected in 87% of the fish samples at concentrations ranging from 2.4 to 38.1 ng/g, lw (overall mean 17.2±10.2 ng/g, lw)

…Biomagnification factors (BMFs) from fish to seals averaged from 17 to 76, indicating that tetra- to hexa-BDEs are highly biomagnified in this marine foodweb." (Shaw 2009b)

"U.S. and EU regulatory bodies have concluded that nanoparticles—especially those smaller than 30 nm—have the potential to pose an entirely new health risk and that it is necessary to carry out an extensive analysis of such risk." (Sherman 2012)

"Just one computer can contain hundreds of chemicals, including lead, mercury, cadmium, brominated flame retardants (BFRs) and polyvinyl chloride (PVC)." (Silicon Valley Toxics Coalition 2015)

"Exposure to phthalates in the United States is widespread. We found measurable concentrations of MEP, MBP, and MBzP in > 97% of the samples tested." (Silva 2004)

"Many of the engineered nanomaterials assessed were found to cause genotoxic responses, such as chromosomal fragmentation, DNA strand breakages, point mutations, oxidative DNA adducts and alterations in gene expression profiles." (Sing 2009)

"Fetal exposure to environmental oestrogens may play a role in the increased incidence of breast cancer… There is widespread human exposure to bisphenol A, an oestrogenic compound that leaches from dental materials and consumer products." (Soto 2008)

"Aside from plastic which has been incinerated, some scientists believe it is plausible that all the plastic ever created since its invention in the late 1940s still exists on the planet, either buried in landfills, buried on shorelines, floating in the ocean, or on the ocean floor." (Stevenson 2011)

"Certain chemicals leached from plastic may have contributed to the huge die-off of American lobster in western Long Island Sound in the last decade. Researchers found that lobsters in the western Long Island Sound, the south shore of Massachusetts and Cape Cod Bay are contaminated with alkyphenols, used commonly in plastic and rubber manufacturing." (Stevenson 2011)

"Mercury intake from food is .03-1.5 ug/kg body weight/week. The food with the largest amount of methyl mercury is fish and seafood products, and methyl mercury is absorbed 95-100% from the intestinal tract." (Takagi 2013)

"Components used in plastics, such as phthalates, bisphenol A (BPA), polybrominated diphenyl ethers (PBDE) and tetrabromobisphenol A (TBBPA), are detected in humans…BPA is one of the highest production volume chemicals in commerce, with over 6 billion pounds produced in 2003." (Talsness 2009)

"TBBPA is the classical halogenated flame retardant chemically bonded to epoxy and polycarbonate resins. It is present in printed circuit boards and casings used in personal computers, printers, fax machines and copiers." (Talsness 2009)

"Model calculations and experimental observations consistently show that polyethylene accumulates more organic contaminants than other plastics." (Teuten 2009)

"Around 4 percent of the world oil production is used as a feedstock to make plastics and a similar amount is used as energy in the process. Yet over a third of current production is used to make items of packaging, which are then rapidly discarded…this linear use of hydrocarbons, via packaging and other short-lived applications of plastic, is simply not sustainable." (Thompson 2009a)

"Plastic items fragment in the environment because of exposure to UV light and abrasion, such that smaller and smaller particles form…but the resulting material does not necessarily biodegrade…Microplastics have a relatively large surface area to volume ratio and are therefore have greater capacity to facilitate the transport of contaminants…but due to limitations in analytical methods, the abundance of smaller fragments is unknown." (Thompson 2013)

"Po-210 is relatively long-lived fallout from the decay of radon in…uranium-contaminated calcium phosphate fertilizer used on tobacco fields…Uranium has a very long half-life and will accumulate in the soil with repeated applications of fertilizer. As a result, modern cigarettes may contain higher levels of Po-210 than those measured 40 years ago." (Tidd 2008)

"Relatively high concentrations of PCBs, CHLs, HCHs, and HCB were also observed in [skipjack tuna] samples collected from some locations in the middle of the Pacific Ocean, indicating the expansion of OC contamination on a global scale." (Ueno 2003)

"Dioxins are produced primarily during the incineration or burning of waste; the bleaching processes used in pulp and paper mills; and the chemical syntheses of trichlorophenoxyacetic acid, hexachlorophene, vinyl chloride, trichlorophenol, and pentachlorophenol." (US CDC 2009)

"Ground water responds more slowly than stream water to changes in pesticide use. A persistent pesticide or degradate can remain in ground water long after its use is discontinued because of the slow rates of ground-water flow and the resulting long residence time of water and pesticides in ground-water flow systems." (US Geological Survey 2006)

"There appears to be a considerable proportion of the manufactured plastic that is unaccounted for in surveys tracking the fate of environmental plastics." (Woodall 2015)

"Certain polychlorinated biphenyls, organochlorine pesticides, PFCs, phenols, PBDEs, phthalates, polycyclic aromatic hydrocarbons, and perchlorate were detected in 99–100% of pregnant women." (Woodruff 2011)

"Embryonic exposure to certain marine toxins or toxicants alters epigenetic programming, leading to long-term effects on gene expression in adult tissues and ultimately contributing to altered neurobehavioral function in adults." (Woods Hole Oceanographic Institution 2014)

The progeny of pyrotechnic-petrochemical-nuclear technology is a global consumer society that wages chemical warfare on the biosphere. "Microplastics are liable to concentrate hydrophobic persistent organic pollutants (POPs), which have a greater affinity for the hydrophobic surface of plastic compared to seawater. Due to their large surface area to volume ratio, microplastics can become heavily contaminated up to six orders of magnitude greater than ambient seawater with waterborne POPs." (Wright 2013)

"Some commercially important fish and their larvae are visual predators, preying on small zooplankton, and may feed on microplastics which most resemble their prey i.e. white, tan and yellow plastic… Microplastic ingestion due to food resemblance may also be applicable to pelagic invertebrates, which are visual raptorial predators." (Wright 2013)

"As plastic continues to fragment, the potential for it to accumulate within the circulatory fluid and phagocytic cells of an organism is likely to increase, as the smaller the microplastics, the greater the abundance available for translocation." (Wright 2013)

"The widespread occurrence of neuro-active pharmaceuticals and metabolites in Minnesota effluents and surface waters indicate that this is likely a global environmental issue." (Writer 2013)

"Due to its unique chemical composition, the upper organic film of the SML [Sea surface microlayer] represents both a sink and a source for a range of pollutants including chlorinated hydrocarbons, organotin compounds, petroleum hydrocarbons, polycyclic aromatic hydrocarbons (PAH) and heavy metals…These pollutants can be enriched in the SML by up to 500 times relative to concentrations occurring in the underlying bulk water column…The SML is also a unique ecosystem, serving as an important habitat for fish eggs and larvae…Due to its unique chemical composition—in particular, its high content of lipids, fatty acids and protein…the SML plays an important role in the fate of persistent organic pollutants (POPs) in aqueous ecosystems…Sources of SML contaminants in the marine environment can be mainly attributed to terrestrially derived wastewater discharges, agricultural and industrial runoff, atmospheric deposition of combustion residues, and shipping activities." (Wurl 2004)

"Hazardous materials that are released as the result of a technologic malfunction precipitated by a natural event are referred to as natural-technologic or na-tech events...Disaster-associated hazardous material releases are of concern, given increases in population density and accelerating industrial development in areas subject to natural disasters. These trends increase the probability of catastrophic future disasters and the potential for mass human exposure to hazardous materials released during disasters." (Young 2004)

The imposition of human ecology on natural ecology has resulted in a biosphere in crisis. The number of self-sustaining ecologically integrated human communities are now declining rapidly in a world of social and political turmoil, declining economic opportunities, infrastructure collapse, mass migration and refugee flight.

Urban Areas as Reservoirs of Environmental Chemicals

If we return to our GSA satellite perch and take another look at our 454 urban areas with over 1 million inhabitants, we observe many of these urban areas are shrouded in vast clouds of effluents derived from industrial activities and transportation networks, especially diesel trucks and the ubiquitous automobile. Urban environments are the subject of widespread monitoring with respect to air pollution particulate size (PM), the reporting units for which range from PM_{10} (10 μm) to ultrafine $PM_{0.1}$ (±100 nm). As noted, associated sorbed chemicals are not often mentioned in media reports on smog pollution events, which are growing in duration and intensity in countries like India, China, and elsewhere. Growing populations, more affluent lifestyles, and access to automobiles are now resulting in increasing levels of air pollution in most urban areas.

Tianjin, China factory explosion, satellite image, 2015-08-13. Image by EPA.

From our GSA satellite overview we can also observe vast landscapes including massive landfills contaminated with in excess of 100,000 types of anthropogenic (manmade) environmental chemicals. Most of these urban areas are located along rivers that drain into the marine environment. In an age of increasing storm intensity, only a few drought stricken areas will not experience washover-washout events of increasing frequency. Such events have included hurricanes Katrina and Sandy, as well as the recent el Nino derived flooding in South Carolina, Texas, Missouri, Northern England, and other locations. Every washover (flooding) event is characterized by remobilized environmental chemicals and other water-born contaminants that are then transported by water in our hemispheric water cycle. The primary destination of these ecotoxins is the marine environment; terrestrial aquatic environments such as the Great Lakes are secondary destinations. Underground aquifers are an unseen tertiary destination now being resupplied with contaminated surface water. The remobilization of environmental chemicals by washover-washout events in highly contaminated urban and industrial landscapes is a key component of ecotoxin transport throughout the biosphere. Nanoparticles, including plastic nanoparticles, are also present in washover-washout events, also eventually reaching all microbiomes where they are components of ever-growing anthropogenic debris fields and are then recycled to higher trophic levels of the biosphere.

There is now a broad public awareness of the unfortunate environmental changes occurring in our round world biosphere in crisis, few of which can be mitigated by well-intentioned meetings such as the recent agreement in Paris (November 30th to December 12th 2015) to limit CO_2 emissions.

Airspace over Mumbai, September 5th, 2010. Image by P.P. Yoonus.

The Legacy of Human Ecology

The following biosphere models are sketches of the imposition of human activities on natural ecosystems. They begin with biogeochemical transport pathways and include sketches of important anthropogenic ecotoxins and the environmental, social and political context of their dispersal. Over 200,000 types of petrochemical effluents are now produced at the rate of tens of billions of tons per year and are continuously dispersed in the atmospheric, terrestrial and aquatic environments of a finite, vulnerable biosphere. These round world biosphere sketches illustrate the cumulative impact of 5,000 years of human civilization, the legacy of humanity in the anthropocene.

Part 1: Pathways of Nanotoxin Translocation

Part 2: Dynamics and Petrochemistry of Biocatastrophe

Part 3: Socio-political Context of Biocatastrophe

Part 1: Pathways of Nanotoxin Translocation

Model 1: Pathway analysis for radionuclides to human consumption, circa 1965.

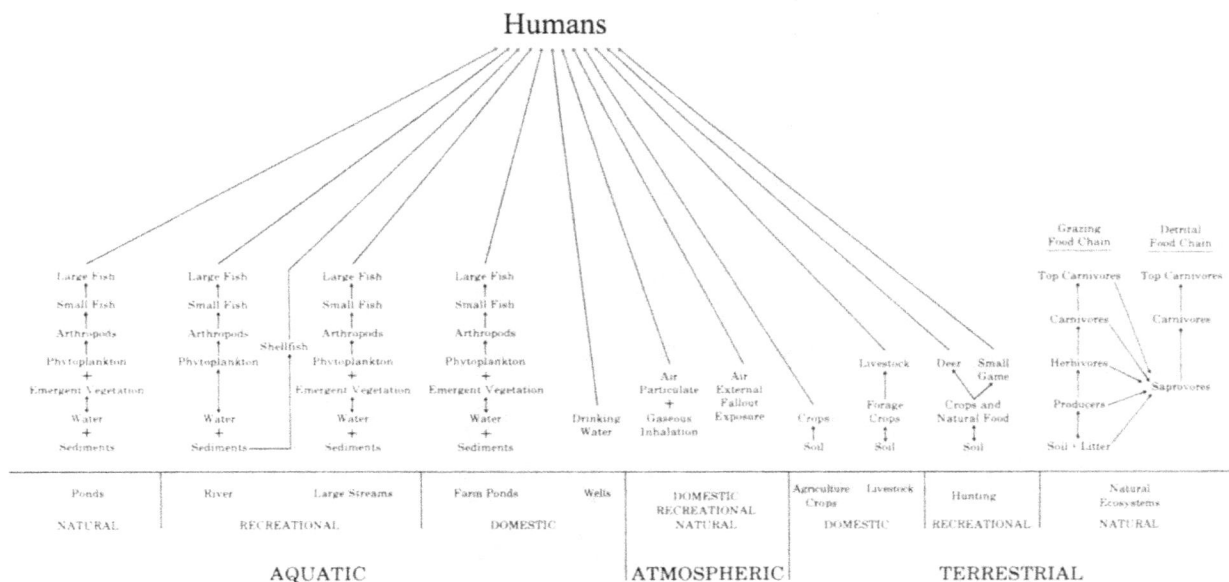

1 Adapted from Platt, et al. "Empirical benefits derived from an Ecosystem Approach to Environmental Monitoring of a Nuclear Fuel Reprocessing Plant" IAEA-SM-172/31, B-268, p.678.

Model 2: The Biosphere and its Environments

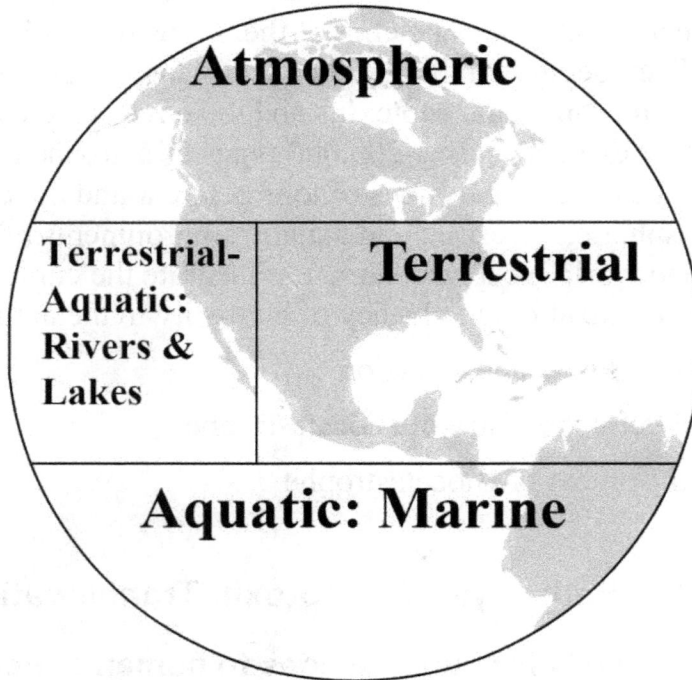

Atmospheric

| Terrestrial-Aquatic: Rivers & Lakes | Terrestrial |

Aquatic: Marine

Model 3: Biogenesis

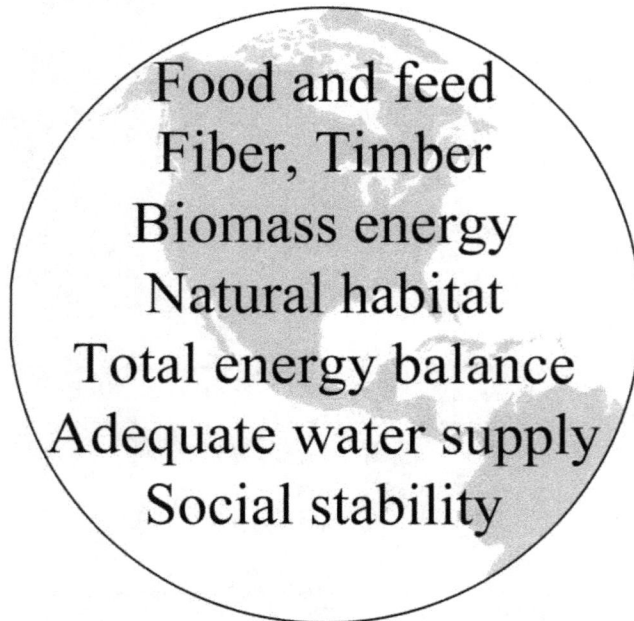

Food and feed
Fiber, Timber
Biomass energy
Natural habitat
Total energy balance
Adequate water supply
Social stability

Model 4: Terrestrial Ecosystems Dynamics

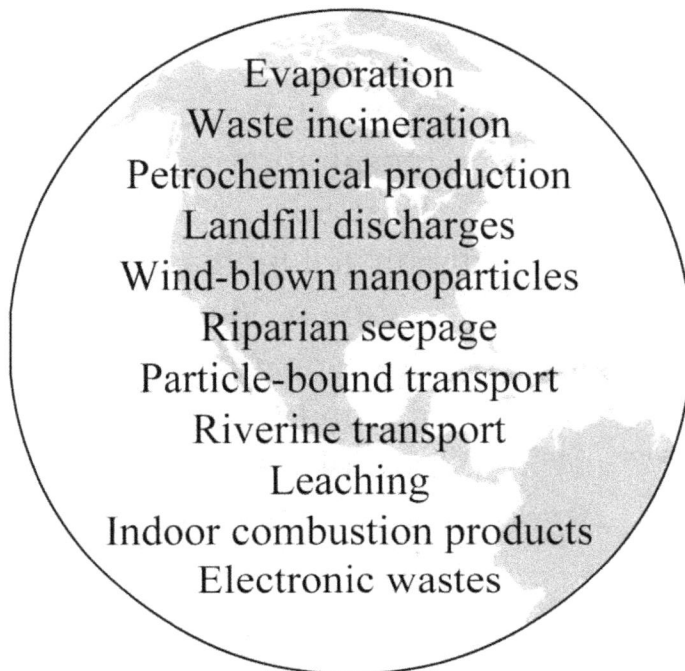

Evaporation
Waste incineration
Petrochemical production
Landfill discharges
Wind-blown nanoparticles
Riparian seepage
Particle-bound transport
Riverine transport
Leaching
Indoor combustion products
Electronic wastes

Model 5: Chemical fallout Pathways

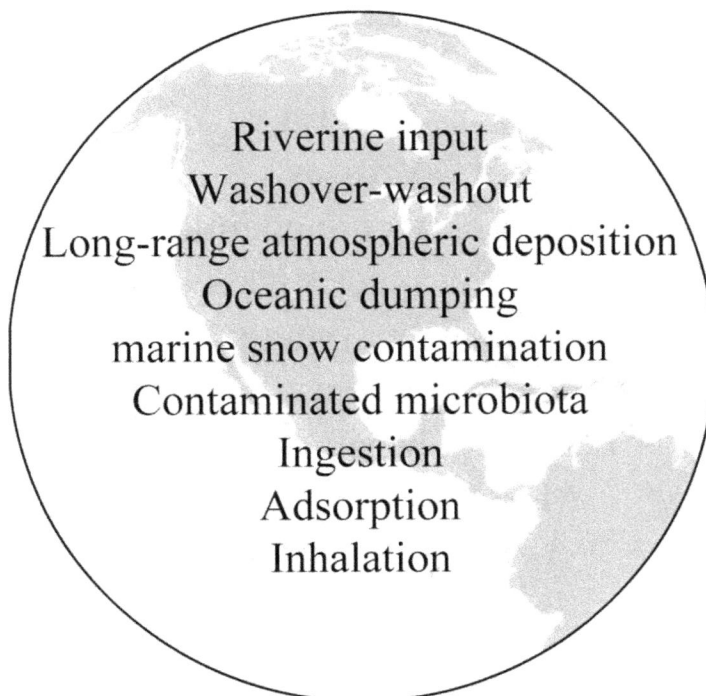

Riverine input
Washover-washout
Long-range atmospheric deposition
Oceanic dumping
marine snow contamination
Contaminated microbiota
Ingestion
Adsorption
Inhalation

Model 6: Atmospheric Transport Dynamics

Evaporation
Transpiration
Dust/Soot
Contaminated biomass
Combustion products
Long range atmospheric transport
Wind driven transport
Hydrolysis
Air-water-snow-gas exchange

Model 7: Contaminants in Terrestrial Water Supplies

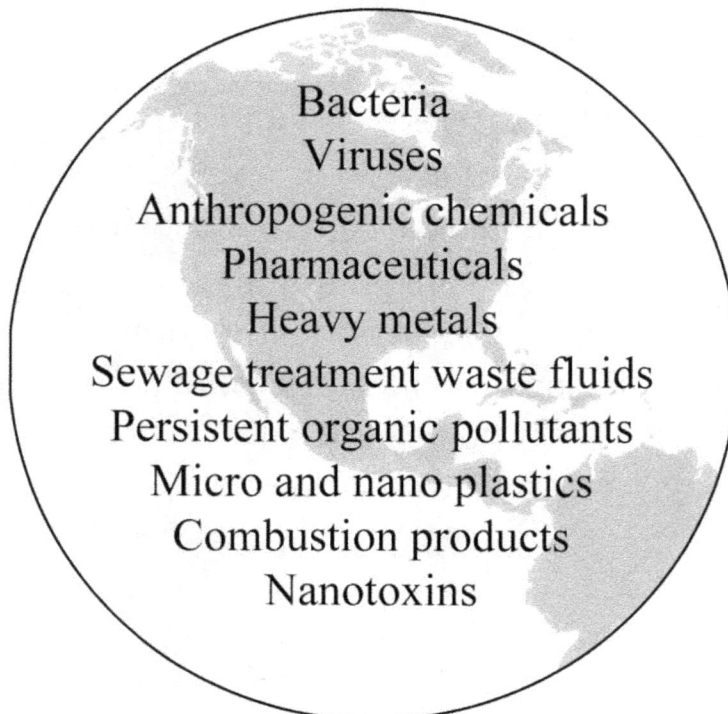

Bacteria
Viruses
Anthropogenic chemicals
Pharmaceuticals
Heavy metals
Sewage treatment waste fluids
Persistent organic pollutants
Micro and nano plastics
Combustion products
Nanotoxins

Model 8: The Marine Environment: Microbial Ecosystems in the Sea Surface Microlayer

MARINE ECOSYSTEM - LEVEL 1 - MICROBIAL FOOD WEB

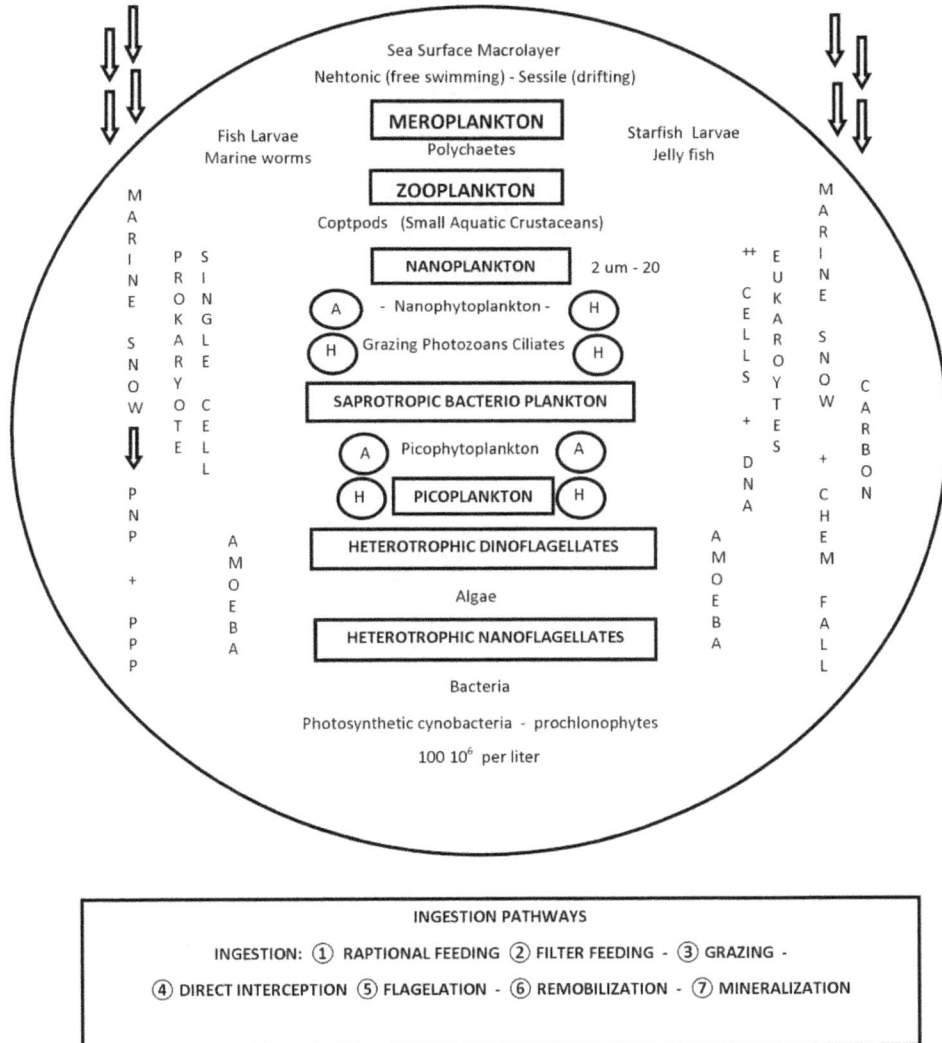

Sea Surface Macrolayer

Nehtonic (free swimming) - Sessile (drifting)

Fish Larvae
Marine worms

Starfish Larvae
Jelly fish

MEROPLANKTON

Polychaetes

ZOOPLANKTON

Coptpods (Small Aquatic Crustaceans)

NANOPLANKTON   2 um - 20

(A)   - Nanophytoplankton -   (H)

(H) Grazing Photozoans Ciliates   (H)

SAPROTROPIC BACTERIO PLANKTON

(A)   Picophytoplankton   (A)

(H) **PICOPLANKTON** (H)

HETEROTROPHIC DINOFLAGELLATES

Algae

HETEROTROPHIC NANOFLAGELLATES

Bacteria

Photosynthetic cynobacteria - prochlonophytes

100 10^6 per liter

Left margin: MARINE SNOW / PNP + PPP

PROKARYOTE

SINGLE CELL

AMOEBA

Right margin: ++ CELLS + DNA / EUKAROYTES / AMOEBA / MARINE SNOW + CHEM FALL / CARBON

INGESTION PATHWAYS

INGESTION: (1) RAPTIONAL FEEDING (2) FILTER FEEDING - (3) GRAZING -

(4) DIRECT INTERCEPTION (5) FLAGELATION - (6) REMOBILIZATION - (7) MINERALIZATION

Judy Weed, Beach Point Shack, Clearwater, FL

Part 2: The Dynamics and Petrochemistry of Biocatastrophe

Model 9: Hydrocarbon Intermediates used for Petrochemical Products

Methane
Ethane, Ethylene, Acetylene
Propane, Propylene, Butane
n-Butenes, Isobutene, Butadiene
Pentanes, Isopentanes, Isoamylenes
Isoprene, Hexanes, Methylpentenes
Benzene, Cyclohexane, Hexanes
Mixed heptenes, Toluene, Di-isobutylene
Xylenes, Ethylbenzene, Styrene
Cumene, Propylene, tetramer
Tri-isobutylene, Dodecylbenzene
n-Olefines, *n*-Paraffins

Model 10: List of Stockholm Convention Chemicals

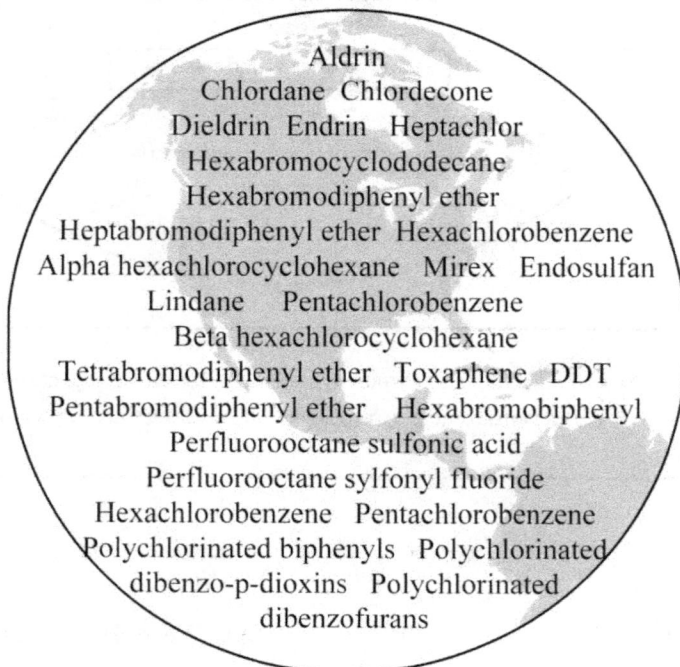

Aldrin
Chlordane Chlordecone
Dieldrin Endrin Heptachlor
Hexabromocyclododecane
Hexabromodiphenyl ether
Heptabromodiphenyl ether Hexachlorobenzene
Alpha hexachlorocyclohexane Mirex Endosulfan
Lindane   Pentachlorobenzene
Beta hexachlorocyclohexane
Tetrabromodiphenyl ether Toxaphene DDT
Pentabromodiphenyl ether   Hexabromobiphenyl
Perfluorooctane sulfonic acid
Perfluorooctane sylfonyl fluoride
Hexachlorobenzene Pentachlorobenzene
Polychlorinated biphenyls Polychlorinated
dibenzo-p-dioxins Polychlorinated
dibenzofurans

Model 11: Chemicals of Concern: HDCs, Carcinogens, etc.

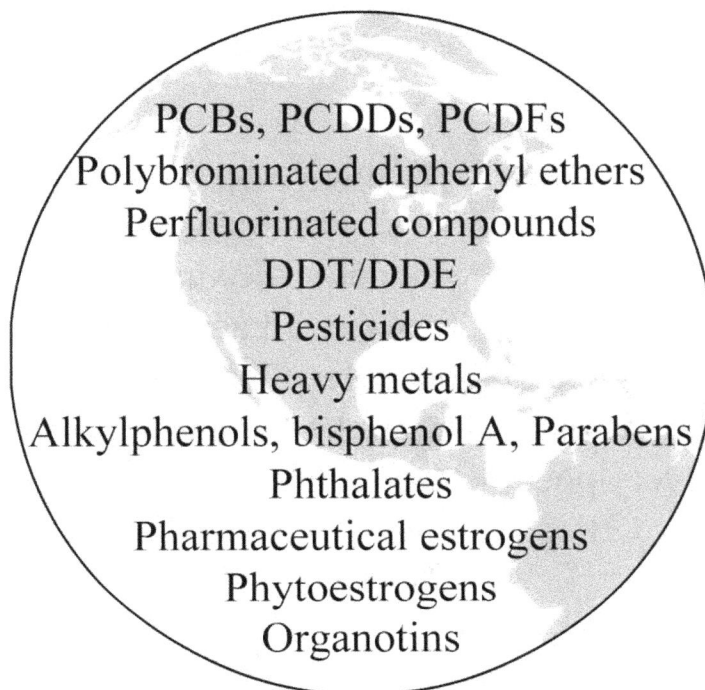

PCBs, PCDDs, PCDFs
Polybrominated diphenyl ethers
Perfluorinated compounds
DDT/DDE
Pesticides
Heavy metals
Alkylphenols, bisphenol A, Parabens
Phthalates
Pharmaceutical estrogens
Phytoestrogens
Organotins

(adapted from World Health Organization 2012)

Model 12: Environmental Working Group's "Dirty Dozen"

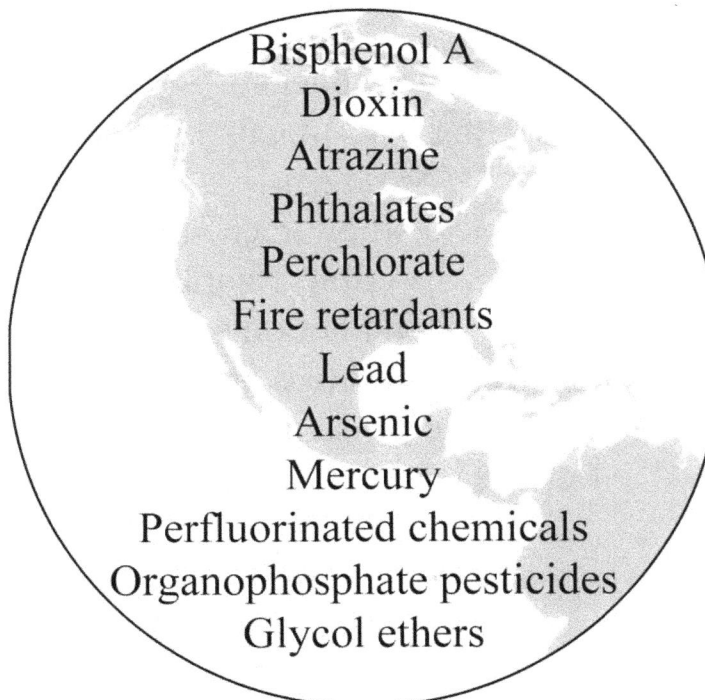

Bisphenol A
Dioxin
Atrazine
Phthalates
Perchlorate
Fire retardants
Lead
Arsenic
Mercury
Perfluorinated chemicals
Organophosphate pesticides
Glycol ethers

Model 13: Contaminants in Aquifers

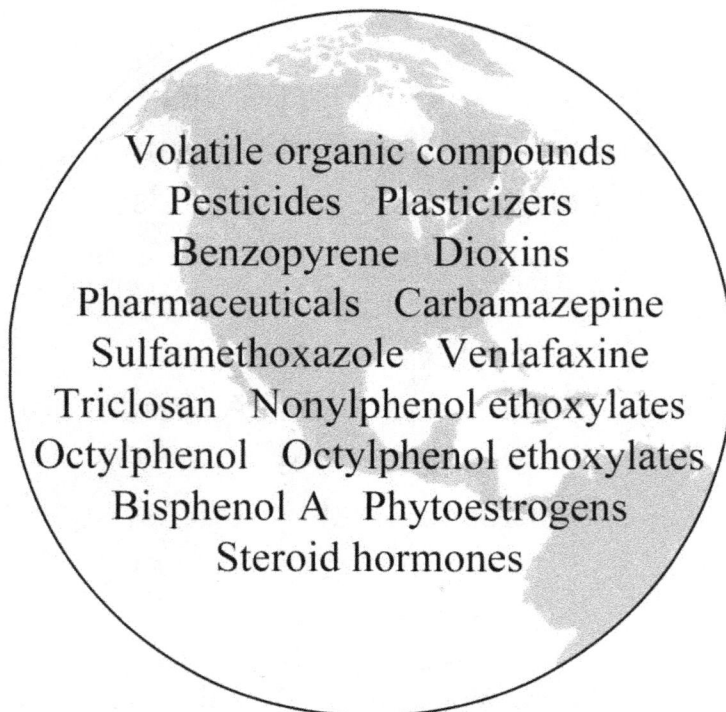

Volatile organic compounds
Pesticides   Plasticizers
Benzopyrene   Dioxins
Pharmaceuticals   Carbamazepine
Sulfamethoxazole   Venlafaxine
Triclosan   Nonylphenol ethoxylates
Octylphenol   Octylphenol ethoxylates
Bisphenol A   Phytoestrogens
Steroid hormones

Model 14: Human Health

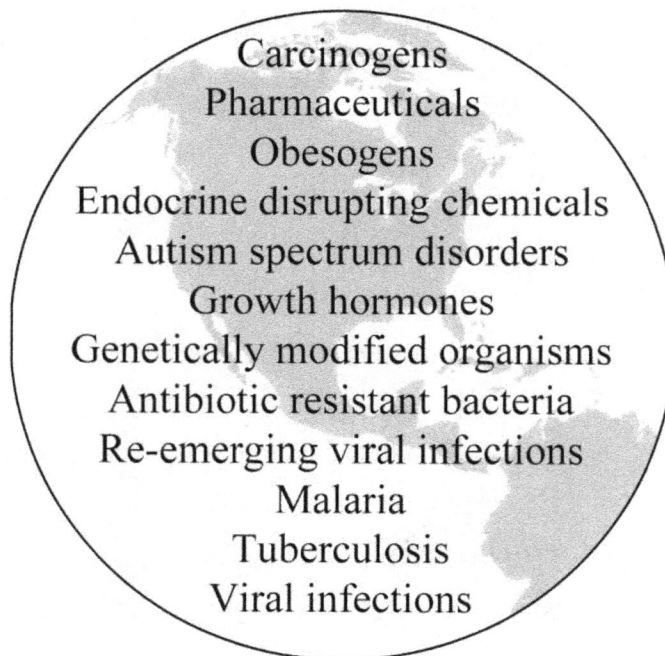

Carcinogens
Pharmaceuticals
Obesogens
Endocrine disrupting chemicals
Autism spectrum disorders
Growth hormones
Genetically modified organisms
Antibiotic resistant bacteria
Re-emerging viral infections
Malaria
Tuberculosis
Viral infections

Model 15: Pharmaceuticals

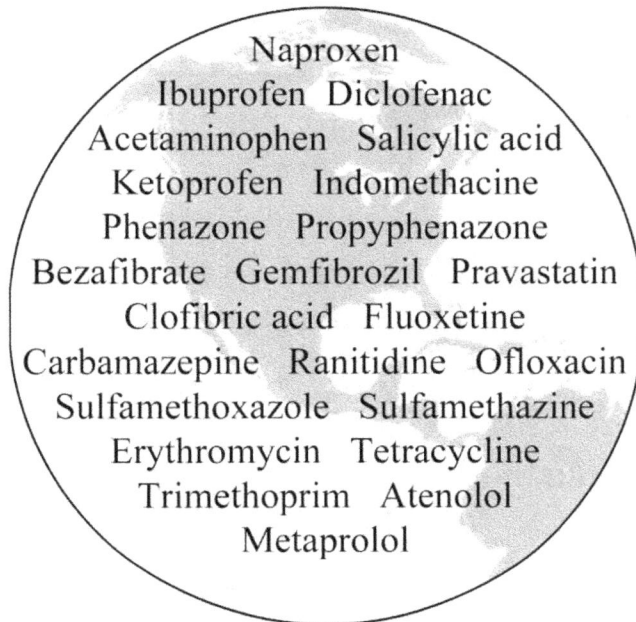

Naproxen
Ibuprofen Diclofenac
Acetaminophen Salicylic acid
Ketoprofen Indomethacine
Phenazone Propyphenazone
Bezafibrate Gemfibrozil Pravastatin
Clofibric acid Fluoxetine
Carbamazepine Ranitidine Ofloxacin
Sulfamethoxazole Sulfamethazine
Erythromycin Tetracycline
Trimethoprim Atenolol
Metaprolol

Most ubiquitous pharmaceuticals detected in the Llobregat River
(adapted from Ginebreda 2009)

Model 16: Known and Suspected Obesogens

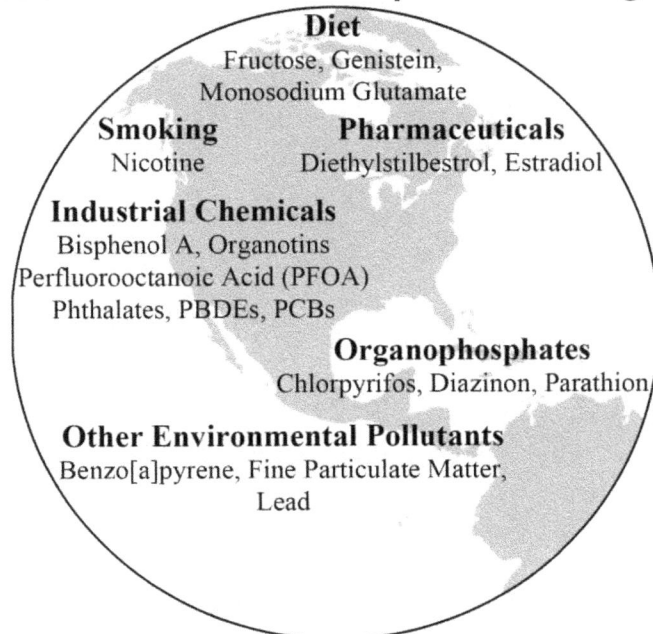

Diet
Fructose, Genistein,
Monosodium Glutamate

Smoking   **Pharmaceuticals**
Nicotine   Diethylstilbestrol, Estradiol

Industrial Chemicals
Bisphenol A, Organotins
Perfluorooctanoic Acid (PFOA)
Phthalates, PBDEs, PCBs

Organophosphates
Chlorpyrifos, Diazinon, Parathion

Other Environmental Pollutants
Benzo[a]pyrene, Fine Particulate Matter,
Lead

(Adapted from Holtcamp 2012)

Model 17: Former End Uses for Various Aroclors (PCBs)

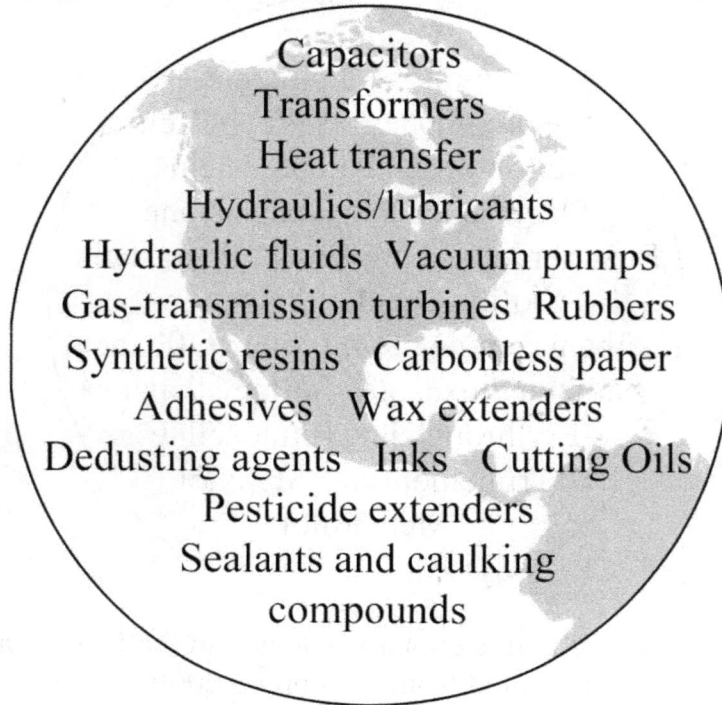

Capacitors
Transformers
Heat transfer
Hydraulics/lubricants
Hydraulic fluids Vacuum pumps
Gas-transmission turbines Rubbers
Synthetic resins   Carbonless paper
Adhesives   Wax extenders
Dedusting agents Inks Cutting Oils
Pesticide extenders
Sealants and caulking
compounds

(Adapted from ATSDR 2015, Table 5-1)

Part 3: Socio-political Context of Biocatastrophe

Model 18: Geosphere as Bank Account

Gold
Silver
Copper
Nickel
Lead
Aluminum
Iron

Rare Earth Elements

Rare Earth Elements

Model 19: Biosphere as Bank Account: Assets

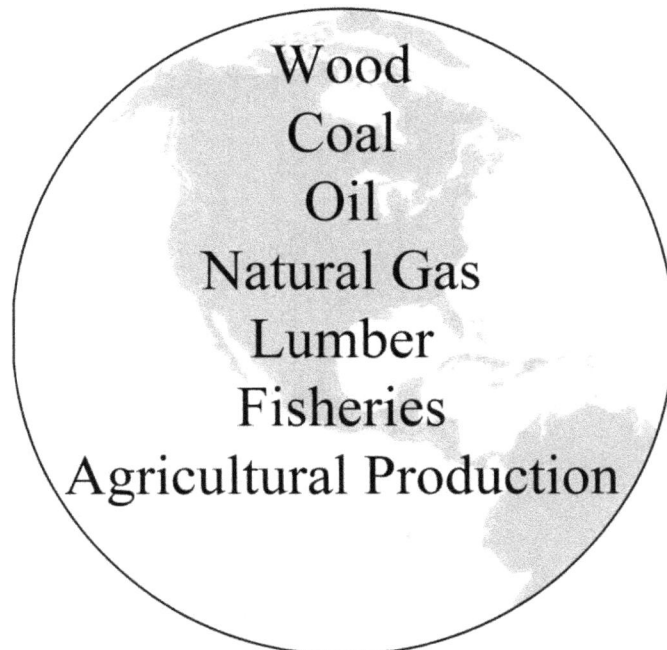

Wood
Coal
Oil
Natural Gas
Lumber
Fisheries
Agricultural Production

Model 20: Biosphere as Bank Account: Withdrawals

Slash
Cut Burn
Hoe Dig Mine
Dam Blast
Frack Shoot
Bomb

Model 21: Phenomenology of Technology Part 1: Age of Pyrotechnology

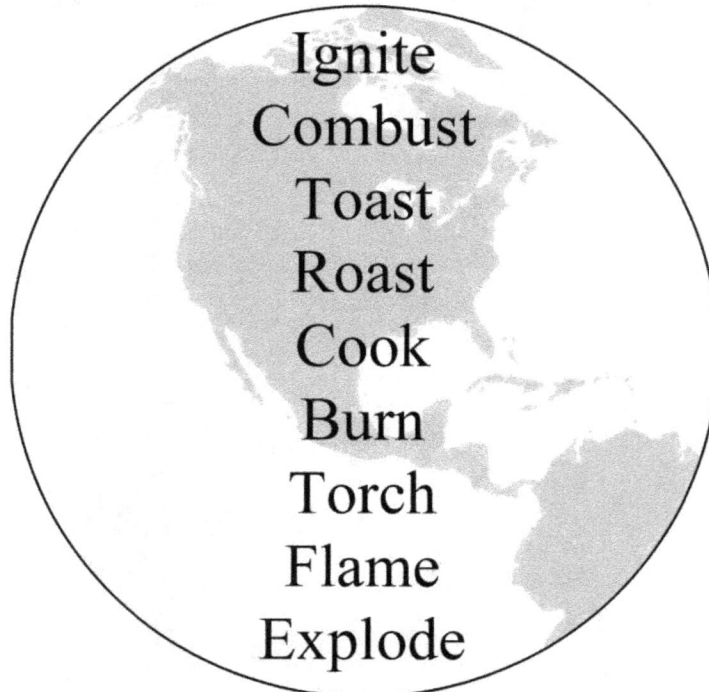

Ignite
Combust
Toast
Roast
Cook
Burn
Torch
Flame
Explode

Model 22: Phenomenology of Technology Part 2: Age of Pyrotechnology Continued

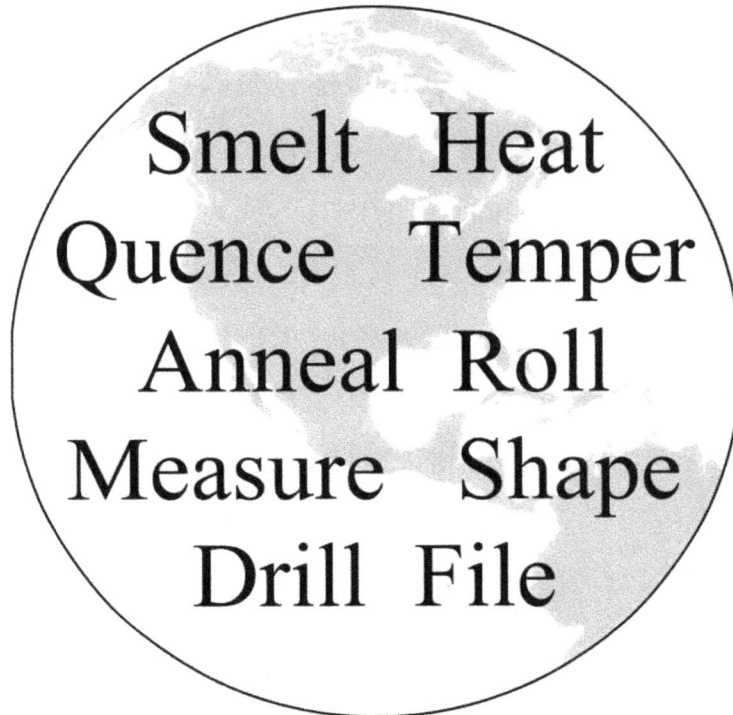

Smelt   Heat
Quence   Temper
Anneal Roll
Measure   Shape
Drill File

Model 23: Phenomenology of Technology Part 3: Urbanization

Dig Pave
Pour Fabricate
Construct
Seal Paint
Use Demolish
Crush Spill
Discard

Model 24: Phenomenology of Technology Part 4: Cascading Industrial Revolutions

Steam engines
Gasoline engines
Petrochemistry
Telecommunications
The age of plastic
Nuclear weapons
Age of information technology
Nanotechnology

Model 25: Politics in the Age of Biocatastrophe

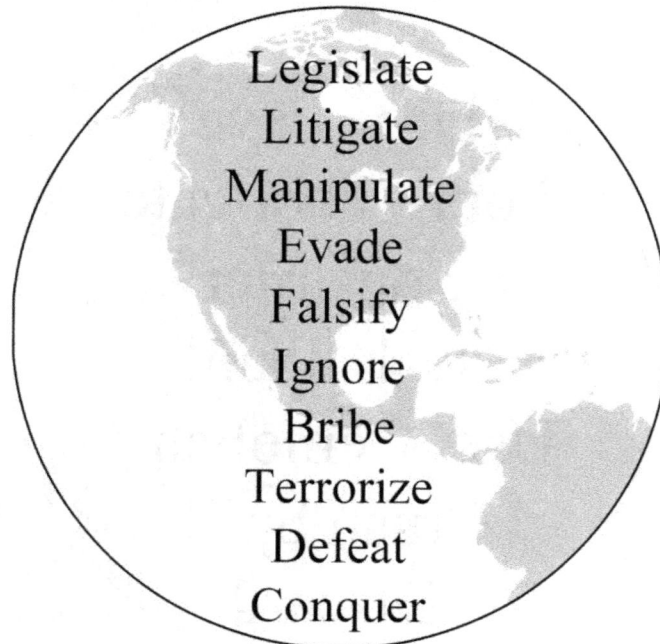

Legislate
Litigate
Manipulate
Evade
Falsify
Ignore
Bribe
Terrorize
Defeat
Conquer

Model 26: Sociology of the Age of Biocatastrophe

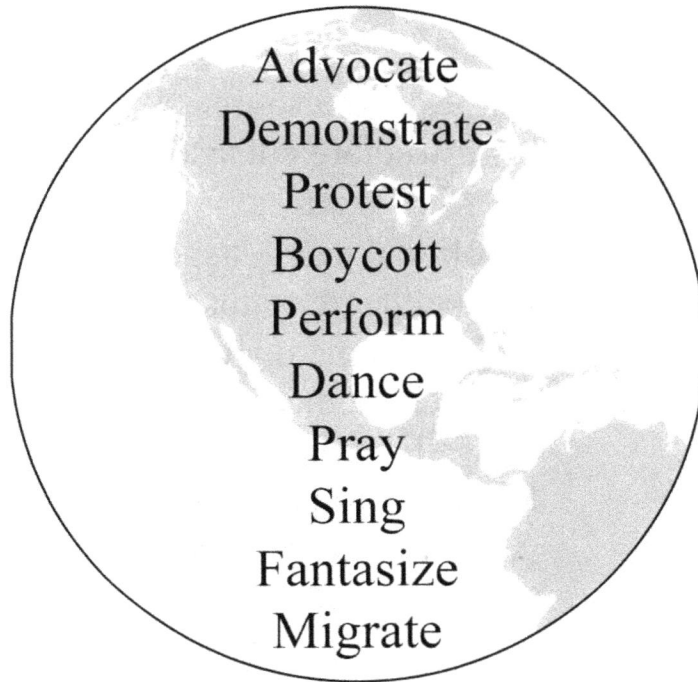

Advocate
Demonstrate
Protest
Boycott
Perform
Dance
Pray
Sing
Fantasize
Migrate

Model 27: Parasitic Shadow Banking Network

Global warfare
Worldwide chemical fallout
Global consumer society
Shadow banking network
Bankrupt pension funds
Income inequality
World indebtedness
Declining public resources

Model 28: Sociopolitical Instability

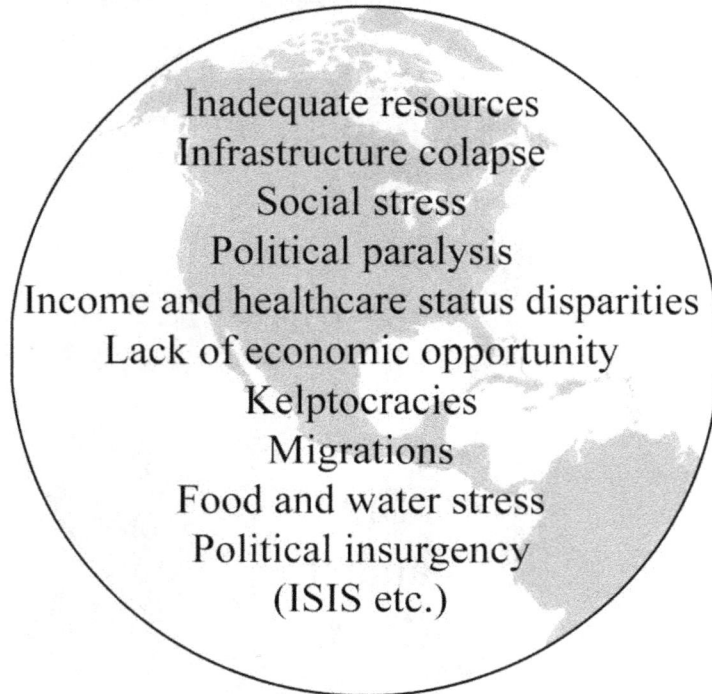

Inadequate resources
Infrastructure colapse
Social stress
Political paralysis
Income and healthcare status disparities
Lack of economic opportunity
Kelptocracies
Migrations
Food and water stress
Political insurgency
(ISIS etc.)

Model 29: Indebtedness

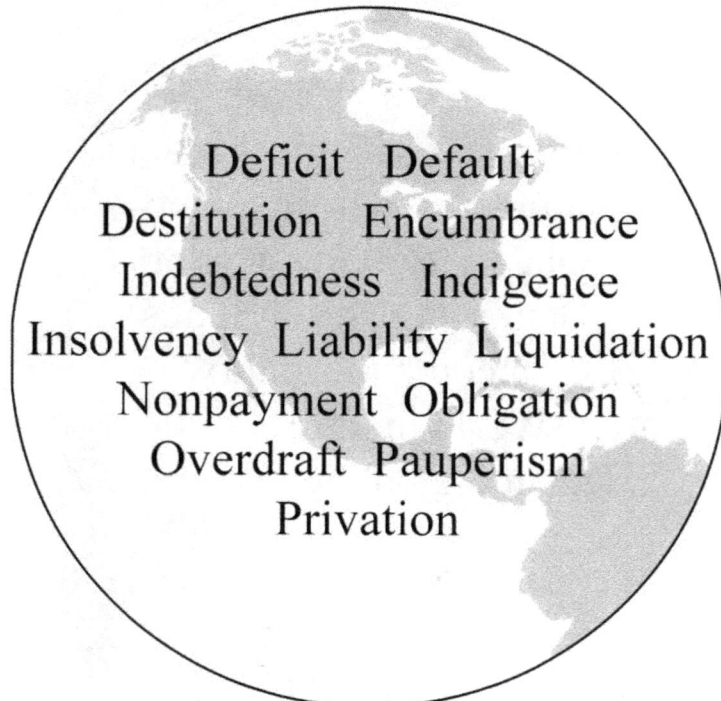

Deficit   Default
Destitution   Encumbrance
Indebtedness   Indigence
Insolvency Liability Liquidation
Nonpayment Obligation
Overdraft Pauperism
Privation

Model 30: Biocatastrophe: The Imposition of Human Ecosystems on Natural Ecosystems

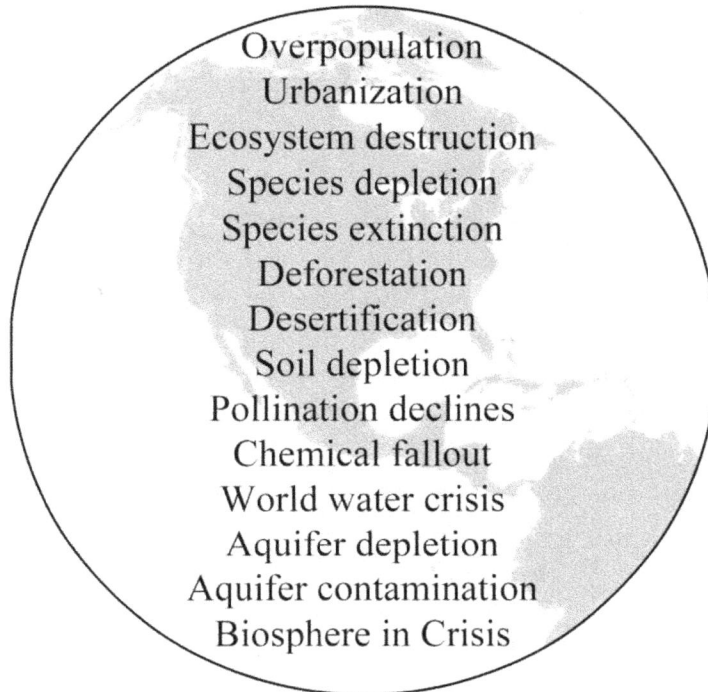

Overpopulation
Urbanization
Ecosystem destruction
Species depletion
Species extinction
Deforestation
Desertification
Soil depletion
Pollination declines
Chemical fallout
World water crisis
Aquifer depletion
Aquifer contamination
Biosphere in Crisis

Model 31:
Footnote to Biocatastrophe: Cataclysmic Climate Change

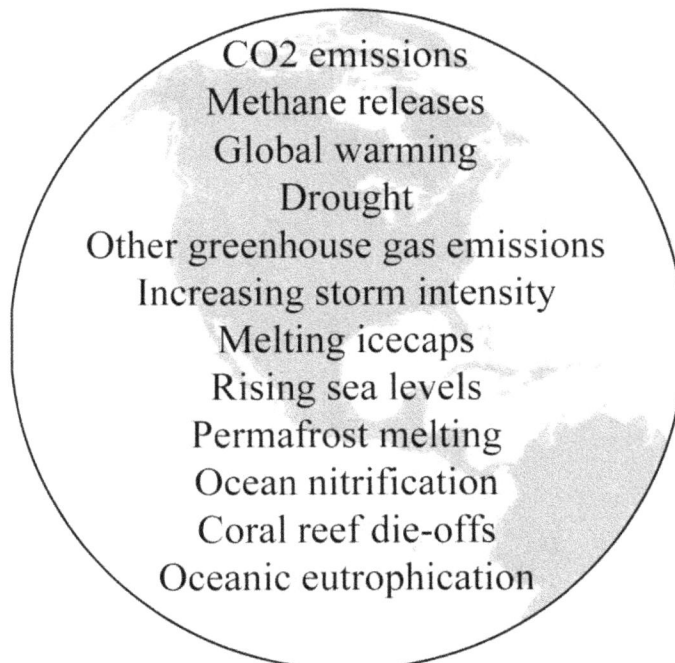

CO2 emissions
Methane releases
Global warming
Drought
Other greenhouse gas emissions
Increasing storm intensity
Melting icecaps
Rising sea levels
Permafrost melting
Ocean nitrification
Coral reef die-offs
Oceanic eutrophication

Postscript

A massive variety of anthropogenic environmental chemicals are now moving throughout the biosphere. All trophic levels of both terrestrial and marine ecosystems are now contaminated with ecotoxins derived from our pyrotechnic-petrochemical lifestyles. The National Security Agency (NSA) has compiled massive classified databases documenting environmental chemicals in pathways to human consumption in all media – air, water, soil, agricultural produce, meat, fish, and drinking water. Nanoparticles are only one of many vectors for their movement throughout the biosphere. The president, Congress, and the NSA have a responsibility to declassify this vast database on environmental chemicals so that individuals and local communities can be alerted to the public safety implications of a biosphere now under attack by the effluents of our pyrotechnic-petrochemical lifestyles. Mitigation of the health physics impact of our accelerating chemical war on the biosphere must begin with community awareness of the persistence and pathways of environmental chemicals, the necessity of increased biological monitoring, and the urgent need for lifestyle changes in the context of sustainable economic activities that minimize the spread of chemical effluents in the biosphere.

Glossary

Additive: A material added to a polymer to enhance processability, performance, or aesthetics. The major types of plastic additives include: antioxidants, antistatic agents, blowing agents, colorants, flame retardants, impact modifiers, lubricants, plasticizers, and heat and ultraviolet (UV) light stabilizers.

Adsorption: "The adhesion of atoms, ions or molecules to a surface; the opposite is desorption; commonly confused with absorption in which a substance permeates another substance." (Stevenson 2011)

Amphipod: "Amphipoda is an order of malacostracan crustaceans with no carapace and generally with laterally compressed bodies. Amphipods range in size from 1 to 340 millimeters (0.039 to 13 in) and are mostly detritivores or scavengers. There are more than 9,500 amphipod species so far described. They are mostly marine animals, but are found in almost all aquatic environments." (http://en.wikipedia.org/wiki/Amphipoda)

Anthropogenic: Effects, processes or materials derived from humans; not natural. (Stevenson 2011)

Autotroph: An organism that is able to form nutritional organic substances from simple inorganic substances such as carbon dioxide. Compare with heterotroph. (Oxford 2010)

Bakelite: "The first fiber-reinforced plastic." (Wikipedia 2015)

Benthic: On the ocean bottom. (Stevenson 2011)

Bioaccumulate: The phenomena of a substance becoming concentrated inside the bodies of living things.

Biogenic compound: Chemical compound having a natural origin.

Biokinetics: 1) "The science of the movements within organisms." (Dorlands Medical Dictionary 2014) 2) Including "…the movements of tissue and related phenomena that occur during the development of organisms" (Miller-Keane Encyclopedia 2014). With respect to this publication, biokinetics is the science of the movements of nanoplastics and nanotoxins within organisms.

Biomagnification: "*Biomagnification* is the sequence of processes in an ecosystem by which higher concentrations of a particular chemical, such as the pesticide DDT, are reached in organisms higher up the food chain, generally through a series of prey-predator relationships." (http://toxics.usgs.gov/definitions/biomagnification.html)

Biome: Regions of the world characterized by common climates with similar plants and animals adapted to those climates.

Biomimetics: Relating to or denoting synthetic methods that mimic biochemical processes. (Oxford 2010) Also called **bionics**, re. the application of biological processes to engineered systems.

Bisphenol A (BPA): 1) A building block of polycarbonate (#7 is often polycarbonate) plastic that is used in thousands of consumer products, including food packaging. 2) "BPA is employed to make certain plastics and epoxy resins. BPA-based plastic is clear and tough, and is made into a variety of common consumer goods, such as water bottles, sports equipment, CDs, and DVDs. Epoxy resins containing BPA are used to line water pipes, as coatings on the inside of many food and beverage cans and in making thermal paper such as that used in sales receipts." (Wikipedia 2015). 3) BPA exposure may disrupt normal breast development in ways that predispose women for later life breast cancer. (Breast Cancer Fund 2014).

Carbon 60 (CO_{60})/Buckyball (BB_{60})/C_{60}: A subset of fullerenes. The term "buckyball" only refers to the spherical fullerenes and is derived from the word "Buckminsterfullerene," which is the geodesic dome/soccer ball shaped C_{60} molecule nanostructure. C_{60} was the first buckyball to be discovered and remains the most common and easy to produce. (Cassone 2007)

Carbon nanotubes: Nanotubes have two forms: single walled (SWCNT) and multi-walled (MWCNT). They are manufactured by thermal stripping carbon atoms from carbon bearing compounds. Their unique electrical, mechanical, and thermal properties put them at the forefront of scientific research and technological development (electronics, computers, aerospace, etc.) They have the strongest tensile strength of any synthetic fiber. Superconducting at low temperatures, they are 1,000 times longer than wide. Carbon nanotubes deteriorate by mass wasting (wear and tear) and by incineration.

Catalyst: Chemical substance that causes or accelerates a chemical reaction without itself being affected. (Rossi 2014)

Chemical fallout: Wet and dry deposition of anthropogenic and remobilized naturally occurring chemicals. Chemicals are often sorbed on particulate matter during dry deposition; chemicals are usually incorporated in rainfall during wet deposition. Radioactive particles in dry deposition are often greatly magnified in wet deposition; during Chernobyl fallout events, radioactive fallout was often two orders of magnitude greater during rainfall events as compared to dry deposition.

Chemical feedstock: For plastic production, chemical feedstocks can be derived from fossil fuels (crude oil, natural gas, and coal) or bio-based resources. Crude oil feedstocks are derived from the "cracking" and distillation (separation) of the feedstock. Feedstocks from natural gas are derived from the processing or separation of that raw material. These processes yield the feedstocks ethane, propane, butane, methane and others. Sources of biobased chemicals include algae, corn, sugarcane, sugar beets, potatoes, and other biological feedstocks. (Rossi 2014)

Chemical footprint: The measure by number and mass of chemicals of high concern, as determined by hazard level, in products and supply chains. (Rossi 2014)

Chemical of high concern (CoHC): Substance that has any of the following properties: 1) persistent, bioaccumulative and toxic (PBT); 2) very persistent and very bioaccumulative (vPvB); 3) very persistent and toxic (vPT); 4) very bioaccumulative and toxic (vBT); 5) carcinogenic; 6) mutagenic; 7) reproductive or developmental toxicant; 8) endocrine disruptor; or 9) neurotoxicant. "Toxic" (T) includes both human toxicity and ecotoxicity. (Rossi 2014)

Compounded plastic product: A product or material consisting of a polymer and a package of additives (for example, colorants, softeners, and flame retardants). (Rossi 2014)

Colloid: Hydrophilic substance that does not dissolve.

Dioxin: Unintentional by-products of high-temperature processes, such as incomplete combustion and pesticide production and are typically emitted from the burning of hospital waste, municipal waste, and hazardous waste, along with automobile emissions, peat, coal, and wood. (Wikipedia)

Endocrine disruption: "The endocrine system is made up of glands throughout the body (e.g., the hypothalamus, pituitary, thyroid, pancreas, adrenals, testes, ovaries) and the hormones that are made by the glands. These hormones (e.g., estrogen, testosterone, growth hormone, insulin, epinephrine and many more) travel through the bloodstream, acting as chemical messengers, regulating many critical bodily functions such as metabolism, blood sugar levels, reproductive function, development, and growth. Mammals, fish, birds, and many other living organisms have endocrine systems. An endocrine disrupting chemical is one that affects the normal functioning of the endocrine system by mimicking the behavior of normal hormones or blocking the effects of normal hormones. Endocrine disruption, especially over long periods of time, can have a broad range of consequences from abnormal growth, to delayed or inhibited reproductive function, to cancer. Regarding endocrine disrupters, the U.S. Environmental Protection Agency (EPA) states that 'there is strong evidence that chemical exposure has been associated with adverse developmental and reproductive effects on fish and wildlife in particular locations.' There are known instances of human endocrine disruption; however, in general, the field of human endocrine disruption caused by exposure to environmentally released chemicals is not well understood." (Stevenson 2011)

Endocrine disruptor: An exogenous substance that causes adverse health effects in an intact organism, or its progeny, consequent to changes in endocrine function. (Lintelmann 2003)

Endocrine system: "The collection of glands of an organism that secrete hormones directly into the circulatory system to be carried towards distant target organs." (Wikipedia 2015)

Endocytosis: "A form of active transport in which a cell transports molecules (such as proteins) into the cell by engulfing them in an energy-using process." (Wikipedia 2015)

Entropy: A thermodynamic quantity representing the unavailability of a system's thermal energy for conversion into mechanical work, often interpreted as the degree of disorder or randomness in the system. (Oxford 2010)

Epigenetics: "The study of mitotically heritable alterations in gene expression potential that are not caused by changes in DNA sequence. Epigenetic mechanisms are established during prenatal and early postnatal development and function throughout life to maintain the diverse gene expression patterns of different cell types within complex organisms… Epigenetics confers an extra level of information which is layered above the DNA sequence information and which, like the sequence, is replicated during cell division." (McAllister 2009)

Estrogenic chemicals: Steroid hormones responsible for feminization. Environmental chemicals that mimic estrogenic steroid hormones.

Eukaryote: "Any organism whose cells contain a nucleus and other organelles enclosed within membranes." (Wikipedia 2015).

Eutrophication: The depletion of oxygen in a nutrient rich environment, such as a marine ecosystem, by overly abundant microbial organisms.

Exogenous agonist: A ligand that can bind to a receptor like the natural substrate and "turn it on." The activation of the hormone receptor then finally leads to the same effects that can be caused by endogenous hormone action. (Lintelmann 2003)

Fiber-reinforced plastic (FRP): "A composite material made of a polymer matrix reinforced with fibers. The fibers are usually glass, carbon, aramid, or basalt. Rarely, other fibers such as paper or wood or asbestos have been used. The polymer is usually an epoxy, vinylester or polyester thermosetting plastic, and phenol formaldehyde resins are still in use… In the late 1970s world polymer production surpassed that of steel, making polymers the ubiquitous material that it is today. Fiber-reinforced plastics have been a significant aspect of this industry from the beginning." (Wikipedia 2015)

Fullerene: Pure carbon, cage-like molecules composed of at least 20 atoms of carbon. (Cassone 2007). See Carbon 60.

Genome: "The genome is the genetic material of an organism. It consists of DNA (or RNA in RNA viruses)." (Wikipedia 2015)

Graphene: "Graphene is an atomic-scale honeycomb lattice made of carbon atoms; pure carbon in the form of a very thin, nearly transparent sheet, one atom thick. It is remarkably strong for its very low weight (100 times stronger than steel) and it conducts heat and electricity with great efficiency...it was first produced in the lab in 2003... it is virtually two-dimensional, it interacts oddly with light and with other materials. Researchers have identified the bipolar transistor effect, ballistic transport of charges and large quantum oscillations." (Wikipedia 2014). Also considered as a PAH.

Global distillation or the grasshopper effect: A geochemical process wherein volatile chemicals vaporize in warmer regions of Earth and are deposited in colder regions. This concentrates persistent organic pollutants at the poles and mountaintops. (Simonich 1995)

Heterotroph: An organism deriving its nutritional requirements from complex organic substances [such as CO_2, especially by photosynthesis]. Compare with autotroph. (Oxford 2010)

Hydrophilic: Having an affinity for water, or causing water to adhere.

Hydrophilic effect: The tendency of polar molecules to be dissolved by water or other polar substances.

Hydrophobic: Having no affinity for water, or repelling water.

Hydrophobic effect: The segregation and repulsion between water and non-polar substances, as in protein folding, lipid bi-layer formation, and insertion of membrane proteins into non-polar lipid environments.

Hydrophobic pollutants: Chemicals that, when in water, preferentially adhere to other substances like plastic, sediment, or fatty tissues.

Hypospadias: "A birth defect of the urethra in the male where the urinary opening is not at the usual location on the head of the penis. It is the second most common birth abnormality in boys, affecting approximately 1 of every 250." (Wikipedia 2015)

Intermediate chemical: Chemical produced by the chemical conversion of primary chemicals to more complicated derivative products such as ethylbenzene, ethylene dichloride, and lactic acid. (Rossi 2014)

Lipophilic: Tending to combine with or dissolve in lipids or fats.

Macrophage: A phagocytic tissue cell of the reticuloendothelial system that is derived from the blood monocyte. The monocyte migrates from the blood into tissues where it transforms into a macrophage. Macrophages are present in most tissues. Macrophages ingest and process degenerated cells and foreign invaders, such as viruses, bacteria, and particles. The long-lived macrophages are reservoirs of HIV. (Buzea 2007)

Marine ecosystem: "They include oceans, salt marshes, intertidal zones, estuaries, lagoons, mangroves, coral reefs, the deep sea, and the sea floor… Marine waters cover two-thirds of the surface of the Earth." (Wikipedia 2015)

Marine microbes: Tiny, single-celled organisms that live in the ocean and account for more than 98 percent of ocean biomass. They include microalgae, bacteria, archaea, protozoa fungi, and viruses. They are only $1/8000^{th}$ the volume of a human cell and span about $1/100^{th}$ the diameter of a human hair. Up to a million of them live in just one milliliter of seawater. (http://oceanexplorer.noaa.gov/)

Marine snow: Marine snow consists of egested zooplankton fecal matter, atmospheric carbon deposited at the sea surface, anthropogenic chemical fallout deposited by long range atmospheric transport and plastic nanoparticles incorporated in fecal matter or sorbed during transport. Particulate matter in marine ecosystems derived from microbial community systems, marine debris, and atmospheric carbon fallout, including sorbed environmental chemicals.

Mesopelagic: A zone in the ocean between 200 meters and 1000 meters depth. (Stevenson 2011)

Mesoplastic: Plastic particles ±10 mm in diameter or length.

Metabolite: "A chemical alteration of the original compound produced by body tissues." (US CDC 2012)

Microbiomes: Microbial communities which are the basis of all biotic communities, as typified by the seasurface microlayer consisting of multiple microbial communities in the top 1 mm (1,000 um) of water.

Mixotroph: "An organism that can use a mix of different sources of energy and carbon, instead of having a single trophic mode on the continuum from complete heterotrophy at one end to autotrophy at the other." (Wikipedia 2015)

Monomer: The molecular unit from which polymers are prepared. REACH Article 3(6) defines a monomer as a substance "which is capable of forming covalent bonds with a sequence of additional like or unlike molecules under the conditions of the relevant polymer-forming reaction used for that particular process." (Rossi 2014)

Moulin: A sinkhole on an ice sheet, as on the Greenland ice sheet, in which surface water in ice sheet rivers cascades into the sink holes and eventually reaches the ocean by traveling underneath the ice sheet.

Nanoaerosol: A collection of nanoparticles suspended in a gas. (NIOSH 2009)

Nanobeads: Nanoparticles that range in size from "0.004 mm to 1.24 mm." (Schneiderman 2014)

Nanoparticle: An aggregate of atoms bonded together with a radius between 1 and 100 nm. It typically consists of 10–105 atoms. (Bhushan 2007). Nanoparticles behave like neither solids, liquids, nor gases, and exist in the topsy-turvy world of quantum physics, which governs those denizens small enough to have escaped the laws of Newtonian physics. This allows them to perform their almost magical feats of conductivity, reactivity, and optical sensitivity, among others. (Hood 2004)

Nanoplankton: Plankton whose size is between 2 to 20 μm. (Wikipedia 2015)

Nanotechnology: The ability to observe and manipulate matter at the nanoscale size range of 1 to 100 nanometers. "The National Nanotechnology Initiative… defines nanotechnology as the manipulation of matter with at least one dimension sized from 1 to 100 nanometers… [taking advantage of] the special properties of matter that occur below the given size threshold… Nanotechnology as defined by size is naturally very broad, including fields of science as diverse as surface science, organic chemistry, molecular biology, semiconductor physics, microfabrication, and molecular self-assembly." (http://en.wikipedia.org/wiki/Nanotechnology). "The creation, manipulation, and application of materials at the nanoscale—involves the ability to engineer, control, and exploit the unique chemical, physical, and electrical properties that emerge from the infinitesimally tiny man-made particles." (Hood 2004)

Nanotoxicology: "Nanotoxicology is defined as the study of the nature and mechanism of toxic effects of nanoscale materials/particles on living organisms and other biological systems. It also deals with the quantitative assessment of the severity and frequency of nanotoxic effects in relation to the exposure of the organisms." (Zhao n.d.)

Naturally occurring nanotoxin: Nanotoxins produced as a result of forest fires and volcanoes, usually in the form of ultrafine particles (UFPs).

Nekton: Aquatic animals that are able to swim and move independently of water currents. Often contrasted with plankton.

Neuston: Small aquatic organisms inhabiting the surface layer or moving on the surface film of water. (Oxford 2010).

NHANES: Administered by the Centers for Disease Control and Prevention (CDC), it is a cross-sectional, nationally representative survey of the health and nutritional status of the US population (Grindler 2015).

Nonylphenol: "A toxic xenobiotic compound classified as an endocrine disrupter capable of interfering with the hormonal system of numerous organisms. It originates principally from the degradation of nonylphenol ethoxylates which are widely used as industrial surfactants." (Soares 2008)

Nurdles: Thermoplastic pre-production resin pellets; the pre-cursor raw material to plastic products, usually <5 mm. (Stevenson 2011)

Obesogen: Chemical agents that inappropriately regulate and promote lipid accumulation and apidogenesis. (Grun 2009). They induce weight gain via endocrine disrupting activity.

Organic Solvents: Solvents used in the computer manufacturing industry and in cleaning products, cosmetics and dry cleaning, including toluene, methylene chloride, and trichloroethylene.

Organotin: "Organotin compounds or stannanes are chemical compounds based on tin with hydrocarbon substituents. Organotin chemistry is part of the wider field of organometallic chemistry." (Wikipedia 2015)

Ozone: "Ground-level ozone, which is the type that is harmful to human health, is formed by the reaction of nitrogen oxides (NOx) and volatile organic chemicals (VOCs) in the presence of sunlight. Motor vehicle exhaust and industrial emissions, gasoline vapors, and chemical solvents as well as natural sources emit NOx and VOC[s] that help form ozone." (www.epa.gov/glo). The stratospheric ozone layer protects the biosphere from intense ultraviolet radiation. Exhaust from the supersonic transport planes (SSTs) provided the potential to deplete the ozone layer, which can also be destroyed by chlorofluorocarbon emissions.

Pacific decadal oscillation (PDO): A recurring pattern of ocean and atmosphere climate variability centered over the mid-latitude Pacific basin, measured as warm or cool surface waters in the Pacific Ocean north of 20 degrees latitude.

Particulate matter (PM): A mixture of solid particles and liquid droplets found in the air and can be composed of many types of materials and chemicals.

(http://www.epa.gov/ncer/science/pm/). Particulate matter often sorbs environmental chemicals, which then can be transported in marine ecosystems by marine snow. See the three categories of PM ($PM_{0.1}$, PM_{10}, and $PM_{2.5}$).

Perfluorinated chemicals (PFCs): Active ingredients or breakdown products of Teflon, Scotchgard, fabric and carpet protectors, and food wrap coatings. A global contaminant, PFCs often accumulate in the environment and the food chain; linked to cancer, birth defects and more. (EWG 2009)

Photolysis: Also known as photodecomposition, photolysis "is a chemical reaction in which a chemical compound is broken down by photons. It is defined as the interaction of one or more photons with one target molecule… Photolysis is part of the light-dependent reactions of photosynthesis." (Wikipedia 2015)

Phthalates: "Phthalates are a group of chemicals used to make plastics more flexible and harder to break. They are often called plasticizers. Some phthalates are used as solvents (dissolving agents) for other materials. They are used in hundreds of products, such as vinyl flooring, adhesives, detergents, lubricating oils, automotive plastics, plastic clothes (raincoats), and personal-care products (soaps, shampoos, hair sprays, and nail polishes). Phthalates are used widely in polyvinyl chloride plastics, which are used to make products such as plastic packaging film and sheets, garden hoses, inflatable toys, blood-storage containers, medical tubing, and some children's toys." (CDC Factsheet 2015)

Picoplankton: "Plankton composed of cells between 0.2 and 2 μm." (Wikipedia 2015)

Plastic: Generically, a polymer and/or the product made from the polymer through its entire life cycle. "The term 'plastics' is used to describe plastic polymers with additives to enable processing and/or give the properties needed for a desired application . . . [including] ...polymer substances, ...polymer substances in mixtures and final articles." (Rossi 2014)

Plastic compounding: The process of preparing plastic materials with desired properties by mixing or blending polymers and additives in a molten state. (Rossi 2014)

Polybrominated dibenzodioxins and furans (PBDD/F): Contaminants in brominated flame retardants; pollutants and byproducts from plastic production and incineration; accumulate in food chain; toxic to developing endocrine (hormone) system (EWG 2009)

Polybrominated diphenyl ethers (PBDEs): The toxic flame retardants added to the plastic cases of televisions, electronic equipment, and the fibers in draperies, furniture, and other textiles. (Schmitt 2007)

Polychlorinated biphenyl (PCB): "A synthetic organic chemical compound of chlorine attached to biphenyl, which is a molecule composed of two benzene rings... 130 of the different PCB arrangements and orientations are used commercially... Polychlorinated biphenyls were widely used as dielectric and coolant fluids, for example in electrical apparatus, cutting fluids for machining operations, carbon paper and in heat transfer fluids... Their destruction by chemical, thermal, and biochemical processes is extremely difficult, and presents the risk of generating extremely toxic dibenzodioxins and dibenzofurans through partial oxidation... Volatilization of PCBs in soil was thought to be the primary source of PCBs in the atmosphere, but recent research suggests ventilation of PCB-contaminated indoor air from buildings is the primary source of PCB contamination in the atmosphere." (Wikipedia 2015)

Polychlorinated naphthalenes (PCNs): Wood preservatives, varnishes, machine lubricating oils, waste incineration; common PCB contaminant; contaminate the food chain; cause liver and kidney damage

Polymer: Long chain of molecules made from repeating parts, called monomers, which are a product of a polymerization reaction. A polymer can be natural or synthetic. In relation to a "compounded plastic product", "polymer" is the stage prior to the addition of performance additives. REACH Article 3(5) defines "polymer substance" as "a substance consisting of molecules characterized by the sequence of one or more types of monomer units." (Rossi 2014)

Polymeric materials: A special kind of formulated mixture made of polymers and typically containing additives to improve performance (e.g., compounded plastics, adhesives, foams, and resins). Polymeric material is a broad term used to describe plastics, resins, adhesives, foams, etc.

Primary chemicals: Fossil fuel-based primary chemicals derived from petroleum, which are the building block chemicals that the vast majority of other chemicals and plastics are manufactured from, e.g. ethylene, propylene and butadiene (olefins); benzene, toluene, and xylene (aromatics); and methanol. Biobased primary chemicals include sugars (glucose) and ethanol from corn. (Rossi 2014)

Quantum dot (Q dot)/(QD): QDs are semiconductor crystals with a diameter of a few nanometers, having many properties resembling those of atoms (Buzea 2007), whose size and shape can be precisely controlled by the duration, temperature, and liquid molecules used in their synthesis. "Researchers have studied applications for quantum dots in transistors, solar cells, LEDs, and diode lasers. They have also investigated quantum dots as agents for medical imaging and as possible qubits in quantum computing. The first commercial release of a product utilizing quantum dots was the Sony XBR X900A series of flat panel televisions released in 2013... The excitation and emission of quantum dots [is] highly tunable. Since the size of a quantum dot may be set when it is made, its conductive properties may be carefully controlled. Quantum dot assemblies consisting of many different sizes, such as gradient multi-layer nanofilms, can be made to exhibit a range of desirable emission properties." (Wikipedia 2015)

Remineralization: The transformation of organic molecules into inorganic forms, which especially occurs in microbiomes. If POPs could be remineralized by microbiomes, they wouldn't be transferred through the food chain to higher trophic levels.

Sea surface microlayer (SML): The top 0.04 millimeter of the sea.

Single-walled carbon nanotube (SWCNT or SWNT): "A long, hollow structure with the walls formed by one-atom-thick sheets of carbon… Most single-walled nanotubes have a diameter of close to 1 nanometer, with a tube length that can be many millions of times longer."

Sorption: Absorption and adsorption considered as a single process. (Oxford 2010)

Styrene: "Also known as ethenylbenzene, vinylbenzene, and phenylethene… it is a colorless oily liquid that evaporates easily and has a sweet smell, although high concentrations have a less pleasant odor. Styrene is the precursor to polystyrene and several copolymers… Various regulatory bodies refer to styrene, in various contexts, as a possible or potential human carcinogen." (Wikipedia 2015)

Synergistic effects: The toxicity of each compound is enhanced by the other compounds in the mixture. When put together, the effects can far exceed the approximated additive effects of, for example, the POP compound mixture. (Wikipedia 2015)

Multi-walled carbon nanotube (MWCNT or MWNT): "Multi-walled nanotubes consist of multiple rolled layers of carbon." (Wikipedia 2015)

Sorbtion: "Plastic debris remains in the environment for a long period of time and sorbs increasing concentrations of a mixture of pollutants over time." (Rochman 2014)

Toxic Equivalents (TEQs): These are used to "report the toxicity-weighted masses of mixtures of PCDDs, PCDFs, and PCBs… TEQs are then used for risk characterization and management purposes, such as prioritizing areas of cleanup." (Wikipedia 2015).

Tributyltin: A highly toxic antifouling compound containing organotins.

Ultrafine particle (UFP): Airborne particles smaller than 100 nm in diameter, frequently occurring in the context of nanometer-diameter particles that have not been intentionally produced but are incidental products of processes involving combustion, welding, or diesel engines. (NIOSH 2009). "The term 'ultrafine particle' is identical with that of nanoparticle." (Hett 2004, 28)

Vinyl chloride: Vinyl chloride is formed in the manufacture of polyvinyl chloride (PVC) or #3 plastic. It was one of the first chemicals designated as a known human carcinogen by the National Toxicology Program (NTP) and the International Agency for Research on Cancer (IARC). It has also been linked to increased mortality from breast cancer among workers involved in its manufacture.

Xenobiotic: Relating to or denoting a substance, typically a synthetic chemical, that is foreign to the body or to an ecological system. (Oxford 2010)

Zooplankton: Tiny floating marine organisms near the bottom of the marine food chain. (Stevenson 2011)

Keyword Index

This keyword index provides an introduction to the subject matter in the bibliographies. The mission of this text is the exploration of the phenomenon of the degradation of microplastics into plastic nanoparticles (PNPs), which then sorb environmental chemicals, and their similarity to the nanoparticles and nanotoxins intentionally and accidentally produced by nanotechnological industries. The key question, which may not be answered for decades, is: will the waste stream produced by nanotechnological industries also sorb environmental chemicals in the same manner as the plastic nanoparticles derived from the degradation of microplastics in the environment? The following keyword index is intended as a guide to the biogeochemical cycling of nanotoxins in our finite round world biosphere.

We have divided the keyword list into five categories:

>Biological
>
>Nano
>
>Chemical
>
>Medical
>
>Ecological

Biological

Antibacterial Properties

Antimicrobial

Bioaccumulation

Bioaccumulation of nanoparticles

Bioactive nanoparticles

Bioavailability

Biobricks

Biocybernetics

Biodegradation

Biodevices

Biodistribution

Bioenergy

Biogeographic connectivity

Biofilm

Biofilm food

Biofilm lattice

Biofouling

Biofusion

Bioinformatics

Biokinetics

Biological application

Biological behavior

Biological microenvironment

Biological pump

Biological self assembly

Biomagnification

Biometrics

Biomimetics

Bionanoscience

Biopersistence

Biosensor

Biotransformation

Biouptake

Coaxial silicon

Ecotoxicological

Ecotoxicological impact

Egestion

False satiation

Global microbiome research

High resolution optical imaging

Impact assessment

Interaction between abiotic and biotic environments

Interaction of nanomaterials with microbes

Interactions with subcellular structures

Intracellular uptake of nanomaterial

In vitro

In vivo

Leaching of constituent contaminants

Microbial synthesis of nanomaterials

Microbiomes

Mitogen

Molecular kinetics

Mulching

Nanophytoplankton

Nano-zooplankton

Nonpolar lipid environments

Novel toxicological risks

Oceanic biogeochemical processes

Picoplankton

Primary emission

Root-soil-microbe interface

Scats contained microplastics

Sea-surface microlayer

Toxicological analysis

Xenobiotic

Nano

Aged nanomaterials

Airborne nanoparticles

Airborne nanoaerosols

Biochip device

Bimetallic nanoparticles

Colloidal semiconductor nanocrystal

Complex nanodevices

Dendrimers

Edible nanocoatings

Engineered plastic nanoparticles

Environmental chemistry of nanometals

Exposure to nanoparticles

Fungal-derived nanoparticles

Inherently ecotoxic nanoform

Intradermal infiltration

Lipid nanotechnology

Metallofullerene

Nanoagriculture

Nanoagrochemicals

Nanoarrays

Nanobacteria

Nano-based antimicrobial packaging

Nano-biointerface

Nanobiotechnology

Nanoboxes

Nano-carrier

Nanocellulose

Nanochemistry

Nanoclay

Nanoclusters

Nanocomposites

Nanocrystal

Nanodevices

Nanodivide

Nanoelectric power

Nanoelectronics

Nanoemulsion

Nanoenhanced

Nanoexposure

Nanofillers (inorganic)

Nanofluidics

Nanofood

Nanofood nonstick lining

Nanofoundaries

Nanohybrid

Nanolithography

Nanomachine

Nanomaterial

Nanomaterial byproduct

Nanomaterial durability

Nanomaterials: diversity of engineered techniques

Nanometaloxides

Nanomedicine

Nanometals

Nanometer scale

Nanometric

Nanonutrient inhibitor

Nano-oncology

Nanoparticle application

Nanoparticle interaction with cells

Nanoparticle production: by chemical manufacture, combustion processes, and environmental transformation

Nanoparticle toxicology

Nanoparticle tracking analysis

Nanoparticle translocation

Nanoparticles (anthropogenic)

Nanoparticles (incidental)

Nanoparticles (manufactured)

Nanoparticles (naturally occurring)

Nanopathology

Nanopesticides

Nanopharmaceuticals

Nanoplastic fragments

Nanoplastics

Nanoplatform

Nanoplanktonic protozoa

Nanoporous membrane

Nanoproducts (accidental)

Nanoproducts (engineered)

Nanoproducts (naturally occurring)

Nano-object

Nanoremediation technologies

Nanorisk

Nano-organism

Nanoscale

Nanoscale hole formation

Nanoscience

Nanoscopic pollution

Nanosensors

Nanosensor packages

Nanosilver

Nanosilver antibacterials

Nanosized particles (NSP)

Nanostigma

Nanosystems

Nanosphere

Nanotechnologies for packaging

Nanotechnology-specific safety laws

Nanotoxic additives

Nanotrees

Nanotribology

Nanotube integrated circuits

Non-degradable nanoparticles

NSP-induced inflammatory and oxidative stress response

NSP: intrinsic toxic properties

NSP lifecycle: manufacture - use – disposal

NSPs: wires, rings, particles, tubes, shells, dots, etc.

Natural nanocomposites

Persistent nanosubstances

Powders of nanomaterials

Self assembling nanotubes

Self assembled quantum dots (5 to 50 nm)

Semiconductor nanoplastics

Semiconductor quantum dots

Synergy of nanotechnology

Chemical

Affinity of POPs with soil organic matter

Agglomeration

Biophysico chemical interaction

Black Carbon

Catalytic behavior

CdSe (Cadmium Selenium)

CdTe (Cadmium Tellurium)

Chemical constituent declaration

Chemicals of high concern (CoHC)

Chemicenteric colloids

Clumping (particle aggregation)

Colorants

Combustion-derived organic volatile compounds

Crystal structure dependent toxicity

Degradation

Degradation products

Diversity of chemical exposure routes

Diversity and versatility of polymers

Gas exchange

Hopane chemical biomarker

Hydrophobic organic contaminant (HOC)

Flame retardants

Floating properties

Glancing angle deposition

High oxygen-moisture barrier property

Insoluble organic polymer

Kinetics of NSP

Lipophilic chemicals (e.g. POPs)

Mass concentration

Neustonic plastic

Nonstick compounds

Macromolecule

Methylsiloxanes

Microbiocide

Microelectronic fabrication

Microfabrication

Microfabricator

Microinstruments

Micromachinery

Microplastics < 1 mm

Microsensor

Molecular assemblies

Nanoparticle metal oxides (TiO_2, SiO_2)

Optoelectronic

Organochlorines

Organic solvent

Organophosphate pesticides

Particle surface adhesion

Persistent free radicals

Physicochemical characterization

Physicochemical properties

Phthalates (DEHP and MEHP)

Plastic microspheres

Plastic scrubbers

Plasticizers

Polycyclic aromatic hydrocarbons (PAHs)

Polyacrylate nanoparticle

Polyfilament nylon

Polymer degradates

Polymer degradation

Polymers of high concern

Polytetrafluoroethylene fume

POPs (Persistent organic pollutants)

Priority pollutants

Post-combustion POP formation

Pyrotechnic society

Semiconductor metal compounds

Silicon revolution

Size dependent toxicity

Size range

Superhydrophobicity

Surface area to volume ratio

Thermal formation of persistent organic pollutants

Toxic volatilization product

Toxicity increases with decreased particle size

Toxicokinetics

Trace concentrations

Transformation of nanomaterials to more toxic metabolites

Triclosan in PVC

Ultrafine diesel exhaust nanoparticles

Unknown and unintended byproducts

UV Stabilizers

Virgin resin pellets as mesoplastics

Volatilization

ZnS

Medical

Biomarkers of toxicity

Blood brain barrier

Carcinogenic trihalomethanes

Cell imaging

Cellular internalization of nanoscale particles

Chronic effects

Chronic neuroinflammation

Chronic toxicity

Cryptorchidism (non-descending testes)

Cyototoxicity

Direct exposure

Disruption of endocrine function

Endocytosis reactive oxygen species

Endocrine disruption

Endometrics

Epigenetic signature of PBTs

Epithelial boundaries

Epithelial injuries

Estrogenic environmental chemicals

Fibrosis

Foreign body granulomas

Genotoxicity

Gynecomastia

Hypoxemia

In situ blood-brain barrier transport of nanoparticles

Inspiration

Intradermal translocation

Life cycle risk assessment

Lipoxidation

Lipophilicity

Low dose chronic exposure

Medbiotech (toxicology, molecular biology)

Mitochondrial damage

Neurogenesis

Neurodegeneration

Obesogenic

Particle inhalation toxicology

Pathogenesis

Personal hygiene products and cosmetics

Phagocytosis

Photothermal therapy

Pleural effusions

Protein engineering nanoform

Protein folding

Pulmonary fibrosis

Risk benefit analysis

Risk profile

Risperdal

RoHS directives

Shape related toxicity of nanoparticles

Synergism of endocrine disruption impacts

Testicular dysgenesis syndrome (TDS)

Toxic assessment

Toxicity threshold

Toxicological endpoint

Toxicological analysis

Toxicology of nanoscaled materials

Toxological profile

Unopposed estrogen

Unusual toxicological effects

Uterine leiomyomas

Xenoestrogens

Ecological

Adhering persistent contaminant

Air nanopollution

Airborne plastic nanoparticles

Air-surface exchange

Anthropogenic chemical fallout

Anthropogenic deforestation

Anthropogenic infrastructure collapse

Aqueous environment

Atmospheric aerosol loading

Atmospheric degradation pathway

Atmospheric deposition

Attenuation mechanism

Automobile exhaust soot

Bacteria

Bacterioplankton

Benthos entanglement

Bioaccumulation of HOC

Bioavailability of contaminants associated with microplastics

Biosolid application

Biosphere in crisis

Bioturbation

Carriers of pollutants

Cause effect relationship

Chemical footprint

Chemical weathering process

Colloidal matter

Coprophagy

Coral Reef Bleaching

Creation-exposure pathways

Cumulative atmospheric global emissions

Defaunation

Degradation in sediment: aerobic and/or anaerobic conditions

Depth transfer process

Disposal-to-emissions pathways

Dissolution

Drifting pollutants

Ecolotoxological impact

Ecophysiological function

Ecotoxicology

End of life destination

Emerging contaminants

Evapotranspiration

Environmental concentration

Environmental fate

Environmental risk assessment

Exposure routes for a diversity of chemicals

Extreme miniaturization

Fate and transport

Finite biosphere

Food chain transport

Food web transfer

Fouling as a transport mechanism

Fragmentation of microplastics

Global distribution mechanisms for plastic

Global source point

Global transport: natural and commercial

Global warming potential of hydrofluorocarbons

Green chemistry

Growth hormone pathways

Habitat alteration

Habitat degradation

Habitat fragmentation

Heterotrophic nanoflagellates (HNF)

Human impact on the atmospheric water cycle

Human interference with the climate system

Human interference with the nitrogen cycle

Indicator species

Invisible fog

Leaching

Life cycle analysis

Life cycle of microplastics in marine environments

Life cycle risk assessment

Long range atmospheric transport

Macrozooplankton

Marine biosphere

Marine debris

Marine pelagic ecosystem

Mass wasting abrasive

Mediating POP transport

Mesozooplankton

Microlitter

Microplasticizing of zooplankton

Natural-technologic event (NA-TECH)

Newston net

Organic micropollutants

Organochlorine soup

Oxidation

Palm oil disaster

Pelagic microplastics

Phytodetritis

Plankton fecal pellet flux

Plankton vs nekton community

Planktonic vector

Plastic concentration vector

Plastic exfoliating microbead

Polylactic acid (PLA) from GEM corn

Post consumer recycled (PCR)

Primary emissions

Progressive fragmentation

Rainfall events

Rainfall washout

Razorbills

Riverine input

Safe operating space

Sedimentary habitat

Sewage sludge

Sink areas

Sinking phytoplankton bloom

Soil microbes

Sorption

Surface erosion

Synergistic biotic/abiotic interaction

Technological convergence

Temporal plastic distribution

Teixobactin

Transfer to depth

Translocation processes

Translocation volatilization

Tree bark concentrations

Tribology

Trophic ecology

Trophic interaction

Trophic level translocation

Trophic transfer

Trophodynamic behavior of emerging pollutants

Ultrafine particle

Unsustainable land use policies

Upper trophic level birds

Washover – washout

Waste stream

Weathering surface photooxidation

Windblown ecotoxins

Windblown nanoplastics

Reporting Units

An appropriate introduction to nanoparticles as vectors of environmental chemicals are the reporting units used by the CDC to document environmental chemicals in human urine, blood, and serum. Some of these chemicals are so toxic that the CDC measures them in parts per quadrillion, e.g. femtograms. Most chemicals, however, are measured in picograms (pg), parts per trillion. Many other chemicals and nanoparticles are measured in nanograms or parts per billion. Some ubiquitous ecotoxins, such as methylmercury in higher trophic levels (e.g. mammals) are measured in micrograms, i.e. parts per million.

Unit	Abbreviation	Value
liter	L	
deciliter	dL	10^{-1} liters
milliliter	mL	10^{-3} liters
gram	g	
milligram	mg	10^{-3} grams
microgram	µg	10^{-6} grams
nanogram	ng	10^{-9} grams
picogram	pg	10^{-12} grams
femtogram	fg	10^{-15} grams

Reporting units table from the Center for Disease Control's *Fourth Annual Report on Human Exposure to Environmental chemicals* (2012).

Other Reporting Units

picometer: pm. 1/1,000,000,000,000[th] of a meter. (one trillionth of a meter)

Angstrom: Å. 1/10,000,000,000[th] of a meter. (one ten billionth of a meter)

nanometer: nm. 1/1,000,000,000[th] of a meter. (one billionth of a meter)

micrometer: µm. 1/1,000,000[th] of a meter. (one millionths of a meter)

millimeter: mm. 1/1000th of a meter. (one thousandth of a meter)

ng/L: nanograms/liter: parts per trillion

µg/L: micrograms/liter: parts per billion

mg/L: milligrams/liter: parts per million

pg/g: picogram per gram dry weight = ppt (parts per trillion)

mg/kg dry weight: milligrams per kilogram

Definitions from the US EPA's Drinking Water Contaminants List

Source: US Environmental Protection Agency. (2009). *Drinking Water Contaminants.* US EPA, Washington, DC. http://water.epa.gov/drink/contaminants/index.cfm

- Maximum Contaminant Level Goal (MCLG) - The level of a contaminant in drinking water below which there is no known or expected risk to health. MCLGs allow for a margin of safety and are non-enforceable public health goals.
- Maximum Contaminant Level (MCL) - The highest level of a contaminant that is allowed in drinking water. MCLs are set as close to MCLGs as feasible using the best available treatment technology and taking cost into consideration. MCLs are enforceable standards.
- Maximum Residual Disinfectant Level Goal (MRDLG) - The level of a drinking water disinfectant below which there is no known or expected risk to health. MRDLGs do not reflect the benefits of the use of disinfectants to control microbial contaminants.)
- Treatment Technique (TT) - A required process intended to reduce the level of a contaminant in drinking water.
- Maximum Residual Disinfectant Level (MRDL) - The highest level of a disinfectant allowed in drinking water. There is convincing evidence that addition of a disinfectant is necessary for control of microbial contaminants.

Examples of relative size

Biological Media

Human hair: 80 μm; range: 17 to 181 μm. (Ley 1999)

Human red blood cell: 7,000 nm (7 μm); range of 6-8 μm

Protozoa like Giardia, Cryptosporidium: 5 microns or larger

Bacteria like holera, E. coli, Salmonella: 0.2-0.5 microns

Viruses like Hepatitis A, rotavirus, Norwalk virus: 0.004 microns or 4 nm (Curtis 1998)

Human Immunodeficiency Virus: 100 nm diameter (University of Wisconsin Madison 2014)

DNA strand: 2.5 nm wide

Protein molecule: 5 nm

Smallest discovered prokaryotes: 2-9 nm^3 (Luef 2015)

Nanoparticles

Nanocrystal clusters (e.g. quantum dots): range of 2-20 nm diameter

Ceramic oxide nanoparticles: range of 2-200 nm diameter

DNA nanobiorods: 5 nm diameter

Nanomaterial: 0.2 – 100 nm

Nanobiomaterials, photosynthetic reaction center (membrane of proteins and pigments that photosynthesizes): 10-20 nm diameter

Nanowires: 1-100 nm diameter

Nanotubes: 1-100 nm diameter (Gogotsi 2006)

Carbon nanotubes – 1-4 nm diameter. (University of Wisconsin Madison 2014)

Plastic nanoparticles: 20-200nm

Plastic nanoparticle fragments: 2-20nm

Chemical Media

Volcanic Ash: 1μm to 2mm (USGS 2009)

Food plastic wrap: 12.5 μm (Dow Industrial 2014)

Naturally occurring colloidal matter in water: 1 nm-1 μm (Nowack 2007)

Ultrafine particles (UFP): < 100 nm in diameter

Ultrafine Soot Particles: <1 μm

Soot particles: < 2.5 μm

Benzene ring corner to corner: 280 pm

Smallest and biggest size of PCB cogeners (~500-600 pm to ?)

Radius (Van der Waals) of a Carbon atom: 1.7 Angstroms or 0.17 nm

Distance between H and O molecules in water: 0.9584 Angstroms or 0.09504 nm

Conversion Factors

Multiply	By	To obtain
Length		
centimeter (cm)	0.3937	inch (in.)
millimeter (mm)	0.03937	inch (in.)
micrometer (µm)	0.00003937	inch (in.)
meter (m)	3.281	foot (ft)
kilometer (km)	0.6214	mile (mi)
Area		
square kilometer (km²)	247.1	acre
square kilometer (km²)	0.3861	square mile (mi²)
square meter (m²)	10.76	square foot (ft²)
square centimeter (cm²)	0.1550	square inch (ft²)
Volume		
cubic meter (m³)	6.290	barrel (petroleum, 1 barrel = 42 gal)
liter (L)	33.82	ounce, fluid (fl. oz)
liter (L)	2.113	pint (pt)
liter (L)	1.057	quart (qt)
liter (L)	0.2642	gallon (gal)
cubic meter (m³)	264.2	gallon (gal)
cubic meter (m³)	0.0002642	million gallons (Mgal)
liter (L)	61.02	cubic inch (in³)
cubic meter (m³)	35.31	cubic foot (ft³)
cubic meter (m³)	1.308	cubic yard (yd³)
Flow rate		
cubic meter per second (m³/s)	70.07	acre-foot per day (acre-ft/d)
cubic meter per second (m³/s)	35.31	cubic foot per second (ft³/s)
cubic meter per second (m³/s)	22.83	million gallons per day (Mgal/d)
meter per second (m/s)	3.281	foot per second (ft/s)
Mass		
gram (g)	0.03527	ounce, avoirdupois (oz)
Pressure		
kilopascal (kPa)	0.1450	pound per square inch (lb/ft²)

(USGS 2011)

Abbreviations/Acronyms

ABRB: Antiobiotic Resistant Bacteria

ABS: Acrylonitrile butadiene styrene

ACD: Anthropogenic climate disruption

ADME: Absorption, distribution, metabolism, elimination

AEGL: Acute exposure guideline level

AF: Accumulation factor

AGE: Anogenital distance

AgNP: Silver Nanoparticle

AhR: Aryl hydrocarbon receptor

ANP: Anthropogenic nanoparticle

ASA: Acrylonitrile styrene acrylate

ASD: Autism spectrum disorder

ASR: Automotive shredder residues

BAF: Bioaccumulation factor

BCF: Bioconcentration factor

BCS: Biochemical soup

BEI: Biological exposure index

BFR: Brominated flame retardant

BL: Blood lead

BMI: Body mass index

BPA: Bisphenol A

BRFSS: Behavioral Risk Factor Surveillance System

BSE: Black swan event

CAFO: Confined animal feeding operation

CBI: Confidential business information

CDC: Centers for Disease Control

CDNP: Combustion derived nanoparticle

CEL: Cancer effects level

CEMP: Community environmental monitoring program

CMCA: Combined maximum covariance analysis

CMR: Carcinogenic, mutagenic, or toxic for reproduction

CNF: Carbon nanofibers

CNT: Carbon nanotube

COD: Chemical oxygen demand

CoHC: Chemicals of high concern

COPC: Contaminants of potential concern

CPP: Central precocious puberty

CTD: Characteristic travel distance

CWAB: Chemical Warfare Against the Biosphere

CWAEG: Cyber Warfare Against the Electronic Grid

DBS: Dried blood spot

DEHP: di-(2-ethylhexyl) phthalate

DES: Diethylstilbestrol

DOC: Dissolved organic carbon

DOHAD: Developmental origins of human adult disease

EA: Endocrine Activity

EAC: Environmental assessment criteria; Endocrine active chemicals

EBA: Ethyl-butyl acrylate

EDC: Endocrine disrupting chemical

EED: Environmental endocrine disruptors

EEDC: Estrogenic endocrine disrupting chemical

EDSP: Endocrine disruptor screening program (EPA 1996)

EDSTAC: Endocrine disruptors screening testing advisory committee

EN: Engineered nanoparticle

ENM: Engineered nanomaterial

ENSO: El Nino Southern Oscillation

ENSP: Engineered nanosized particle

EPCRA: Emergency Planning and Community Right-to-know Act

ESEM: Environmental scanning electron microscope

FCM: Food contact materials

FMD: Floating marine debris

FQPA: Food quality protection act

FRP: Fiber-reinforced plastic

FT-IR: Fourier transform infrared spectroscopy

GC-MS: Gas chromatography – mass spectrometry

GEM: Gaseous elemental mercury

GEOINT: Geospatial Intelligence

GESAMP: Group of Experts on the Scientific Aspects of Marine Environmental Protection

GHG: Greenhouse gas

GHS: Globally harmonized system of classification and labeling of chemicals

GIC: Geomagnetically induced current

HAB: Harmful algal bloom

HBCD: Hexabromocyclododecane

HCB: Hexachlorobenzene

HCH: Hexylcyclohexane

HDPE: High density polyethylene

HF: Heterotrophic flagellate

HFC: Hydrofluorocarbon

HFCS: High fructose corn syrup

HHW: Household hazardous waste

HNF: Heterotrophic nanoflagellate

HOC: Hydrophobic organic contaminant

HRT: Hydraulic retention time

IMI: International Microbiome Initiative

IMNP: Intentionally manufactured nanoplastic

K_{OA}: Octanol-air partition coefficient

K_{OW}: Octanol-water partition coefficient

IARC: International Agency on Research on Cancer

LD50: Lethal dose, 50% (median lethal dose)

LDPE: Low density polyethylene

IFCS: Intergovernmental forum on chemical safety

IUPAC: International Union of Pure and Applied Chemistry

IWGN: Interagency Working Group on Nanoscience

LCA: Life cycle assessment

LC50: Lethal concentration 50%

LDPE: Low density polyethylene

LOAEL: Lowest observed adverse effect level

LOD: Limit of detection

LRAT: Long range atmospheric transport

LSGM: Least square geometric mean

MCL: Maximum contaminant level

MEHP: Mono-(2-ethylhexyl) phthalate

MBR: Membrane bioreactors

MISR: Multi-angle imaging spectroradiometer

MNM: Manufactured nanomaterial

MNP: Manufactured nanoparticle

MNP: Metal nanoparticle

MODIS: Moderate resolution imaging spectroradiometer

MOL: Maximum contaminant level

MPP: Microplastic particles

MQL: Method quantization limit

MSW: Municipal solid waste

MSWC: Municipal solid waste combustion (plant)

MWCNT: Multiwalled carbon nanotube

MWS: Municipal waste stream

NADP: National Atmospheric Deposition Monitoring Network

NAPAP: National Acide Precipitation Asessment Program

NATA: National air toxics assessment

NCCOS: National Center for Coastal Ocean Science

NEI: National emission inventory

NEMS: Nanoelectromechanic systems

NHANES: National Health and Nutrition Examination Survey

NIMA: National Imagery & Mapping Agency

NMDMP: National Marine Debris Monitoring Program

NNI: National nanotechnology initiative

NOAEL: No observed adverse effect level

NOM: Natural organic matter

NONP: Naturally occurring nanoparticle

NP: Nonylphenol

NPF: Nanoparticle formation

NPL: National priorities list

NSAID: Nonsteroidal anti-inflammatory drug

NSP: Nanosized particles

NTP: National Toxicology Program

OC: Organic carbon

OCP: Organochlorine pesticide

OMI: Ozone monitoring instrument

OPP: Office of Pesticide Programs (EPA)

OPP: Organophosphorus pesticide

OWC: Organic wastewater compound

PA: Polyamide

PAH: Polycyclic aromatic hydrocarbon

PBB: Polybrominated biphenyl

PBDE: Polybrominated diphenylether

PBT: Persistent bioaccumulative and toxic (chemicals)

PBM: Polymer-based material

PC: Polycarbonate

PCB: Polychlorinated biphenyl

PCDD: Polychlorinated dibenzo-p-dioxin

PCDF: polychlorinated dibenzofuran

PCP: Personal care products

PD: Pharmacodynamic

PDO: Pacific Decadal Oscillation

PE: Polyethylene

PE-HD: Polyethylene, high density

PE-LD: Polyethylene, low density

PE-MD: Polyethylene, medium density

PEL: Probable effect level

PET or PETE: Polyethylene terephthalate

PFOA: Perfluorooctanoic acid

PhAC: Pharmaceutically active compounds

PHAH: Polyhalogenated aromatic hydrocarbon

PK: Pharmacokinetic

PMD: Plastic marine debris

PMMA: Polymethyl methylacrylate

PNP: Plastic nanoparticle

POP: Persistent organic pollutant

POSI: Perfluorooctane sulfonate

PP: Polypropylene

PPA: Pollution Prevention Act

PPAR: Peroxisome proliferator-activated receptors

PPCP: Pharmaceutical and personal care product

PPCS: Pyrotechnic Petrochemical Consumer Society

PPNS: Pyrotechnic-Petrochemical-Nuclear Society

PRTR: Pollution release and transfer register

PS: Polystyrene

PS-E: Polystyrene, expandable

PTWI: Provisional tolerable weekly intake

PUR: Polyurethane

PVC: Polyvinylchloride

RDF: Refuse derived fuel

RGM: Reactive gaseous mercury

REACH: Registration, evaluation, and authorization of chemicals

REL: Recommended exposure limit

RoC: Report on cancer (by the National Toxicology Program)

RoHS: Registry of Hazardous Substances

ROS: Reactive oxygen species

SAN: Styrene-acrylonitrile

SDWA: Safe drinking water act

SeaWiFS: Sea-viewing wide field-of-view sensor

SEM: Scanning electron microscope

SL: Soil lead

SML: Sea surface microlayer

SMM: Sustainable materials management

SMX: Sulfamethoxazole

SOG: Soil organic matter

SRT: Solid retention time

SSL: Soil screening limit

SST: Sea surface temperature. Also Supersonic transport.

STEM: Scanning transmission electron microscope

STP: Sewage treatment plant

SVOC: Semivolatile organic compound

SWNT: Single walled carbon nanotubes

TBBPA: tetrabromobisphenol A

TBT: Tributylin compounds

TCS: Triclosan

TDS: Testicular dysgenesis syndrome

TEF: Toxic equivalency factor

TEL: Threshold effect level

TEQ: Toxic equivalent

TK: Toxicokinetic

TOC: Total organic carbon

TOMP: Toxic organic micro pollutants

ToxPF: Toxicological profile

TP: Transformation product

TRI: Toxics release inventory

TT: Treatment Technique

UMI: Unified Microbiome Initiative

UFP: Ultra fine particles

vZVI: Nanoscale zero-valent iron

VOC: Volatile organic compound

WEEP: Waste electronic & electrical products

WTE: Waste to energy

WWTP: Wastewater treatment plant

Links

Suggestions, corrections and updates welcomed: curator@davistownmuseum.org

US Government

Agency for Toxic Substances and Disease Registry (ATSDR)

- Toxicological Profiles and ToxFAQs: http://www.atsdr.cdc.gov/toxpro2.html
- Toxic Substances Portal: http://www.atsdr.cdc.gov/substances/index.asp

Centers for Disease Control and Prevention (CDC) Resources:

- National Biomonitoring Program: http://www.cdc.gov/biomonitoring/
- National Center for Health Statistics (NCHS): http://www.cdc.gov/nchs
  - National Health and Nutrition Examination Survey (NHANES): http://www.cdc.gov/nchs/nhanes.htm
  - Fourth National Report on Human Exposure to Environmental chemicals: http://www.cdc.gov/exposurereport
- National Institute for Occupational Safety and Health (NIOSH)
  - Databases and Information Resources: http://www.cdc.gov/niosh/database.html
  - Registry of Toxic Effects of Chemical Substances (RTECS): http://www.cdc.gov/niosh/rtecs

Lawrence Livermore National Laboratory

- Global Security: https://www-gs.llnl.gov/

National Institutes of Health (NIH)

- National Institute for Environmental Health Sciences (NIEHS): http://www.niehs.nih.gov
- National Toxicology Program (NTP): http://ntp.niehs.nih.gov
- National Library of Medicine (NLM), Toxicology Data Network: http://toxnet.nlm.nih.gov

NOAA

- Marine Debris Program: http://marinedebris.noaa.gov/
- Sea Grant Program: http://seagrant.noaa.gov/Home.aspx

U.S. Environmental Protection Agency (EPA)

- Endocrine Disruptor Screening Program (EDSP): http://www.epa.gov/endo/index.htm

- Ground level ozone: http://www.epa.gov/glo/

- Integrated Risk-Information System (IRIS): http://www.epa.gov/iris

- Office of Prevention, Pesticides, and Toxic Substances (OPPTS): http://www.epa.gov/opptsmnt/index.htm

- Particulate matter: http://www.epa.gov/ncer/science/pm/

- Toxics release inventory: http://www2.epa.gov/toxics-release-inventory-tri-program

U.S. Food and Drug Administration (FDA)

- Center for Food Safety and Applied Nutrition: http://www.cfsan.fda.gov

- National Center for Toxicological Research: http://www.fda.gov/nctr

U. S. Geological Survey (USGS)

- Environmental Health – Toxic Substances: http://toxics.usgs.gov/

- Ground-water Contaminant Transport: http://toxics.usgs.gov/topics/gwcontam_transport.html

- National Water Information System: http://waterdata.usgs.gov/nwis

International

Food and Agriculture Organization (FAO) of the United Nations: http://www.fao.org/home/en/

International Agency for Research on Cancer (IARC): www.iarc.fr

- Monographs on the Evaluation of Carcinogenic Risks to Humans: http://monographs.iarc.fr/ENG/Monographs/allmonos90.php

International Occupational Safety and Health Information Center

- International Chemical Safety Cards: http://www.ilo.org/public/english/protection/safework/cis/products/icsc/dtasht/index.htm

National Institute for Public Health and the Environment (Netherlands): http://www.rivm.nl/en/

World Health Organization

- International Programme on Chemical Safety (IPCS): http://www.who.int/pcs

- Monographs of the Joint FAO/WHO Meeting on Pesticide Residues: http://www.inchem.org/pages/jmpr.html

- WHO Global Alert and Response: http://www.who.int/csr/en/

Other Links

Adventurers and Scientists for Conservation (ASC) Microplastics Project: http://www.adventurescience.org/microplastics.html

Beat the Microbead: http://www.beatthemicrobead.org/en/

BioDiversity Research Institute: http://www.briloon.org/

Breast Cancer Research Foundation: http://www.bcrfcure.org/

Center for Responsible Nanotechnology (CRN): http://crnano.org/

Centre for the Study of Existential Risk (CSER): http://cser.org/

Coral Reef Monitoring Network: http://www.icriforum.org/gcrmn

Endocrine Society: http://www.endocrine.org/

Foresight Institute: http://www.foresight.org/

Future Earth: Research for Global Sustainability: http://www.futureearth.org/

Future of Humanity Institute: http://www.fhi.ox.ac.uk/

Future of Life Institute (FLI): http://futureoflife.org/

Global Catastrophic Risk Institute: http://gcrinstitute.org/

GoodGuide Scorecard, suspected endocrine toxicants: http://scorecard.goodguide.com/health-effects/chemicals-2.tcl?short_hazard_name=endo

Greenpeace: http://www.greenpeace.org/international/en/

Intergovernmental Forum on chemical safety (IFCS): http://www.who.int/ifcs/en/

International Conference on Chemical Management (ICCM):

International Pellet Watch (IPW): http://www.pelletwatch.org/

International Program on Chemical Safety (IPCS):

International Union of Pure and Applied Chemistry (IUPAC): http://www.iupac.org/

Interagency Working Group on NanoScience, Engineering and Technology (IWGN): http://www.wtec.org/loyola/nano/toc.htm

Katriana sampling results: Bywater/Marigny including agriculture street landfill: http://www.nrdc.org/health/effects/katrinadata/bywater.asp.

Lifeboat Foundation: Safeguarding Humanity: http://lifeboat.com/ex/main

Machine Intelligence Research Institute (MIRI): https://intelligence.org/

Marine & Environmental Research Institute (MERI): http://www.meriresearch.org/

Millennium Alliance for Humanity & the Biosphere (MAHB): http://mahb.stanford.edu/

National Nanotechnology Initiative (NNI): http://www.nano.gov/

New Orleans environmental quality test results: http://www.nrdc.org/health/effects/katrinadata/contents.asp.

Professor Zhon L. Wang's Nano Research Group: http://www.nanoscience.gatech.edu/zlwang/research.html

Project on Emerging Nanotechnologies (PEN): http://www.nanotechproject.org/

Red List Index (RLI): http://www.iucnredlist.org/

Schoodic Institute: http://www.schoodicinstitute.org/

Silicon Valley Toxics Coalition: http://www. http://svtc.org/

Stockholm Convention: http://chm.pops.int/

Strategic Approach to International Chemicals Management (SAICM): http://www.saicm.org/

Union of Concerned Scientists: http://www.ucsusa.org/

Worldwatch Institute: http://www.worldwatch.org/

Bibliography

5 Gyres Institute. (2007). *Baseline sampling for plastic pollution in the Gulf of Mexico.* 5 Gyres. http://5gyres.org/wp-content/uploads/2013/02/Post-KatrinaMarinePollutionSurveyApril2007.pdf

5 Gyres Institute. (2013). *Microplastics in consumer products and in the marine environment.* 5 Gyres. http://5gyres.org/media/5_Gyres_Position_Paper_on_Microplastics.pdf

Abu-Hilal, Ahmad and Al-Najjar, Tariq. (2009). Plastic pellets on the beaches of the northern Gulf of Aqaba, Red Sea. *Aquatic Ecosystem Health & Management.* 12. pg. 461-470. http://www.tandfonline.com/doi/abs/10.1080/14634980903361200?journalCode=uaem20

- "The occurrence of plastic pellets on the Jordanian beaches along the northeastern side of the Gulf of Aqaba (Red Sea) is being reported for the first time."

- "When compared with other beaches of other parts of the world, the Jordanian beaches on the Gulf of Aqaba are considered heavily polluted with these pellets."

Adams, C., Walker, K., Obare, S. and Docherty, K. (2014). Size-dependent antimicrobial effects of novel palladium nanoparticles. *PLoS ONE.* 9(1). http://www.plosone.org/article/fetchObject.action?uri=info:doi/10.1371/journal.pone.0085981&representation=PDF

- "Multidrug resistant pathogenic outbreaks are one of the most important global health issues we face today (e.g. extensively drug-resistant tuberculosis, multi-drug resistant cholera and methicillinresistant *Staphylococcus aureus*)…"

- "Palladium Nanoparticles are highly antimicrobial, and that fine-scale (<1 nm) differences in size can alter antimicrobial activity."

Aitken, R., Creely, K. and Tran, C. (2004). *Nanoparticles: An occupational hygiene review.* Institute of Occupational Medicine. http://www.hse.gov.uk/research/rrpdf/rr274.pdf

- "There are four main groups of nanoparticle production processes (gas-phase, vapour deposition, colloidal and attrition) all of which may potentially result in exposure by inhalation, dermal or ingestion routes."

- "Knowledge gaps:…1. The nanoparticle nomenclature is not sufficiently well described or agreed…2. There are no convenient methods by which exposures to nanoparticles in the workplace can be measured or assessed… 3. Insufficient knowledge concerning nanoparticle exposure is available… 4. The effectiveness of control approaches has not been evaluated… 5. Knowledge concerning nanoparticle risks is inadequate for risk assessments."

Alaee, M., Arias, P., Sjodin, A. and Bergman, Å. (2003). An overview of commercially used brominated flame retardants, their applications, their use patterns in different countries/ regions and possible modes of release. *Environ Int.* 29. pg. 683-9.

Alfred-Wegener Institut. (2015). *The role of protozooplankton in the pelagic ecosystem.* Alfred-Wegener Institut. http://www.awi.de/de/forschung/fachbereiche/biowissenschaften/polare_biologische_oz eanographie/research_themes/phytoplankton_protists/protists/

- "Pelagic ecosystems are driven by the interaction of organisms that fall into four functional components: the autotrophic phytoplankton and the heterotrophic bacteria, protozoa and metazoa."

- "Despite their heterogeneity they share common features that identify them as a distinct component: they tend to be of the same size class as their prey and, being unicellular and heterotrophic, can potentially have higher growth rates than the phytoplankton."

- "There are various large-celled protozoa capable of feeding on bloom-forming phytoplankton using various feeding modes."

- "Dinoflagellates…are able to graze on a large variety of different size classes depending on feeding behaviour."

Aliani, S., Griffa, A. and Molcard, A. (2003). Floating debris in the Ligurian Sea, north-western Mediterranean. *Marine Pollution Bulletin.* 46. pg. 1142-9. http://www.resodema.org/publications/publication28.pdf

Alivisatos, A., Blaser, M., Brodie, E., et. al. (2015). A unified initiative to harness Earth's microbiomes. *Science.* 350(6260). Pg. 507-508. http://www.sciencemag.org/content/350/6260/507.full.pdf

- "Despite their centrality to life on Earth, we know little about how microbes interact with each other, their hosts, or their environment."

- "The scientific community must also integrate ethicists, social scientists, regulators, and legal professionals at an early stage to ensure that risks associated with microbiome research are accurately assessed and proactively addressed."

- No mention is made of the "billions of tons of human-made toxic chemicals [that] have overwhelmed the degrading and recycling capacity of microbiomes" cited in the *Nature* proposal to formulate a international microbiome initiative." (Dubilier 2015)

Allen, J., McClean, M. and Webster, T. (2007). Personal exposure to polybrominated diphenyl ethers (PBDEs) in residential indoor air. *Environmental Science and Technology*. 41. pg. 4574-9. http://www.researchgate.net/publication/6143494_Personal_exposure_to_polybrominat ed_diphenyl_ethers_(PBDEs)_in_residential_indoor_air

- "We used personal air samplers to measure indoor air exposure to polybrominated diphenyl ethers (PBDEs) for 20 residents of the Greater Boston Area (Massachusetts)."

- "PBDEs are persistent in the environment, highly hydrophobic, and bioaccumulate in biota and humans."

- "The highest PBDE concentrations in humans and the environment have been reported in the United States."

- "Vast reservoirs of PBDEs remain in existing consumer products, potentially contributing to environmental and human burdens of PBDEs for decades."

- "The personal cloud effect we observed is consistent with previous studies found that specific household activities (e.g., folding blankets, walking, dry dusting) increased personal exposure to particulate matter 1.4-1.6 times compared to a stationary area monitor."

Allison, D. B., Mentore, J. L., Heo, M., et al. (1999). Antipsychotic-induced weight gain: A comprehensive research synthesis. *Am. J. Psychiatry*. 156. pg. 1686-96.

Allsopp, M., Erry, B., Stringer, R., et al. (2000). A recipe for disaster: A review of persistent organic pollutants in food. *Greenpeace Research Laboratories*. http://www.greenpeace.org/international/en/publications/reports/recipe-for-disaster-a-review/

Almeda, R., Wambaugh, Z., Chai, C., et al. (2013a). Effects of crude oil exposure on bioaccumulation of polycyclic aromatic hydrocarbons and survival of adult and larval stages of gelatinous zooplankton. *PLoS ONE*. 8(10). http://www.plosone.org/article/fetchObject.action?uri=info:doi/10.1371/journal.pone.00 74476&representation=PDF

- "Some of the most toxic PAHs of crude oil can be bioaccumulated in gelatinous zooplankton and potentially be transferred up the food web and contaminate apex predators."

Almeda, R., Wambaugh, Z., Wang, Z., et al. (2013b). Interactions between zooplankton and crude oil: Toxic effects and bioaccumulation of polycyclic aromatic hydrocarbons. *PLoS ONE*. 8(6). http://www.plosone.org/article/fetchObject.action?uri=info:doi/10.1371/journal.pone.0067212&representation=PDF

- "The negative impact of oil spills on mesozooplankton may be increased by the use of chemical dispersant and UV radiation."

- "Mesozooplankton and protozoans may play an important role in fate of PAHs in marine environments."

- "Despite their importance in marine environments, our knowledge of the interactions between zooplankton and anthropogenic pollutants is very limited."

- "Protozoans in the water reduced the toxic effects of crude oil and the bioaccumulation of PAHs in copepods."

Alvarez-Roman, R., Naik, A., Kalia, Y. (2004). Skin penetration and distribution of polymeric nanoparticles. *Journal of Control Release*. 99(1). pg. 53-62. http://www.ncbi.nlm.nih.gov/pubmed/15342180

- "Encapsulation using nanoparticulate systems is an increasingly implemented strategy in drug targeting and delivery."

- "Non-biodegradable, fluorescent, polystyrene nanoparticles (diameters 20 and 200 nm)… accumulated preferentially in the follicular openings."

- "This distribution increased in a time-dependent manner, and the follicular localization was favoured by the smaller particle size."

Anderson, T. D. and MacRae, J. D. (2006). Polybrominated diphenyl ethers in fish and wastewater samples from an area of the Penobscot River in Central Maine. *Chemosphere*. 62. pg. 1153-60. http://www.ncbi.nlm.nih.gov/pubmed/16084563

- "Fish tissue samples were collected from sites along the Penobscot River in central Maine. The total concentration of tetra- to hepta-PBDEs in these samples were calculated and generally increased from upstream to downstream locations ranging from 800 to 1810 ng/g lipid at the northernmost site to 5750-29000 ng/g at the downstream sampling site."

- "PBDE congeners from a WWTP at Orono, Maine…was concentrated in the biosolids."

Anderson, H., Imm, P., Knobeloch, L., et. al. (2008). Polybrominated diphenyl ethers (PBDE) in serum: Findings from a US cohort of consumers of sport-caught fish. *Chemosphere.* 73(2). pg. 187-94. http://www.ncbi.nlm.nih.gov/pubmed/18599108

Andersson, B., Babrauskas, V., Holmstedt, G., et al. (1999). *Simulated fires in substances of pesticide type.* http://lup.lub.lu.se/luur/download?func=downloadFile&recordOId=525800&fileOId=1266586

Andersson, Berit. (2003). *Combustion products from fires: Influence from ventilation conditions.* Lund University, Sweden, Department of Fire Safety Engineering. http://lup.lub.lu.se/luur/download?func=downloadFile&recordOId=642023&fileOId=642045

Andersson, Elin. (2014). *Microplastics in the oceans and their effect on the marine fauna.* SLU, Dept. of Biomedical Sciences and Veterinary Public Health. Uppsala, Sweden. http://stud.epsilon.slu.se/6634/7/andersson_e_140904.pdf

- "Different kinds of plastics accumulate different kinds of POPs and in different quantities."

Andersson, P., Blom, A., Johannisson, A., et al. (1999). Assessment of PCBs and hydroxylated PCBs as potential xenoestrogens: In vitro studies based on MCF-7 cell proliferation and induction of vitellogenin in primary culture of rainbow trout hepatocytes. *Environmental Contamination Toxicology.* 37. pg. 145-50. http://link.springer.com/article/10.1007%2Fs002449900499

Andrady, Anthony L. (1994). Assessment of environmental biodegradation of synthetic polymers: a review. *Journal of Macromolecular Science.* 34(1). pg. 25-75. http://www.tandfonline.com/doi/abs/10.1080/15321799408009632#preview

Andrady, Anthony L. (2003). *Plastics and the environment.* John Wiley and Sons. New York, NY.

- The index has no reference to microplastics or nanoplastics. The text has no discussion of microplastics with the exception of once on page 384.

- It contains extensive discussion of problems associated with the thermal destruction of waste, including plastic waste.

- It contains extensive information on the growing use of polymers in automobile applications as of 2000.

Andrady, Anthony L. (2009). *Proceedings of the international research workshop on the occurrence, effects and fate of micro-plastic marine debris*. National Oceanic and Atmospheric Administration. http://marinedebris.noaa.gov/sites/default/files/Microplastics.pdf

Andrady, Anthony L. (2010). *Measurement and occurrence of microplastics in the environment. Presentation at the 2nd research workshop on microplastic debris, Tacoma, WA*. National Oceanic and Atmospheric Administration. http://marinedebris.noaa.gov/sites/default/files/microplastics_workshop_2012.pdf

Andrady, Anthony L. (2011). Microplastics in the marine environment. *Marine Pollution Bulletin*. 62(8). http://ac.els-cdn.com/S0025326X11003055/1-s2.0-S0025326X11003055-main.pdf?_tid=e5b36fda-f4e1-11e4-b599-00000aab0f6b&acdnat=1431021358_6fa1067d335a442793615a60486a4aad

- "Weathering degradation of plastics on the beaches results in their surface embrittlement and microcracking, yielding microparticles that are carried into water by wind or wave action. Unlike inorganic fines present in sea water, microplastics concentrate persistent organic pollutants (POPs) by partition. The relevant distribution coefficients for common POPs are several orders of magnitude in favour of the plastic medium."

- "Land-based sources including beach litter contributes about 80% of the plastic debris."

- "About 18% of the marine plastic debris found in the ocean environment is attributed to the fishing industry."

- Any toxicity associated with plastics in general, including meso- or microplastics, can be attributed to one or more of the following factors: (a) Residual monomers from manufacture present in the plastic or toxic additives used in compounding of plastic may leach out of the ingested plastic...) (b) Toxicity of some intermediates from partial degradation of plastics. For instance, burning polystyrene can yield styrene and other aromatics and a partially burnt plastic may contain significant levels of styrene and other aromatics. (c) The POPs present in sea water are slowly absorbed and concentrated in the microplastic fragments. Plastics debris does 'clean' the sea water of the dissolved pollutant chemicals. On being ingested, however, these can become bioavailable to the organisms."

- "Sea water typically contains low levels of a host of chemical species such as insecticides, pesticides and industrial chemicals that enter the ocean via waste water and runoff. POPs such as polychlorinated biphenyls (PCBs), polybrominated diphenyl ethers (PBDEs), and perfluorooctanoic acid (PFOA) have a very large water-polymer distribution coefficient, KP/W [L/kg], in favour of the plastic… The microparticles laden with high levels of POPs can be ingested by marine biota… Microparticles and nanoparticles fall well within the size range of the staple phytoplankton diet of zooplanktons such as the Pacific Krill. There is little doubt that these can be ingested."

- "A recent study by Rios and Moore (2007) on plastic mesoparticles on four Hawaiian, one Mexican and five California beaches showed very significant levels of pollutants in the particles. The ranges of values reported were: P PAH = 39–1200 ng/g: P PCB = 27–980 ng/g: P DDT = 22–7100 ng/g. These are cumulative values for 13 PCB congeners and 15 PAHs. The cumulative levels found in plastic pellets collected from locations near industrial sites were understandably much higher. Highest values reported were P PAH = 12,000 ng/g and DDT = 7100 ng/g. A 2009 study reported data for 8 US beaches (of which 6 were in CA) as follows: P PCB = 32–605 ng/g; P DDT = 2–106 ng/g; and P HCH(4 isomers) = 0–0.94 ng/g."

- "Recent work has suggested that micro- and mesoplastic debris may also concentrate metals in addition to the POPs. This is an unexpected finding as the plastics are hydrophobic but the oxidized surface could carry functionalities that can bind metals."

- "The critical ecological risk is not due to low-levels of POPs in water but from the bioavailability of highly concentrated pools of POPs in microplastics that can potentially enter the food web via ingestion by marine biota."

- "Engineered plastic nanoparticles derived from post-consumer waste as well as from meso-/microplastics via degradation pose a specific challenge to the ecosystem."

- "Small Eukaryotic protists, Diatoms and Flagellates that measure in the range of 200 nm to a couple of microns are abundant in the oceans."

Andrady, A. and Neal, M. (2009). Applications and societal benefits of plastics. *Philosophical Transactions of the Royal Society B: Biological Sciences*. 364(1526). pg. 1977-1984. http://www.ncbi.nlm.nih.gov/pmc/articles/PMC2873019/

Anguissola, S., Garry, D., Salvati, A., et al. (2014). High content analysis provides mechanistic insights on the pathways of toxicity induced by amine-modified polystyrene nanoparticles. *PLoS ONE*. 9(9). http://www.plosone.org/article/fetchObject.action?uri=info:doi/10.1371/journal.pone.0108025&representation=PDF

Antignac, J., Cariou, R., Maume, D., et al. (2008). Exposure assessment of fetus and newborn to brominated flame retardants in France: preliminary data. *Molecular Nutrition Food Research*. 52. pg. 258–265.
http://www.ncbi.nlm.nih.gov/pubmed/18186099

- "Preliminary results obtained on 26 individuals (mother/newborn pairs) mainly demonstrated the presence of polybromodiphenylethers (PBDE) and tetrabromobisphenol A both in maternal and fetal matrices, and a possible risk of overexposure of newborns through breastfeeding."

Arnot, Jon and Gobas, Frank. (2006). A review of bioconcentration factor (BCF) and bioaccumulation factor (BAF) assessments for organic chemicals in aquatic organisms. *Environmental Reviews*. 14(4). pg. 257-97.
http://www.nrcresearchpress.com/doi/abs/10.1139/a06-005?src=recsys&journalCode=er#.VQXnYI7F-Jc

- "A review of 392 scientific literature and database sources includes 5317 bioconcentration factor (BCF) and 1656 bioaccumulation factor (BAF) values measured for 842 organic chemicals in 219 aquatic species… 45% of BCF values are subject to at least one major source of uncertainty."

- "Field BAFs tend to be greater than laboratory BCFs… only 0.2% of current use organic chemicals have BAF measurements."

Arthur, Courtney, Baker, Joel and Bamford, Holly, Eds. (2009). *Proceedings of the International Research Workshop on the occurrence, effects, and fate of microplastic marine debris: September 9-11, 2008*. Technical memorandum NOS-OR&R-30. NOAA, Tacoma, WA.

- This is a ±528 page report sponsored by NOAA on microplastic debris. The terms nanoplastic or plastic nanoparticles (PNP) do not appear in this report, nor are they analyzed or discussed with respect to the report title "effects and fate of microplastic debris."

- The following quotations are from the workshop session abstracts by authors cited in this bibliography.

- Dr. Joel Baker, University of Washington Tacoma: "The overall impact and consequences of marine microplastics has not yet been articulated."

- Dr. R. C. Thompson, University of Plymouth, Plymouth, UK: "There is also concern that ingestion of small items of plastic debris could facilitate the transport of toxic chemicals to marine organisms… but more work will be required to reach firm conclusions."

- Dr. Thompson is famous for posing the unanswered question "where have all the plastics gone?" an issue studiously avoided in this depressing NOAA report.

- Dr. Emma Teuten, University of Edinburgh, Edinburgh, UK: "While it has been unequivocally demonstrated that contaminants concentrate on plastics in the environment, little evidence exists supporting their subsequent transfer to animals."

- Dr. Miriam Doyle, University of Washington, Seattle, WA: "The plastic particles were assigned to three plastic product types: product fragments, fishing net and line fibers, and industrial pellets; and five size categories: <1 mm, 1-2.5 mm, 2.5-5 mm, 5-10 mm, and >10 mm… Their ubiquity in the plankton samples and predominance of particles <2.5 mm, implies persistence in these pelagic ecosystems as a result of continuous breakdown from larger plastic debris fragments, and widespread distribution by ocean currents."

- So, what about plastics <1 mm, e.g. ±250 um, and especially, plastic nanoparticles (PNP) <250 nm?

- Dr. Upal Ghosh, University of Maryland: "The majority of hydrophobic contaminants such as PAHs and PCBs are strongly bound to carbonaceous particles…anthropogenic organic particulates in impacted sediments (coal, soot, charcoal, wood, coal tar pitch, and humic materials."

- Rainen Lohman, University of Rhode Island: "Based on available information, it seems unlikely that the amount of microplastics in the marine environment is currently large enough to be an important geochemical reservoir for POPs, as research was presented that pointed to a much stronger binding of organic pollutants to the more abundant black carbon that to plastic polymers."

- Dr. Alan J. Mearns, NOAA: "It remains unclear to what extent microplastics represent a source for food chain POP accumulation compared to other sources."

- Dr. Anthony L. Andrady, Research Triangle Institute, Durham, NC: "The plastic microparticles well known to be present in the oceans, likely originate from both the slow deterioration of the floating or submerged plastics, as well as the fragmentation of plastics degraded to embrittlement in the beach environment."

- The evasion of discussing the fate of microplastics as they continue to evolve into smaller particles and their increasing uptake of ecotoxins of every description as a result of their increasing surface to volume ratio makes this NOAA report a classic example of governmental and institutional evasion of one of the most important environmental issues of the 21st century.

Asakura, H., Matsuto, T. and Tanaka, N. (2004). Behavior of endocrine-disrupting chemicals in leachate from MWS landfill sites in Japan. *Waste Management.* 24. pg. 613-22. http://wastegr2-er.eng.hokudai.ac.jp/home_old/publish/09.pdf

- "In Japan, 3.18 million tons (33% of annual production) of plastics were landfilled in 1999 (Plastic Waste Management Institute, 1999). Therefore, EDCs in landfills may be released to the environment with leachate."

- "Concentrations of BPA and DEHP were almost constant regardless of season."

- "Concentration of BPA in raw leachate tends to decrease as the years go by, but the concentration of DEHP was observed to remain at a constant level."

- "BPA ranged widely from 0.07 to 228 lg/l and was 2000 times higher than in surface water. Among the four phthalates, DEHP was over 30 times higher than in surface water."

Ashton, K., Holmes, L. and Turner, A. (2010). Association of metals with plastic production pellets in the marine environment. *Marine Pollution Bull.* 60. pg. 2050-5.

Asmatulu, R. (2011). *Toxicity of nanomaterials and recent developments in lung disease.* Department of Mechanical Engineering, Wichita State University, Fairmount, KS, USA. http://cdn.intechopen.com/pdfs-wm/17355.pdf

- "Nanomaterials, including nanoparticles, nanotubes, nanofibers, and nanocomposites, in the forms of metals and alloys, ceramics, polymers, and composites are all produced by nanotechnology processes and are considered to be the next generation of materials for manufacturing faster cars and planes, more powerful computers and satellites, more sensitive sensors, stronger materials for structural applications, and better micro- and nanochips and batteries… Nanomaterials have outstanding mechanical, electrical, optical, magnetic, quantum mechanic, and thermal properties. Nanomaterials are already found in more than a thousand different products, including bacteria-free cloth, concrete, filtration units, sunscreen, car bumpers, tooth paste, polymeric coatings, solar and fuel cells, lithium-ion batteries, tennis rackets, wrinkle-resistant clothing, and optical, electronic, and sensing devices."

Atlantic States Marine Fisheries Commission. (2012). The Expanded Multispecies Virtual Population Analysis (MSVPA-X). ASMFC. http://www.asmfc.org/uploads/file/2012MSPVA_Update.pdf

ATSDR. (1994). *Toxicological profile for chlordane.* Agency for Toxic Substances and Disease Registry. Atlanta, GA. http://www.atsdr.cdc.gov/toxprofiles/tp31-c4.pdf

ATSDR. (1995a). *Toxicological profile for diethyl phthalate (DEP).* Agency for Toxic

Substances and Disease Registry. Atlanta, GA.
http://www.atsdr.cdc.gov/toxprofiles/tp73.pdf

ATSDR. (1995b). *Toxicological profile for Mirex and Chlordecone*. Agency for Toxic Substances and Disease Registry. Atlanta, GA.
http://www.atsdr.cdc.gov/toxprofiles/tp66-c4.pdf

ATSDR. (1996). *Toxicological profile for Endrin*. Agency for Toxic Substances and Disease Registry. Atlanta, GA.
http://www.atsdr.cdc.gov/toxprofiles/tp.asp?id=617&tid=114

ATSDR. (2001). *Toxicological profile for di-n-butyl phthalate (DBP)*. Agency for Toxic Substances and Disease Registry. Atlanta, GA.
http://www.atsdr.cdc.gov/toxprofiles/tp135.pdf

ATSDR. (2002a). *Toxicological profile for 2-di(ethylhexyl)phthalate (DEHP)*. Agency for Toxic Substances and Disease Registry. Atlanta, GA.
http://www.atsdr.cdc.gov/toxprofiles/tp9.pdf

ATSDR. (2002b). *Toxicological profile for DDT, DDE, and DDD*. US Agency for Toxic Substances and Disease Registry. http://www.atsdr.cdc.gov/toxprofiles/tp35.pdf

ATSDR. (2002c). *Toxicological profile for Aldrin/Dieldrin*. US Agency for Toxic Substances and Disease Registry. Atlanta, GA.
http://www.atsdr.cdc.gov/toxprofiles/tp.asp?id=317&tid=56

ATSDR. (2004a). *Final interaction profile for persistent chemicals found in breast milk (chlorinated dibenzo-p-dioxins, hexachlorobenzene, p,p'-DDE, methylmercury, and polychlorinated biphenyls)*. US Agency for Toxic Substances and Disease Registry.
http://www.atsdr.cdc.gov/interactionprofiles/ip03.html

ATSDR. (2004b). *Toxicological profile for polybrominated biphenyls and polybrominated diphenyl ethers (PBBs and PBDEs)*. Agency for Toxic Substances and Disease Registry. Atlanta, GA. http://www.atsdr.cdc.gov/toxprofiles/tp68.pdf

ATSDR. (2005). *Curtis Bay Coast Guard Yard public health assessment final release*. U.S. Department of Health and Human Services, Agency for Toxic Substances and Disease Registry. Atlanta, GA.
http://www.atsdr.cdc.gov/HAC/pha/CurtisBay121504/CurtisBay-pt1.pdf and
http://www.atsdr.cdc.gov/HAC/pha/CurtisBay121504/CurtisBay-pt2.pdf

ATSDR. (2007a). *Health consultation: Great Kills Park, Richmond County Gateway National Recreation Area, National Park Service, Staten Island, New York*. U.S. Department of Health and Human Services, Agency for Toxic Substances and Disease Registry. Atlanta, GA.

http://www.atsdr.cdc.gov/HAC/pha/GatewayNatlRecreationArea/GatewayNationalRec AreaHC053107.pdf

ATSDR. (2007b). *Toxicological profile for heptachlor and heptachlor epoxide*. Department of Health and Human Services, Agency for Toxic Substances and Disease Registry. Atlanta, GA. http://www.atsdr.cdc.gov/toxprofiles/tp.asp?id=746&tid=135

ATSDR. (2013). *Toxicological profile for hexachlorobenzene*. U.S. Department of Health and Human Services, Agency for Toxic Substances and Disease Registry. Atlanta, GA. http://www.atsdr.cdc.gov/toxprofiles/tp.asp?id=627&tid=115

ATSDR. (2015). *Toxicological profile for polychlorinated biphenyls (PCBs)*. U.S. Department of Health and Human Services, Agency for Toxic Substances and Disease Registry. Atlanta, GA. http://www.atsdr.cdc.gov/toxprofiles/tp.asp?id=142&tid=26

Aulakh, Rabinder, S., Gill, Jatinder Paul S., Bedi, Jasbir S., et al. (2006). Organochlorine pesticide residues in poultry feed, chicken muscle and eggs at a poultry farm in Punjab, India. *Journal of the Science of Food and Agriculture*. 86(5). pg. 741-4. http://onlinelibrary.wiley.com/doi/10.1002/jsfa.2407/abstract

- "The results indicated that poultry feed could be one of the major sources of contamination for chicken and eggs. These residues are present despite complete ban on the use of technical HCH and DDT for agricultural purposes in India."

Ausman, K., Carter, J., Karn, B., Kreyling, W., Lai, D., Olin, S., Monteiro-Riviere, N., Warheit, D. and Yang, H. (2005). Principles for characterizing the potential human health effects from exposure to nanomaterials: Elements of a screening strategy. *Particulate Fibre Toxicology*. 2(8). pg. 1-35. http://www.particleandfibretoxicology.com/content/2/1/8

- "Physicochemical properties that may be important in understanding the toxic effects of test materials include particle size and size distribution, agglomeration state, shape, crystal structure, chemical composition, surface area, surface chemistry, surface charge, and porosity."

Austin, L., Dunford, J., Milley, M. and Bostick, T. (2013). *SIGAR 14-13 inspection report: Forward operating base Sharana: Poor planning and construction resulted in $5.4 million spent for inoperable incinerators and continued use of open-air burn pits*. Special Investigator General for Afghanistan Reconstruction.

Azarmi, S., Roa, W. H. and Lobenberg, R. (2008). Targeted delivery of nanoparticles for the treatment of lung diseases. *Advanced Drug Delivery Reviews*. 60(8). pg. 863-75. http://www.ncbi.nlm.nih.gov/pubmed/18308418

- "By developing colloidal delivery systems such as liposomes, micelles and nanoparticles a new frontier was opened for improving drug delivery. Nanoparticles with their special characteristics such as small particle size, large surface area and the capability of changing their surface properties have numerous advantages…for treatment or diagnostic purposes."

Backhurst, M. K. and Cole, R. G. (2000). Subtidal benthic marine litter at Kawau Island, north-eastern New Zealand. *Journal of Environmental Management.* 6. pg. 227-37. http://www.sciencedirect.com/science/article/pii/S0301479700903815

Bae, B., Jeong, J. and Lee, S. J. (2002). The quantification and characterization of endocrine disruptor bisphenol-A leaching from epoxy resin. *Water Science Technology.* 46. pg. 381–387. http://www.ncbi.nlm.nih.gov/pubmed/12523782

- "Bisphenol-A (BPA), a known endocrine disruptor, is a main building block of epoxy resin which has been widely used as a surface coating agent on residential water storage tanks."

- "BPA leashing increased as the water temperature increases. In addition, microbial growth, measured by colony forming units, in epoxy coated water tanks was higher than that in a stainless steel tank."

Bagulayan, A., Bartlett-Roa, J., Carter, A., et al. (2012). Journey to the center of the gyre: The fate of the Tohoku Tsunami debris field. *Oceanography.* 25(2). pg. 200-7. http://www.tos.org/oceanography/archive/25-2_bagul.pdf

- "The 9.0 magnitude Tohoku earthquake that struck off the coast of Japan on March 11, 2011, was the fourth largest earthquake in recorded history and the largest ever to hit a densely populated region."

- "The ensuing tsunami inundated an area of about 561 km^2, washing away an estimated 24.9 million tonnes of debris, including wood, sediments, plastics, industrial chemicals, and structural components."

- "Two weeks following the tsunami, the meltdown of the Fukushima Daiichi nuclear reactors released radioactive elements into the atmosphere and coastal waters."

- "We predict that the Tohoku debris field will create a rare perturbation for ecosystems interconnected across the North Pacific, exacerbating the accumulating human impacts on the world ocean."

- "Long-lived radioactive isotopes could be incorporated into sediment, phytoplankton, and brown algae, eventually bioaccumulating in higher trophic levels such as copepods, molluscs, polychaetes, and fishes."

- "Chemicals and small floating plastics typically concentrate within a microlayer at the air-sea interface where the majority of buoyant eggs and larvae occur. As pollutant exposed larvae and juveniles are preyed upon, chemical concentrations may bioaccumulate and compound deleterious effects."

- "While this is an unprecedented influx of debris, the widespread short-term effects will only be a pulse in the rising levels of anthropogenic impacts to the ocean."

Baillie-Hamilton, P. (2002). Chemical toxins: A hypothesis to explain the global obesity epidemic. *Journal of Alternative Complementary Medicine.* pg. 185-92. http://www.ibarguchi.ca/teaching-CHEO/Baillie-Hamilton-Chems%20and%20obesity.pdf

- "The commonly held causes of obesity—overeating and inactivity—do not explain the current obesity epidemic."

- "Many of these chemicals are better known for causing weight loss at high levels of exposure but much lower concentrations of these same chemicals have powerful weight-promoting actions."

- "The current level of human exposure to these chemicals may have damaged many of the body's natural weight-control mechanisms."

- "In their daily lives, human beings are now exposed to tens of thousands of these chemicals, in the forms of pesticides, dyes, pigments, medicines, flavorings, perfumes, plastics, resins, rubber-processing chemicals, intermediate chemicals, plasticizers, solvents, and surface-active agents."

- "The average person now has many hundreds of industrial chemicals lodged in his or her body, with many of these toxins being transferred across the fetal–maternal blood barrier."

- "These chemicals—which human beings are exposed to quite regularly—include:"

  - Pesticides, for example, organochlorines, such as dichlorodiphenyltrichloroethane (DDT), endrin, lindane, and hexachlorobenzene
  - Organophosphates
  - Carbamates, including dithiocarbamates
  - Polychlorinated biphenyls

117

- Polybrominated biphenyls, which are commonly used as fire retardants
- Plastics, such as phthalates and bisphenol A
- Heavy metals, such as cadmium and lead
- Solvents

- "Many synthetic chemicals [are] being used by the agricultural community to promote animal fattening and growth. These substances, generally known as growth promoters, include such synthetic chemicals as antithyroid drugs, corticosteroids, anabolic steroids, organophosphate pesticides, carbamates, antibacterials, and ionophores."

- "The earth's environment has changed significantly during the last few decades because of the exponential production and usage of synthetic organic and inorganic chemicals...The coincidence of the obesity epidemic with the appearance of these chemicals in the environment indicates the possibility of a causative relationship."

Bailey, R., van Wijk, D. and Thomas, P. (2009). Sources and prevalence of pentachlorobenzene in the environment. *Chemosphere*. 75(5). pg. 555-64. http://www.sciencedirect.com/science/article/pii/S0045653509000770

Bakir, A., Rowland, S. and Thompson, R. (2012). Competitive sorption of persistent organic pollutants onto microplastics in the marine environment. *Marine Pollution Bulletin*. 64. pg. 2782-9. http://www.ncbi.nlm.nih.gov/pubmed/23044032

Balakrishnan, B., Henare, K., Thorstensen, E., et al. (2010). Transfer of bisphenol A across the human placenta. *American Journal of Obstetric Gynecology*. 202(4). pg. 1-7. http://www.ncbi.nlm.nih.gov/pubmed/20350650

Baltic Marine Environment Protection Commission HELCOM. (2014). *Preliminary study on synthetic microfibers and particles at a municipal waste water treatment plant*. BMEPC HELCOM. Helsinki, Finland. http://helcom.fi/Lists/Publications/Microplastics%20at%20a%20municipal%20waste%20water%20treatment%20plant.pdf

- "The objective of this project was to study the amount of microplastic litter arriving at the Central Wastewater Treatment Plant (WWTP) of St. Petersburg...the results of this study show that the WWTPs may operate as a point source of microplastic litter into the aquatic environment."

- "In the sampling process, water was filtered through different mesh-sized filters: 300, 200 and 20 um, respectively."

- "Microplastics settle or are captured into the sludge during the processes [96%], but some of them also pass the treatment and end up in the water environment with the purified wastewater."

- No mention is made of micro- or nanoplastics < 20 um, all of which would be translocated in the purified wastewater.

Barber, L. B., Murphy, S. F., Verplanck, P. L., et al. (2006). Chemical loading into surface water along a hydrological, biogeochemical, and land use gradient—A holistic watershed approach. *Environmental Science and Technology*. 40(2). pg. 475-86. http://toxics.usgs.gov/highlights/pharm_watershed/index.html

Barber, L., Antweiler, Ronald C., Flynn, Jennifer L. et al. (2011a). Lagrangian mass-flow investigatins of inorganic contaminants in wastewater-impacted streams. *Environ. Sci. Technol.* 45(7). pg. 2575-83.

Barber, L., Keefe, S., Kolpin, D., et al. (2011b). *Lagrangian sampling of wastewater treatment plant discharges into Boulder Creek, Colorado and Fourmile Creek, Iowa during the summer of 2003 and spring of 2005--Hydrological and water-quality data.* U.S. Geological Survey Open-File Report 2011. US Geological Survey. http://pubs.usgs.gov/of/2011/1054/report/OF11-1054.pdf

- This USGS study included the following categories: Trace elements and major elements, acidic organic wastewater compounds, antibiotic compounds, pharmaceutical compounds, neutral organic wastewater compounds, steroid and steroidal-hormone compounds, pesticide compounds.

Barber, L., Vajda, A., Douville, C., et al. (2012). Fish endocrine disruption responses to a major wastewater treatment facility upgrade. *Environmental Science and Technology*. pg. 2121-31. http://pubs.acs.org/doi/abs/10.1021/es202880e

Barber, L., Keefe, S., Brown, G., et al. (2013). Persistence and potential effects of complex organic contaminant mixtures in wastewater-impacted streams. *Environmental Science Technology*. 47(5). http://pubs.acs.org/doi/abs/10.1021/es303720g

Barlow, S., Chesson, A., Collins, J., et al. (2009). Scientific opinion of the scientific committee on a request from the European commission on the potential risks arising from nanoscience and nanotechnologies on food and feed safety. *The EFSA Journal*. 958. pg. 1-39. http://www.efsa.europa.eu/en/efsajournal/doc/958.pdf

Barnes, D., Galgani, F., Thompson, R. and Barlaz, M. (2009). Accumulation and fragmentation of plastic debris in global environments. *Philosophical Transcripts of the Royal Society of London Biological Science*. 364(1526). pg. 1985-98. http://www.ncbi.nlm.nih.gov/pmc/articles/PMC2873009/

Barnes, K., Kolpin, D., Meyer, M., et al. (2002). *Water-quality data for pharmaceuticals, hormones, and other organic wastewater contaminants in U.S. streams, 1999-2000*. U.S. Geological Survey Open-File Report 02-94. http://toxics.usgs.gov/pubs/OFR-02-94/index.html

Barr, D., Wang, R. and Needham, L. (2005). Biologic monitoring of exposure to environmental chemicals throughout the life stages: requirements and issues for consideration for the national children's study. *Environmental Health Perspectives*. 113. pg. 1038-91. http://www.ncbi.nlm.nih.gov/pmc/articles/PMC1280353/pdf/ehp0113-001083.pdf

- "The matrices available for analyses include blood, urine, breast milk, adipose tissue, and saliva, among others."

- "Generally, persistent organic chemicals are measured more readily in blood-based matrices or other lipid-rich matrices."

Barrena, R., Casals, E., Colon, J., et al. (2009). Evaluation of ecotoxicity of model nanoparticles. *Chemosphere*. 75. pg. 850-7. http://ictaservidor.uab.es/99_recursos/1240388690024.pdf

Bartrons, M., Grimalt, J., de Mendoza, G. and Catalan, J. (2012). Pollutant dehalogenation capability may depend on the trophic evolutionary history of the organism: PBDEs in freshwater food webs. *PLoS ONE*. 7(7). http://www.plosone.org/article/fetchObject.action?uri=info:doi/10.1371/journal.pone.0041829&representation=PDF

Basi, Christian. (August 22, 2013). Toxic nanoparticles might be entering human food supply, MU study says. *Missouri University News Releases*. http://munews.missouri.edu/news-releases/2013/0822-toxic-nanoparticles-might-be-entering-human-food-supply-mu-study-finds/

- "Farmers have used silver nanoparticles as a pesticide because of their capability to suppress the growth of harmful organisms."

- "More than 1,000 products on the market are nanotechnology-based products," said Mengshi Lin, associate professor of food science in the MU College of Agriculture, Food and Natural Resources. "This is a concern because we do not know the toxicity of the nanoparticles. Our goal is to detect, identify and quantify these nanoparticles in food and food products and study their toxicity as soon as possible."

- "The growing trend to use other types of nanoparticles has revolutionized the food industry by enhancing flavors, improving supplement delivery, keeping food fresh longer and brightening the colors of food. However, researchers worry that the use of silver nanoparticles could harm the human body."

Battaglin, W., Kuivila, K., Winton, K. and Meyer, M. (2008). *Occurrence of Chlorothalonil, its transformaction products, and selected other pesticides in Texas and Oklahoma streams, 2003-2004*. U.S. Geological Survey. http://pubs.usgs.gov/sir/2008/5016/pdf/SIR08-5016.pdf

Bauer, M. and Herrmann, R. (1997). Estimation of the environmental contamination by phthalic acid esters leaching from household wastes. *Sci. Total Environ*. 208. pg. 49-57.

Bauer, M. and Herrmann, R. (1998). Dissolved organic carbon as the main carrier of pthalic acid esters in municipal landfill leachates. *Waste Management Research*. 16. pg. 446-54. http://www.researchgate.net/publication/245383598_Dissolved_organic_carbon_as_the_main_carrier_of_phthalic_acid_esters_in_municipal_landfill_leachates

Baun, A., Sørensen, S., Rasmussen, R., Hartmann, N., and Koch, C. (2008). Toxicity and bioaccumulation of xenobiotic organic compounds in the presence of aqueous suspensions of aggregates of nano-C60. *Aquatic Toxicology*. 86. pg. 379-87. http://www.ncbi.nlm.nih.gov/pubmed/18190976

- "This study is the first to demonstrate the influence of C(60)-aggregates on aquatic toxicity and bioaccumulation of other environmentally relevant contaminants. The data provided underline that not only the inherent toxicity of manufactured nanoparticles, but also interactions with other compounds and characterisation of nanoparticles in aqueous suspension are of importance for risk assessment of nanomaterials."

Bell, Jennifer M. (2014). *Characterization, composition and source identification of Iraqi aerosols*. Dissertation. University of Alaska, Fairbanks. https://scholarworks.alaska.edu/bitstream/handle/11122/4580/Bell_uaf_0006E_10199.pdf?sequence=1

- "A study initiated in 2008 was designed to determine the concentrations and compositions of fine particulate matter in Baghdad, Iraq."

- "A combination of processes (brittle fragmentation, saltation, long-range transport, and midair collisions during high wind conditions) occur that result in excess mechanical grinding to produce ultrafine soil particles during high wind scenarios."

- "The production of these particles are important in that the fine particulate matter concentrations frequently exceed military exposure guidelines of 65 µg m-3 and individual constituents, such as lead, exceed U.S. national ambient air quality standards designed to protect human health."

- No mention is made in this dissertation of any environmental chemicals sorbed by or associated with the UFPs noted in this analysis.

Bellou, N., Papathanassiou, E., Dobretsov, S., et al. (2012). The effect of substratum type, orientation and depth on the development of bacterial deep-sea biofilm communities grown on artificial substrata deployed in the Eastern Mediterranean. *Biofouling*. 28(2). pg. 199-213. http://www.ncbi.nlm.nih.gov/pubmed/22352335

Bennett, Samuel W., Zhou, Dongxu, Mielke, Randall and Keller, Arturo A. (2012). Photoinduced disaggregation of TiO_2 nanoparticles enables transdermal penetration. *PLoS ONE*. 7(11). http://journals.plos.org/plosone/article?id=10.1371/journal.pone.0048719

- "Under many aqueous conditions, metal oxide nanoparticles attract other nanoparticles and grow into fractal aggregates as the result of a balance between electrostatic and Van Der Waals interactions."

- "Ambient light and other light sources can partially disaggregate nanoparticles from the aggregates and increase the dermal transport of nanoparticles, such that small nanoparticle clusters can readily diffuse into and through the dermal profile, likely via the interstitial spaces."

- "Photoinduced disaggregation may have important health implications."

Bergmann, Melanie and Klages, Michael. (2012). Increase of litter at the Arctic deep-sea observatory HAUSGARTEN. *Marine Pollution Bulletin*. 64. pg. 2734-41. http://www.resodema.org/publications/publication29.pdf

- "To quantify litter on the deep seafloor over time, we analysed images… taken in 2002, 2004, 2007, 2008 and 2011 (2500 m depth). Litter increased from 3635 to 7710 items km2 between 2002 and 2011. Plastic constituted the majority of litter (59%)."

- "The changes in litter could be an indirect consequence of the receding sea ice, which opens the Arctic Ocean to the impacts of man's activities."

Bernhoft, A., Wiig, O. and Skaare, J. U. (1997). Organochlorines in polar bears (*Ursus maritimus*) at Svalbard. *Environ Pollut*. 95. pg. 159-75.

Besis, Athanasios and Samara, Constantini. (2012). Polybrominated diphenyl ethers (PBDEs) in the indoor and outdoor environments: A review on occurrence and human exposure. *Environmental Pollution*. 169. pg. 217-29.

http://www.researchgate.net/profile/Constantini_Samara/publication/224948065_Polyb
rominated_diphenyl_ethers_%28PBDEs%29_in_the_indoor_and_outdoor_environment
s--
a_review_on_occurrence_and_human_exposure/links/53d8b4ad0cf2e38c63318c42.pdf

- "Due to their high production volume, widespread usage, and environmental persistence, PBDEs have become ubiquitous contaminants in environmental media, biota and humans. As their levels are rapidly increasing in the environment, these chemicals have evolved from 'emerging contaminants' to globally-distributed organic pollutants."

- "Concentrations in indoor dust are highest in workplaces, particularly in e-waste storage and recycling facilities. Car dust in particular has markedly higher PBDEs content than house dust."

Besseling, E., Wegner, A., Foekema, E., et al. (2013). Effects of microplastic on fitness and PCB bioaccumulation by the lugworm *Arenicola marina*. *Environmental Science and Technology*. 47. pg. 593-600. http://pubs.acs.org/doi/abs/10.1021/es302763x

Besseling, E., Wang, B., Lurling, M. and Koelmans, A. (2014). Nanoplastic affects growth of *S. obliquus* and reproduction of *D. magna*. *Environmental Science and Technology*. 48(20). pg. 12336-43. http://pubs.acs.org/doi/abs/10.1021/es503001d

- "Direct life history shifts in algae and *Daphnia* populations may occur as a result of exposure to nanoplastic."

Betts, K. (2002). Rapidly rising PBDE levels in North America. *Environmental Science and Technology*. 36(3). pg. 50A-52A. http://www.ncbi.nlm.nih.gov/pubmed/11871568

Betts, K. (2008a). Why small plastic particles may pose a big problem in the oceans. *Environmental Science and Technology*. 42. pg. 8995. http://cmore.soest.hawaii.edu/education/teachers/science_kits/materials/Marine_Debris/Day_1-supplemental_materials/Betts_%282008%29_Plastic_particles.pdf

Betts, K. (2008b). New thinking on flame retardants. *Environmental Health Perspectives*. 116(5). pg. A210-A213. http://www.ncbi.nlm.nih.gov/pmc/articles/PMC2367656/pdf/ehp0116-a00210.pdf

Beychok, Milton. (1987). A database for dioxin and furan emissions from refuse incinerators. *Atmospheric Environment*. 21(1). http://dx.doi.org/10.1016%2F0004-6981%2887%2990267-8

Beyer, A., Mackay, D., Matthies, M., et. al. (2000). Assessing long-range transport potential of persistent organic pollutants. *Environmental Science and Technology*. 34(4). pg. 699-703. http://pubs.acs.org/doi/abs/10.1021/es990207w

Beyler, Craig and Hirschler, Marcelo. (2001). Thermal decoposition of polymers. In *SFPE handbook of fire protection engineering.* NFPA, Quincy, MA. http://www.ewp.rpi.edu/hartford/~ernesto/F2012/EP/MaterialsforStudents/Patel/Beyler _Hirschler_SFPE_Handbook_3.pdf

Bhattacharya, P., Lin, S., Turner, J. and Chun Ke, P. (2010). Physical adsorption of charged plastic nanoparticles affects algal photosynthesis. *Journal of Physical Chemistry.* https://www.researchgate.net/profile/Priyanka_Bhattacharya/publication/215980572_P hysical_Adsorption_of_Charged_Plastic_Nanoparticles_Affects_Algal_Photosynthesis/ links/09e415011b2323a446000000.pdf

Bhushan, Bharat. (2002). *Introduction to tribology.* Wiley, New York, NY.

Bhushan, Bharat. (1999). *Handbook of Micro/nanotribology.* CRC Press. Boca Raton, FL.

Bhushan, Bharat. (2008a). *Nanotribology and nanomechanics – an introduction, 2nd edition.* Springer, Berlin. Heidelberg, Germany.

Bhushan, B. and Jung, Y. (2008b). Wetting, adhesion and friction of superhydrophobic and hydrophilic leaves and fabricated micro-/nanopatterned surfaces. *Journal of Physics.* 20(22). http://iopscience.iop.org/0953-8984/20/22/225010/pdf/0953-8984_20_22_225010.pdf

Bhushan, Bharat. (2009a). *Introduction to nanotechnology.* Wiley India Pvt. Ltd. http://link.springer.com/referenceworkentry/10.1007%2F978-3-540-29857-1_1#

- "Nanotechnology encompasses the production and application of physical, chemical, and biological systems at scales ranging from individual atoms or molecules to submicron dimensions, as well as the integration of the resulting nanostructures into larger systems."

- "Science and technology research in nanotechnology promises breakthroughs in areas such as materials and manufacturing, nanoelectronics, medicine and healthcare, energy, biotechnology, information technology, and national security. It is widely felt that nanotechnology will be the next Industrial Revolution."

- "When the dimension of a material is reduced from a large size, the properties remain the same at first, then small changes occur, until finally when the size drops below 100 nm, dramatic changes in properties can occur. If only one length of a three-dimensional nanostructure is of nanodimension, the structure is referred to as a quantum well; if two sides are of nanometer length, the structure is referred to as a quantum wire. A quantum dot has all three dimensions in the nano range."

- "NEMS refers to nanoscopic devices that have a characteristic length of less than 100 nm and that combine electrical and mechanical components."

- "Biologically inspired design, adaptation or derivation from nature is referred to as biomimetics, a term coined by the polymath Otto Schmitt in 1957. Biomimetics is derived from the Greek word biomimesis."

- "For example, one key question is what happens to nanoparticles (such as buckyballs or nanotubes) in the environment and whether they are toxic in the human body, if digested."

Bhushan, B. (2009b). Biomimetics – Lessons from nature, an overview. *Philosophical Transactions of the Royal Society of London*. 367. pg. 1445-86. https://www.mecheng.osu.edu/nlbb/files/nlbb/PhilTrans_Biomimetics_Overview_BB_09.pdf

Bidleman, T. F., Helm, P. A., Braune, B. M. et al. (2010). Polychlorinated naph-thalenes in polar environments: A review. *Science of the Total Environment*. 408. pg. 2919-35. http://www.sciencedirect.com/science/article/pii/S0048969709008572

Biggers, William J. and Laufer, Hans. (2004). Identification of juvenile hormone-active alkylphenols in the lobster *Homarus americanus* and in marine sediments. *Biol. Bull.* 206. pg. 13-24.

Biles, J., McNeal, T., Begley, T. and Hollifield, H. (1997). Determination of bisphenol-A in reusable polycarbonate food-contact plastics and migrations to food simulating liquids. *Journal of Agricultural Food Chemistry*. 45. pg. 3541-4. http://pubs.acs.org/doi/abs/10.1021/jf970072i

- "Bisphenol-A (BPA) is a principal reactant in the preparation of polycarbonate (PC) plastics and has been shown in in vitro cell proliferation studies to exhibit estrogen-like characteristics. Reusable baby bottles, water carboys, and other housewares are often made of PC."

- "Residual amounts of BPA found in PC food contact articles ranged from 7 to 58 μg/g."

Birnbaum, L. S. and Staskal, D. F. (2004). Brominated flame retardants: Cause for concern? *Environ Health Perspect*. 112. pg. 9-17.

Blaser, S.A., Scheringer, M., MacLeod, M. and Hungerbuhler, K. (2008). Estimation of cumulative aquatic exposure and risk due to silver: Contribution of nanofunctionalized plastics and textiles. *Science of the Total Environment.* 390. pg. 396-409. http://www.ncbi.nlm.nih.gov/pubmed/18031795

- "Products with antimicrobial effect based on silver nanoparticles are increasingly used in Asia, North America and Europe. This study presents an analysis of risk to freshwater ecosystems from silver released from these nanoparticles incorporated into textiles and plastics."

Blaxill, Mark F. (2004). What's going on? The question of time trends in autism. *Public Health Reports.* 119. http://www.publichealthreports.org/issueopen.cfm?articleID=1413

- "Increases in the reported prevalence of autism and autistic spectrum disorders in recent years have fueled concern over possible environmental causes."

- "Large increases in prevalence in both the United States and the United Kingdom that cannot be explained by changes in diagnostic criteria or improvements in case ascertainment."

- "Reported rates of autism in the United States increased from 3 per 10,000 children in the 1970s to 30 per 10,000 children in the 1990s."

Blaylock, Russell. (2013). *Impacts of chemtrails on human health. Nanoaluminum: Neurodegenerative and neurodevelopmental effects*. Research on Globalization. http://www.globalresearch.ca/impacts-of-chemtrails-on-human-health-nanoaluminum-neurodegenerative-and-neurodevelopmental-effects/5342624

Boerger, C., Lattin, G., Moore, S. et al. (2010). Plastic ingestion by planktivorous fishes in the North Pacific Central Gyre. *Marine Poll. Bull.* 60. pg. 2275-8. http://www.cleanwateraction.org/files/publications/Fish%20ingestion%20of%20plastic.pdf

- "Debris and zooplankton density were measured at three depths…Density of debris was greatest near the bottom."

- "The mass of plastic collected exceeded that of zooplankton, though when the comparison was limited to the size of most zooplankton, zooplankton mass was three times that of debris."

de Boer, J., Wester, P. G., van der Horst, A. and Leonards, P. E. G. (2003). Polybrominated diphenyl ethers in influents, suspended particulate matter, sediments, sewage treatment plant and effluents and biota from the Netherlands. *Environ Pollut.* 122. pg. 63-74.

de Boer, I., Rue, T., Hall, Y., et al. (2011). Temporal trends in the prevalence of diabetic kidney disease in the United States. *J.of the American Medical Association.*

305(24). pg. 2532-9. http://jama.jamanetwork.com/article.aspx?articleid=646748

- "Prevalence of DKD in the United States increased from 1988 to 2008 in proportion to the prevalence of diabetes. Among persons with diabetes, prevalence of DKD was stable despite increased use of glucose-lowering medications and renin-angiotensin-aldosterone system inhibitors."

Bolton, T. and Havenhand, J. (1998). Physiological versus viscosity-induced effects of an acute reduction in water temperature on microsphere ingestion by trochophore larvae of the serpulid polychaete Galeolaria caespitosa. *Journal of Plankton Research*. 20(11). pg. 2153-64. http://plankt.oxfordjournals.org/content/20/11/2153.full.pdf

- "When 3 and 10 um spheres were supplied to larvae both separately and in combination, reduced water temperature resulted in a 60% decline in the number of microspheres ingested."

- "Data reported here were obtained with an acute change in temperature using polymer microspheres which obviously do not constitute part of the natural diet of these larvae."

Boon, J. P., Lewis, W. E., Tjoen-A-Choy, M. R., et al. (2002). Levels of polybrominated diphenyl ether (PBDE) flame retardants in animals representing different trophic levels of the North Sea food web. *Environ Sci Technol*. 36. pg. 4025-32.

Borm, P. and Berube, D. (2008). A tale of opportunities, uncertainties, and risks. *Nanotoday*. 3(1-2). pg. 56-9. http://www.sciencedirect.com/science/article/pii/S1748013208700161

Borm, P. and Kreyling, W. (2004). Toxicological hazards of inhaled nanoparticles – potential implications for drug delivery. *Journal of Nanoscience and Nanotechnology*. 4(5). pg. 521-31. http://www.protocol-online.org/forums/uploads/monthly_11_2009/msg-6644-1257857797.ipb

- "A large gap is present between research on NP in inhalation toxicology and in nanoscaled drug carrying. This review recommends a closer interaction between both disciplines to gain insight in the role of NP size and properties and their mechanisms of acute and chronic interaction with biological systems."

Borysiewicz, Mieczyslaw. (2008). *Pentachlorophenol*. United Nations Economic Commission for Europe: Institute of Environmental Protection, Warsaw, Poland. http://www.unece.org/fileadmin/DAM/env/lrtap/TaskForce/popsxg/2008/Pentachlorophenol_RA%20dossier_proposal%20for%20submission%20to%20UNECE%20POP%20protocol.pdf

Bostrom, Nick and Cirkovic, Milan, eds. (2008). *Global catastrophic risks*. Oxford University Press, Oxford, UK. http://www.global-catastrophic-risks.com/book.html

Botham, C. and Holmes, P. (2005). *Chemicals purported to be endocrine disruptors: A compilation of existing lists*. MRC Institute for Environment and Health. http://www.cranfield.ac.uk/about/people-and-resources/schools-institutes-research-centres/school-of-applied-sciences/groups-institutes-and-centres/ieh-reports-/endocrine-disruptors/w20.pdf

- A detailed, if outdated, listing of known and suspected EDCs.

- "The largest group are the general anthropogenic chemicals, consisting of 539 individual chemicals, metabolites or degradation products. The second largest category constitutes the biocides, and includes 225 chemicals, metabolites or degradation products. The third category comprises biogenic compounds, that is those of natural origin, that have been suggested as having endocrine activity; this comprises 62 substances. The other categories are: Pharmaceuticals (58 substances); inorganic compounds and organo-metallic complexes (54) and consumer products (28)."

- Information from US governmental sources is minimal (2 citations). Other sited sources include Germany, Japan, the CAIF and ILL EPAs, and environmental organizations such as FOE and the Colburn's "Our Stolen Future."

- No life cycle assessment of the movement of EDCs in the biosphere through biogeochemical cycling, including sorption by plastic nanoparticles, are referenced.

Bouldin, J., Ingle, T., Sengupta, A., et al. (2008). Aqueous toxicity and food chain transfer of quantum dots in freshwater algae and *Ceriodaphnia dubia*. *Environmental Toxicological Chemistry*. 27(9). pg. 1958-63. http://www.ncbi.nlm.nih.gov/pubmed/19086211

- "Fluorescent semiconductor nanocrystals have a protective organic coating…the transfer of core metals from intact nanocrystals may occur at levels well above toxic threshold values, indicating the potential exposure of higher trophic levels."

- "If algal uptake ultimately results in QD breakdown and core metal release, then lethality to primary producers also may lead to food chain disruption."

Bouwmeester, H., Dekkers, S., Noordam, M., et al. (2009). Review of health safety aspects of nanotechnologies in food production. http://ac.els-cdn.com/S0273230008002468/1-s2.0-S0273230008002468-main.pdf?_tid=73553584-b488-11e4-b06e-

Bowmer, T. and Kershaw, P. (2010). *Proceedings of the GESAMP International Workshop on Micro-plastic Particles as a Vector in Transporting Persistent, Bioaccumulating and Toxic Substances in the Oceans, June 10*. UNESCO-IOC, Paris, France. http://www.gesamp.org/data/gesamp/files/media/Publications/Reports_and_studies_82/gallery_1510/object_1670_large.pdf

- "The bioaccumulation of the contaminant load (absorbed and plastic additives), … represent an additional and significant vector for transferring pollutants."

- "Plastics with their accumulated contaminant load are directly ingestible by organisms…microplastics have recently been found in the circulatory systems and other tissues of filter feeding organisms such as the blue mussels."

- "Many aspects of the marine plastic debris problem … are currently poorly known and understood."

- "Any such assessment should aim at providing estimates of plastics inputs to the oceans, describe the rates of fragmentation to micro-plastics, as well as their fate and distribution."

- "According to the US-EPA municipal solid waste statistics for 2008 (US EPA 2008) 30 million tons of plastic waste is produced annually, of which only 7.1% is recovered. A further 19.8 million tons of rubber, leather and textiles, containing a substantial polymer component achieved 15% recovery."

- "Polymers therefore do not biodegrade to any significant extend under natural conditions…organisms with a greater fat content could be a better indicator of bioaccumulation of PBTs."

- "Human health impacts through the food-chain should also be considered as part of an attempt to assess the socio-economic consequences."

- "It has been demonstrated that marine microplastics contain a wide-range of organic contaminants including polychlorinated biphenyls (PCBs), polycyclic aromatic hydrocarbons (PAHs), petroleum hydrocarbons, organochlorine pesticides (DDTs, HCHs), polybrominated diphenylethers (PBDEs), alkylphenols and bisphenol A (BPA)."

- "The impact of micro-plastics on different trophic levels needs further study, e.g. filter feeders, surface benthic feeders, deposit feeders, predators (including sea-birds)."

- This report notes that there are 200 plastics families in production.

- No reference is made to the fate of microplastics after further fragmentation. "Long distance atmospheric transport processes…" provide the chemical fallout that will be sorbed by plastic nanoparticles.

Boyce, D. G., Dowd, M. Lewis, M. R. et al. (2014). Estimating global chlorophyll changes over the past century. *Prog. Oceanogr*. 122. pg. 163-73.

Bradley, Paul and Kolpin, Dana. (2013). Managing the effects of endocrine disrupting chemicals in wastewater-impacted streams. *Intech.* http://cdn.intechopen.com/pdfs-wm/43177.pdf

Brack, H. G. (1984). *Information sampler on long-lived radionuclides: Basic information sources, bibliography with emphasis on pathways and inventories of nuclear effluents, incomplete checklist of biologically significant radionuclides and other appendices*. Pennywheel Press. Hulls Cove, ME.

Brack, H. G. (2010a). *Essays on Biocatastrophe and the Collapse of Global Consumer Society*. Pennywheel Press, Hulls Cove, ME.

Brack, H. G. (2010b). *Biocatastrophe Lexicon: An Epigrammatic Journey Through the Tragedy of our Round-World Commons*. Pennywheel Press, Hulls Cove, ME.

Brack, H. G. (2010c). *Biocatastrophe: The legacy of human ecology: Toxins, health effects, links, appendices, and bibliographies*. Pennywheel Press, Hulls Cove, ME.

Branch, T. A., DeJoseph, B. M., Ray, L. J. et al. (2013). Impacts of ocean acidification on marine seafood. *Trends Ecol. Evol*. 28. pg. 178-86.

Brausch, J. and Rand, G. (2011). A review of personal care products in the aquatic environment: Environmental concentrations and toxicity. *Chemosphere*. 82(11). pg. 1518-32. http://www.ncbi.nlm.nih.gov/pubmed/21185057

- "The primary issues of concern with PCPs are their ability to bioaccumulate to high levels as well as the propensity to cause estrogenic and endocrine effects."

Braydich-Stolle, L., Hussain, S., Schlager, J. and Hofmann, M. (2005). *In Vitro* cytotoxicity of nanoparticles in mammalian germline stem cells. *Toxicological Science*. 88(2). pg. 412-9. http://www.ncbi.nlm.nih.gov/pmc/articles/PMC2911231/pdf/nihms217521.pdf

Breast Cancer Fund. (2014). *Chemicals in Plastics*. Breast Cancer Fund. http://www.breastcancerfund.org/clear-science/environmental-breast-cancer-links/plastics/

- "Plastic is everywhere—it's used in consumer products and packaging of all kinds. And while it solves a lot of problems for manufacturers and can seem convenient to consumers, there are also serious risks to human health and the environment from its widespread use. Three plastics have been shown to leach toxic chemicals when heated, worn or put under pressure: polycarbonate, which leaches bisphenol A; polystyrene, which leaches styrene; and PVC, or polyvinyl chloride, which break down into vinyl chloride and sometimes contains phthalates that can leach."

- "Data indicate that there is a substantial decrease in the amount of dioxin remaining in a woman's breast fat tissue after she has [given birth]; unfortunately, this is because the chemicals have been passed on to her newborn via breast milk…the release of the chemicals from storage in breast fat cells, initiated by the process of milk synthesis, may actually trigger genotoxic (cancer-causing) effects in the breast tissue."

Brede, C., Fjeldal, P., Skjevrak, I. and Herikstad, H. (2003). Increased migration levels of bisphenol A from polycarbonate baby bottles after dishwashing, boiling and brushing. *Food Additive Contamination*. 20. pg. 684-9. http://people.oregonstate.edu/~rochefow/STEPS%20Plastics%20in%20Daily%20Life/Articles%20on%20Plastics/Leaching%20in%20Plastics/BPA%20Leaching/Bisphenol%20A%20Leaching.pdf

- "Migration testing performed with both new and used bottles revealed a significant increase in migration of bisphenol A due to use. This finding might be explained by polymer degradation."

- "None of the bottles released bisphenol A at levels that exceed the recently established provisional tolerable daily intake (0.01 mg kg1 body weight/day) in the European Union."

Breggin, K. L. and Pendergrass, J. (2007). *Where does the nano go?* Woodrow Wilson International Centre for Scholars, Project on emerging nanotechnologies. http://www.nanotechproject.org/file_download/files/NanoEnd-of-Life_Pen10.pdf

Breivik, K., Sweetman, A., Pacyna, J. and Jones, K. (2007). Towards a global historical emission inventory for selected PCB congeners — a mass balance approach: 1. Global production and consumption. *Science of the Total Environment*. 290(1-3). pg. 181-98. http://www.sciencedirect.com/science/article/pii/S0048969701010750

- "In particular, the information on imports and exports for the principal users of PCBs around the time of peak production is considered to be fairly reliable. The estimates account for a reported historical global production of approximately

1.3 million t PCBs, more than 70% of which are tri-, tetra- and pentachlorinated biphenyls. The results further suggest that almost 97% of the global historical use of PCBs have occurred in the Northern Hemisphere."

Producer	Country	Start	Stop	Amount	Reference
Monsanto	USA	1930	1977	641 246	de Voogt and Brinkman (1989)
Geneva Ind.	USA	1971	1973	454	de Voogt and Brinkman (1989)
Kanegafuchi	Japan	1954	1972	56 326	Tatsukawa (1976)
Mitsubishi	Japan	1969	1972	2461	Tatsukawa (1976)
Bayer AG	West Germany	1930	1983	159 062	de Voogt and Brinkman (1989)
Prodelec	France	1930	1984	134 654	de Voogt and Brinkman (1989)
S.A. Cros	Spain	1955	1984	29 012	de Voogt and Brinkman (1989)
Monsanto	U.K.	1954	1977	66 542	de Voogt and Brinkman (1989)
Caffaro	Italy	1958	1983	31 092	de Voogt and Brinkman (1989)
Chemko	Czechoslovakia	1959	1984	21 482	Schlosserová (1994)
Orgsteklo	USSR (Russia)	1939	1990	141 800	AMAP (2000)
Orgsintez	USSR (Russia)	1972	1993	32 000	AMAP (2000)
Xi'an	China	1960	1979	8000	Jiang et al. (1997)
Total		1930	1993	1 324 131	

Breivik, K., Sweetman, A., Pacyna, J. and Jones, K. (2007). Towards a global historical emission inventory for selected PCB congeners — a mass balance approach: 2. Emissions. *Science of the Total Environment*. 290(1-3). pg. 199-224. http://www.sciencedirect.com/science/article/pii/S0048969701010750

Breivik, K., Sweetman, A., Pacyna, J. and Jones, K. (2007). Towards a global historical emission inventory for selected PCB congeners — a mass balance approach: 3. An update. *Science of the Total Environment*. 377(2-3). pg. 296-307. http://www.ncbi.nlm.nih.gov/pubmed/17395248

Brigden, K. and Santillo, D. (2006). *Toxic chemicals in computers exposed*. Greenpeace Research Laboratories, Exeter, UK. http://www.genderchangers.org/docs/200609_Greenpeace_ToxicChemicalsincomputers.pdf

Brinkman, U. A. Th. and De Kok, A. (1980). Production, properties and usage of polychlorinated biphenyls. In: Kimbrough, R. D., Ed. *Halo-genated biphenyls, terphenyls, naphthalenes, dibenzodioxins and related products*. Elsevier, Amsterdam.

Brinton, W., Dietz, C., Bouyounan, A. and Matsch, D. (2011). *New opportunities in recycling and product manufacture eliminate the environmental hazards inherent in the composting of plastic-coated paper products*. Woods End Laboratories & Eco-Cycle. http://ecocycle.org/files/pdfs/microplastics_in_compost_white_paper.pdf

Brooke, D., Crookes, M., Gray, D., and Robertson, S. (2009). *Environmental risk assessment report: Decamethylcyclopentasiloxane*. Environment Agency, Bristol, UK. https://www.gov.uk/government/uploads/system/uploads/attachment_data/file/290561/scho0309bpqx-e-e.pdf

- "D5 meets the screening criteria for a very persistent (vP) and very bioaccumulative (vB) substance."

- "The bioconcentration factor (BCF) for BCF of D5 in fish is 7060 l/kg (determined experimentally). In addition, D5 is accumulated by fish from diet, and a growth-corrected and lipid-normalised biomagnification factor (BMF) of 3.9 is derived from the available experimental data. Thus, D5 meets the vB criterion."

- "D5 shows essentially no acute toxicity to aquatic organisms… However, the available long-term fish toxicity data may not cover all of the relevant toxicological endpoints, so it is not fully established whether or not D5 has the potential to cause effects in fish over long-term exposure."

- "D5 is lost from water by volatilisation to the air, where subsequent degradation occurs."

Brown, T.N. and Wania, F. (2008). Screening chemicals for the potential to be persistent organic pollutants: A case study of Arctic contaminants. *Environmental Science and Technology*. 42(14). pg. 5202-9. http://www.ncbi.nlm.nih.gov/pubmed/18754370

- "Within a data set of more than 100,000 distinct industrial chemicals, the methodology identifies 120 high production volume chemicals which are structurally similarto known Arctic contaminants."

Browne, M., Dissanayake, A., Galloway, T., et al. (2008). Ingested microscopic plastic translocates to the circulatory system of the mussel, *Mytilus edulis*. *Env. Science and Technology*. 42. pg. 5026-31. http://www.ncbi.nlm.nih.gov/pubmed/18678044

- "Particles translocated from the gut to the circulatory system within 3 days and persisted for over 48 days."

- " Smaller particles were more abundant than larger particles and our data indicate as plastic fragments into smaller particles, the potential for accumulation in the tissues of an organism increases."

Browne, M. A., Galloway, T. and Thompson, R. (2007) Microplastic – An emerging contaminant of potential concern? *Integrated Environmental Assessment and*

Management. 3(4). pg. 559-66.

Browne, M., Galloway, T. and Thompson, R. (2010). Spatial patterns of plastic debris along estuarine shorelines. *Environmental Science and Technology*. 44. pg. 3404-9. http://www.researchgate.net/publication/43078614_Spatial_patterns_of_plastic_debris_along_Estuarine_shorelines

- "Our study suggests that most plastic is fragmented and that the main types of polymers present in macroplastic debris are not representative of the mixture of polymers found as microplastic debris."

Browne, M., Crump, P., Niven, S., et al. (2011). Accumulation of microplastic on shorelines worldwide: Sources and sinks. *Environmental Science and Technology*. 45(21). pg. 9175-9. http://pubs.acs.org/doi/abs/10.1021/es201811s

Browne, M., Stewart, N., Galloway, T., et al. (2013). Microplastic moves pollutants and additives to worms, reducing functions linked to health and biodiversity. *Current Biology*. 23(23). pg. 2388-92. http://www.cell.com/current-biology/abstract/S0960-9822(13)01253-0

- "Microplastic transferred pollutants and additive chemicals into gut tissues of lugworms, causing some biological effects, although clean sand transferred larger concentrations of pollutants into their tissues. Uptake of nonylphenol from PVC or sand reduced the ability of coelomocytes to remove pathogenic bacteria by >60%."

Brumfiel, G. (2006). Consumer products leap aboard the nano bandwagon. *Nature*. 440. pg. 262. http://www.nature.com/nature/journal/v440/n7082/full/440262b.html

Buckley, Christine. (August 10, 2010). Lobster dieoffs linked to chemicals in plastics. *UConn Today*. University of Connecticut, Storrs, CT. http://today.uconn.edu/blog/2010/08/lobster-dieoffs-linked-to-chemicals-in-plastics/

Bucheli, T. and Orjan, G. (2003). Soot sorption of non-*ortho* and *ortho* substituted PCBs. *Chemosphere*. 53. pg. 515-522. http://www.ncbi.nlm.nih.gov/pubmed/12948535

- "This strong interaction with soot, particularly of non-ortho substituted PCBs, may fundamentally affect their environmental distribution and bioavailable exposure."

Bundy, A., Shannon, L, Rochet, M., et. al. (2010). The good(ish), the bad, and the ugly: A tripartite classification of ecosystem trends. *Oxford Journals*. 67(4). http://icesjms.oxfordjournals.org/content/67/4/745.full.pdf+html

Buxton, H.T. and Kolpin, D.W. (2002). *Pharmaceuticals, hormones, and other organic wastewater contaminants in U.S. streams*. U.S. Geological Survey Fact Sheet FS-027-02. http://toxics.usgs.gov/pubs/FS-027-02/pdf/FS-027-02.pdf

Buzea, C., Blandino, I. and Robbie, K. (2007). Nanomaterials and nanoparticles: Sources and toxicity. *Biointerphases*. 2(4). pg. MR17-MR172. http://arxiv.org/ftp/arxiv/papers/0801/0801.3280.pdf

- "Humans have always been exposed to tiny particles via dust storms, volcanic ash, and other natural processes, and that our bodily systems are well adapted to protect us from these potentially harmful intruders. The reticuloendothelial system in particular actively neutralizes and eliminates foreign matter in the body, including viruses and non-biological particles. Particles originating from human activities have existed for millennia, e.g. smoke from combustion and lint from garments, but the recent development of industry and combustion-based engine transportation has profoundly increased anthropogenic particulate pollution. Significantly, technological advancement has also changed the character of particulate pollution, increasing the proportion of nanometer-sized particles - "nanoparticles" and expanding the variety of chemical compositions. Recent epidemiological studies have shown a strong correlation between particulate air pollution levels, respiratory and cardiovascular diseases, various cancers, and mortality. Adverse effects of nanoparticles on human health depend on individual factors such as genetics and existing disease, as well as exposure, and nanoparticle chemistry, size, shape, agglomeration state, and electromagnetic properties."

- "Animal and human studies show that inhaled nanoparticles are less efficiently removed than larger particles by the macrophage clearance mechanisms in the lung, causing lung damage, and that nanoparticles can translocate through the circulatory, lymphatic, and nervous systems to many tissues and organs, including the brain."

- "Reduction in fossil fuel combustion would have a large impact on global human exposure to nanoparticles, as would limiting deforestation and desertification."

- "While uncontained nanoparticles clearly represent a serious health threat, fixed nanostructured materials, such as thin film coatings, microchip electronics, and many other existing nanoengineered materials, are known to be virtually benign."

- "The number of publications on the topic of nanomaterials has increased at an almost exponential rate since the early 1990s, reaching about 40,000 in the year 2005, as indicated by a search on ISI Web of Knowledge database. There is also a notable rise in the number of publications discussing their toxicity, particularly in the past two years. The total number of papers on toxicity, however, remains low compared to the total number of publications on nanomaterials, with only around 500 publications in the year 2005."

Byrne, J. and Baugh, J. (2008). The significance of nanoparticles in particle-induced pulmonary fibrosis. *McGill Journal of Medicine*. 11(1). pg. 43-50. http://www.ncbi.nlm.nih.gov/pmc/articles/PMC2322933/

Caballero, Benjamin. (2007). The global epidemic of obesity: An overview. *Epidemiological Review*. 29. pg. 1-5. http://epirev.oxfordjournals.org/content/29/1/1.full.pdf+html

Calafat, A., Ye, X., Wong, L., et al. 2008 exposure of U.S. population to Bisphenol A and 4-tertiary-octylphenol: 2003-2004. *Environmental Health Perspectives*. 116. pg. 39-44. http://www.ncbi.nlm.nih.gov/pmc/articles/PMC2199288/pdf/ehp0116-000039.pdf

- "Bisphenol A (BPA) and 4-*tertiary*-octylphenol (tOP) are industrial chemicals used in the manufacture of polycarbonate plastics and epoxy resins (BPA) and nonionic surfactants (tOP). These products are in widespread use in the United States."

- "We measured the total (free plus conjugated) urinary concentrations of BPA and tOP in 2,517 participants $\geq$ 6 years of age in the 2003–2004 National Health and Nutrition Examination Survey."

- "BPA and tOP were detected in 92.6% and 57.4% of the persons, respectively... Females had statistically higher BPA least square geometric mean (LSGM) concentrations than males ($p = 0.043$). Children had higher concentrations than adolescents ($p < 0.001$), who in turn had higher concentrations than adults ($p = 0.003$)."

- "LSGM concentrations were lowest for participants in the high household income category (> $45,000/year)."

- CDC biomonitoring reports have been diminishing since this publication and may cease entirely in view of the political and congressional restraints on CDC publications, research, and reporting.

Carey, Mark. (2011). Intergenerational transfer of plastic debris by Short-tailed Shearwaters (*Ardenna tenuirostris*). *Emu*. 111. pg. 229-34. http://www.publish.csiro.au/?act=view_file&file_id=MU10085.pdf

- "All birds sampled contained plastic, averaging 7.6 particles per bird. The mean mass of plastic per bird was 113 mg. The most common type of plastic was user plastic, followed by industrial pellets."

Carlsen, E., Giwercman, A., Keiding, N. (1992). Evidence for decreasing quality of semen during past 50 years. http://www.ncbi.nlm.nih.gov/pubmed/1393072

- "Linear regression analysis…showed a significant decrease in mean sperm concentration between 1940 and 1990 from $113x10^6$/ml to $55x10^6$/ml."

Caron, D., Davis, P., Madin, L. and Sieburth, J. (1986). Enrichment of microbial populations in macro-aggregates (marine snow) from surface waters of the North Atlantic. *Journal of Marine Research*. 44. pg. 543-65. http://www.ingentaconnect.com/content/jmr/jmr/1986/00000044/00000003/art00007

Carpenter, E. J. and Smith Jr., K. L. (1972). Plastics on the Sargasso Sea surface. *Science*. 175. pg. 1240-1. http://5gyres.org/media/Carpenter_1972_plastic_in_North_Atlantic_Gyre.pdf

Carpenter, E. J., Anderson, S., Harvey, G., et al. (1972) Polystyrene spherules in coastal waters. *Science*. 178(4062). pg. 749-50. http://www.ncbi.nlm.nih.gov/pubmed/4628343

- One of the first commentators on plastic in the marine environment.

Carpenter, S. R. (2003). *Regime shifts in lake ecosystems: Pattern and variation*. Vol. 15, Excellence in Ecology Series. Ecology Institute, Oldendorf/Luhe, Germany. http://www.int-res.com/book-series/excellence-in-ecology-books/ee15/

Carrick, H. and Fahnestiel, G. (1989). *Biomass, size-structure, and composition of phototrophic and heterotrophic nanoflagellate communities in lakes Huron and Michigan*. http://www.researchgate.net/profile/Hunter_Carrick/publication/237183294_Biomass_Size_Structure_and_Composition_of_Phototrophic_and_Heterotrophic_Nanoflagellate_Communities_in_Lakes_Huron_and_Michigan/links/53ee90d90cf2711e0c4204fb.pdf

Carson, H., Nerheim, M., Carroll, K. and Eriksen, M. (2013a). The plastic-associated microorganisms of the North Pacific Gyre. *Marine Pollution Bulletin*. 75. 126-32. http://www.sciencedirect.com/science/article/pii/S0025326X13004475

- "Microorganisms likely mediate processes affecting the fate and impacts of marine plastic pollution, including degradation, chemical adsorption, and colonization or ingestion by macroorganisms."

- "Bacillus bacteria (mean 1664 ± 247 individuals mm(-2)) and pennate diatoms

(1097 ± 154 mm(-2)) were most abundant, with coccoid bacteria, centric diatoms, dinoflagellates, coccolithophores, and radiolarians present."

Carson, H. S. (2013). The incidence of plastic ingestion by fishes: From the prey's perspective. *Marine Pollution Bulletin*. 74(1). pg. 170-4. http://www.sciencedirect.com/science/article/pii/S0025326X13003779

- "Tooth widths of marks ranged from 1 to 20 mm, suggesting many species ingest plastic."
- The tip of the iceberg of plastic ingestion by marine biota.

Carwile, J., Luu, H., Bassett, L., et al. (2009). Polycarbonate bottle use and urinary bisphenol A concentrations. *Environmental Health Perspectives*. 117(9). pg. 1368-72. http://www.ncbi.nlm.nih.gov/pmc/articles/PMC2737011/

Cassone, A. -L. and Soudant, P., eds. (2014). *Detailed program of MICRO 2014 & presentation abstracts: Fate and impact of microplastics in marine ecosystems: 13-15 January 2014 Plouzane – France*. http://micro2014.sciencesconf.org/conference/micro2014/pages/Book_of_abstracts_MICRO2014.pdf

- "Microplastics are considered a pervasive and widespread pollutant of marine ecosystems across the globe. Over recent years numerous studies have looked to enumerate microplastics within a range of marine habitats, however, the complexities of sampling very small microplastics at dilute concentrations has resulted in the majority of these studies only focusing on microplastics > 333 μm in diameter. As our lab-based research has identified that microplastics within the size range 2 – 230 μm can be ingested by and cause harm to an array of marine life, including zooplankton, mussels, polychaetes and crabs, it becomes increasingly important to identify and quantify microplastics of this size range within the marine environment itself."
- The following is a selection of the abstracts presented at this meeting.
- Marc Long. Can phytoplankton species impact microplastic behavior within water column?
- Won Joon Shim. Fragmentation of polyethylene, polypropylene and expanded polystyrene with an accelerated mechanical abrasion experiment
- Marcus Eriksen. Microplastics and suspected microbeads in the Laurentian Great Lakes of North America
- Natalie Welden. Rope degradation and microfibre formation – A benthic exposure trial
- Amy Lusher. The ubiquitous nature of microplastics in the North Atlantic
- Gunnar Gerdts. Dangerous hitchhikers: Evidence for potentially pathogenic *Vibrio* spp. on microplastic particles
- Carmen Gonz´alez-Fern´andez. Interactive effects of microplastics and fluoranthene on mussels *Mytilus* sp.
- Carlo Giacomo Avio. Microplastics and trophic transfer of polycyclic aromatic

hydrocarbons (PAHs) to marine organisms

- Albert Koelmans. Microplastics as vectors of chemical contaminants
- Sang Hee Hong. Plastic debris as a vector in transporting toxic additive chemicals in the marine environment: hexabromocyclododecanes in expanded polystyrene fragments
- Adil Bakir. Relative importance of microplastics as a pathway for the transfer of persistent organic pollutants to marine life
- Magnus Svendsen Nerheim. The plastic-associated microorganisms of the North Pacific Gyre
- Lisa Devriese. The role of microplastics as a vector for PCBs through the marine trophic levels
- Ellen Besseling. Ecotoxic effects of nano plastic on freshwater plankton (*Scenedesmus obliquus* and *Daphnia magna*)
- Marc Suquet. Microplastics are love-killers for Pacific oysters!
- Lisbeth Van Cauwenberghe. Microplastics are taken up by marine invertebrates living in natural habitats
- Amy Lusher. Microplastics in our food: Ingestion by commercially important fish species
- Rossana Sussarellu. Microplastics: Effects on oyster physiology and reproduction
- Andrew Watts. Two alternate mechanisms for uptake of microplastics into the Shore crab *Carcinus maenas*: Trophic transfer and direct exposure
- Matthew Cole. Improving microplastic detection in plankton-rich samples
- Laurent Colasse. Initial assessment of microplastic on the French coasts: The special case of industrial granules
- Fabienne Lagarde. Investigation of the first stage of polymer degradation by combined Raman and AFM study
- Lisbeth Van Cauwenberghe. Microplastic pollution in deep-sea sediments
- Liv Ascer. Morphological changes in polyethylene abrasives of Brazilian cosmetics caused by mechanical stress
- Matthew Cole. Science under sail: On the hunt for < 333 μm microplastic debris in the Gulf of Maine
- Martin Ogonowski. Ecological and ecotoxicological effects of microplastics and associated contaminants on aquatic biota
- Stephanie Wright. Microscopic PVC as a vector for PAHs: Bioaccumulation and toxicity in a sediment dwelling marine polychaete
- David Mazurais. Impact of polyethylene microbeads ingestion on sea bass larvae development
- Amy Lusher. Ingestion of microplastics by mesopelagic fish from the North Atlantic
- Marina Santana. Intake and size selection of microplastic particles (PVC) by marine invertebrates: a preliminary assessment of biological risks
- Lisa Devriese. Occurrence of synthetic fibres in brown shrimp on the Belgian part of the North Sea
- Paul Farrell. Trophic level transfer of microplastic: *Mytilus edulis* (L.) to *Carcinus maenas* (L.).

Castle, S., Thomas, B., Reager, J., et. al. (2014). Groundwater depletion during drought threatens future water security of the Colorado River Basin. *Geophysical Research Letters*. 41. pg. 5904-11.
http://onlinelibrary.wiley.com/doi/10.1002/2014GL061055/abstract

- "Stream flow of the Colorado River Basin is the most over allocated in the world. Recent assessment indicates that demand for this renewable resource will soon outstrip supply, suggesting that limited groundwater reserves will play an increasingly important role in meeting future water needs."

- "The rapid rate of depletion of groundwater storage (-5.6 ± 0.4 km^3yr) far exceeded the rate of depletion of Lake Powell and Lake Mead. Results indicate that groundwater may comprise a far greater fraction of Basin water use than previously recognized, in particular during drought, and that its disappearance may threaten the long-term ability to meet future allocations to the seven Basin states."

- "At question is the potential impact of solely managing surface water allocations and diversions in the Basin, without regard to groundwater loss, on meeting future water demands."

- One more chapter in the world water crisis.

Castro-Jimenez, J., Ghiani, M., Hanke, G., et al. (2009). *Polychlorinated biphenyls (PCBs) in the Mediterranean Sea atmosphere and seawater*. European Commission-DG Joint Research Centre, Institute for Environment and Sustainability.
http://jcastrojimenez.net/web_documents/mesaep_07.pdf

Cattaneo, A., Gornati, R., Chiriva-Internati, M. and Bernardini, G. (2009). *Ecotoxicology of nanomaterials: The role of invertebrate testing*. University of Insubria, Varese, Italy.
http://www.researchgate.net/publication/26626843_Ecotoxicology_of_nanomaterials_the_role_of_invertebrate_testing/links/0912f50b35f7e77d8a000000

- "Engineered nanomaterials represent a new and expanding class of chemicals whose environmental hazard is actually poorly determined…This review highlights the role of invertebrates as valuable and validated test organisms for assessing ecotoxicity of new and/or untested chemicals."

Cauwenberghe, L, Claessens, M., Vandegehuchte, M., Mees, J. and Janssen, C. (2013a). Assessment of marine debris on the Belgian continental shelf. *Marine Pollution Bulletin*.73. pg. 161-9. http://www.vliz.be/imisdocs/publications/248267.pdf

- "In terms of weight, macrodebris still dominates the pollution of beaches, but in the water column and in the seafloor microplastics appear to be of higher importance: here, microplastic weight is approximately 100 times and 400 times higher, respectively, than macrodebris weight."

- "Hence, per km^2 there are around 9.2×10^9 to 3.5×10^{10} microplastic particles present in the upper 10 cm of the seabed."

Cauwenberghe, L., Vanreusel, A., Mees, J. and Janssen, C. (2013b). Microplastic pollution in deep-sea sediments. *Environmental Pollution*. 182. pg. 495-9. http://onemoregeneration.org/wp-content/uploads/2012/07/Microplastic-pollution-in-deep-sea-sediments.pdf

- "We found plastic particles sized in the micrometre range in deep-sea sediments collected at four locations representing different deep-sea habitats ranging in depth from 1100 to 5000 m."

Cauwenberghe, L. and Janssen, C. R. (2014). Microplastics in bivalves cultured for human consumption. *Environmental Pollution*. 193. pg. 65-70. http://www.sciencedirect.com/science/article/pii/S0269749114002425

- "Microplastics are present throughout the marine environment and ingestion of these plastic particles (<1 mm) has been demonstrated in a laboratory setting for a wide array of marine organisms. Here, we investigate the presence of microplastics in two species of commercially grown bivalves: *Mytilus edulis* and *Crassostrea gigas*. Microplastics were recovered from the soft tissues of both species…the annual dietary exposure for European shellfish consumers can amount to 11,000 microplastics per year. The presence of marine microplastics in seafood could pose a threat to food safety, however."

CBD Technical Series 67. (2012). *Impacts of Marine Debris on Biodiversity*. Secretariat of the Convention on Biological Diversity. http://www.thegef.org/gef/sites/thegef.org/files/publication/cbd-ts-67-en.pdf

- "Impacts of marine debris were reported for 663 species. Over half of these reports documented entanglement in and ingestion of marine debris, representing a 40% increase since the last review in 1997, which reported 247 species."

- "Marine debris is an important contributor among the anthropogenic stresses acting on habitats and biodiversity."

Cedervall, T., Hansson, L., Lard, M., et al. (2012). Food chain transport of nanoparticles affect behavior and fat metabolism in fish. *PLoS ONE*. 7(2). http://www.plosone.org/article/fetchObject.action?uri=info:doi/10.1371/journal.pone.00

32254&representation=PDF

- "The commercial use of nanoparticles in, for example detergents, cosmetics, food, and dental products, is therefore rapidly growing, leading to a rapidly increasing release of possibly very potent particles into the environment. This specifically raises concerns about nanoparticle effects in freshwater and marine ecosystems since many products containing nanoparticles will end up there through sewage systems."

Chalew, Talia and Halden, Rolf. (2009). Environmental exposure of aquatic and terrestrial biota to triclosan and triclocarban. *Journal of American Water Works Association*. 45(1). pg. 4-13. http://www.ncbi.nlm.nih.gov/pmc/articles/PMC2684649/pdf/nihms100867.pdf

- "The synthetic biocides triclosan (5-chloro-2-(2,4-dichlorophenoxy)phenol) and triclocarban (3,4,4'-trichlorocarbanilide) are routinely added to a wide array of antimicrobial personal care products and consumer articles."

- "The highest biocide levels, measured in the mid parts-per-million range, were determined to occur in aquatic sediments and in municipal biosolids destined for land application."

- "Crustacea and algae were identified as the most sensitive species, susceptible to adverse effects from biocide exposures in the parts-per-trillion range."

Chalmers, A., Nilles, M., Krabbenhoft, D., et al. (2003). Wet distribution of mercury in the Boston metropolitan area. *National Atmospheric Deposition Program Meeting*. Washington, DC. http://nh.water.usgs.gov/project/nawqa/ac_nadp03.htm

- " Concentrations of total Hg in wet deposition during 2002 ranged from 2-20 ng/L (nanograms per liter) at the 4 sites."

- "Localized urban emission sources may have significant effects on concentrations of Hg in rainfall in New England, and may result in variable deposition patterns on a sub-regional scale."

Chan, Vivian. (2006). Nanomedicine: An unresolved regulatory issue. *Regulatory Toxicology and Pharmacology*. 46. pg. 218. http://www.sciencedirect.com/science/article/pii/S0273230006000821

Chan, Warren. (2007). *Advances in experimental medicine and biology, volume 620: Bio-applications of nanoparticles*. Springer, New York, NY.

Chanda, Manas. (2012). *Plastics technology handbook, fourth edition (plastics engineering)*. CRC Press, FL.

Chang, M., Lee, D. and Lai, J. (2007). Nanoparticles in wastewater from a science-based industrial park – Coagulation using polyaluminium chloride. *Journal of Environmental Management.* 85. pg. 1009-14. http://www.ncbi.nlm.nih.gov/pubmed/17202026

- "Wastewater from the HSIP contains numerous nano-sized silicate particles whose size distributions peak at 2 and 90 nm."

- "Polyaluminum chloride (PACl) was used in the field to coagulate these particles…nano-particles agglomerated in approximately linear aggregates of sizes 100-300 nm. Prolonged contact between residual PACl and the nano-particles generated large aggregates with sizes of up to 10 microm."

Chang, C. (2010). The immune effects of naturally occurring and synthetic nanoparticles. *Journal of Autoimmunology.* 34(3). pg. J234-46. http://www.ncbi.nlm.nih.gov/pubmed/19995678

- "While the benefits of this new science to human civilization are seemingly immeasurable… these particles can also lead to harmful effects on human health."

Chao, H., Wang, S., Lee, W., et al. (2007). Levels of polybrominated diphenyl ethers (PBDEs) in breast milk from central Taiwan and their relation to infant birth outcome and maternal menstruation effects. *Environment International.* 33. pg. 239-45. http://www.nature.com/pr/journal/v70/n6/pdf/pr20111086a.pdf

Chao, H., Tsou, T., Huang, H., et. al. (2011). Levels of breast milk PBDEs from southern Taiwan and their potential impact on neurodevelopment. *Pediatric Research.* 70(6). pg. 596-600. http://www.nature.com/pr/journal/v70/n6/pdf/pr20111086a.pdf

- "Fourteen PBDEs in 70 breast milk were analyzed using a high-resolution gas chromatograph/high-resolution mass spectrometer." Mean: 5.64 ng/g lipid, range of 1.44-118 ng/g lipid.

- "The median of 14PBDEs (the sum of 14 PBDE congeners) was 2.92 ng/g lipid."

- "Prenatal or postnatal exposure to BDE-209 potentially delays the neurological development."

- "BDE-196 might help language development."

- "It is worth noting that BDE-209 is commonly used as a brominated fire retardant in electronic equipment and that it constitutes approximately 80% of the world market demand for PBDEs."

Chao, H., Shy, C., Huang, H., et al. (2014). Particle-Size Dust Concentrations of Polybrominated Diphenyl Ethers (PBDEs) in Southern Taiwanese Houses and Assessment of the PBDE Daily Intakes in Toddlers and Adults. *Aerosol and Air Quality Research*. 14. Pg. 1299-1309. http://aaqr.org/VOL14_No4_June2014/23_AAQR-12-12-OA-0342_1299-1309.pdf

Charlier, C., Albert, A., Zhang, L., et al. (2004). Polychlorinated biphenyls contamination in women with breast cancer. *Clinica Chimica Acta*. 347(1-2). pg. 177-81. http://www.ncbi.nlm.nih.gov/pubmed/15313156

Chaudhry, Q., Castle, L., Bradley, E., et al. (2006). *A03063: Assessment of current and projected applications of nanotechnology for food contact materials in relation to consumer safety and regulatory implications*. UK FSA. http://www.foodbase.org.uk//admintools/reportdocuments/346-1-648_A03063_Final_Report.pdf

Chen, A., Dietrich, K., Huo, X. and Ho, S. (2011). Developmental neurotoxicants in e-waste: An emerging health concern. *Environmental Health Perspectives*. 119(4). 431-8. http://ehp.niehs.nih.gov/wp-content/uploads/119/4/ehp.1002452.pdf

- "Electronic waste (e-waste) has been an emerging environmental health issue in both developed and developing countries, but its current management practice may result in unintended developmental neurotoxicity in vulnerable populations."

- "E-waste is the fastest-growing stream of municipal solid waste."

- "In developing countries where most informal and primitive e-waste recycling occurs, environmental exposure to lead, cadmium, chromium, polybrominated diphenyl ethers, polychlorinated biphenyls, and polycyclic aromatic hydrocarbons is prevalent at high concentrations in pregnant women and young children."

Chen, C., Anaya, J., Zhang, S., et al. (2011). Effects of engineered nanoparticles on the assembly of exopolymeric substances from phytoplankton. *PLoS ONE*. 6(7). http://www.plosone.org/article/fetchObject.action?uri=info:doi/10.1371/journal.pone.0021865&representation=PDF

- "The unique properties of engineered nanoparticles (ENs) that make their industrial applications so attractive simultaneously raise questions regarding their environmental safety."

- "Polystyrene nanoparticles (23 nm) were used in our study."

- "Our results clearly demonstrate how nanowaste (e.g. nanoparticles) can potentially disturb the marine carbon cycle and ecosystem... That indirect influences from ENs can potentially pose greater environmental threats than those of direct toxicity."

Chen, Eric Y. T., Garnica, Maria, Wang, Yung-Chen, et al. (2011). Mucin secretion induced by titanium dioxide nanoparticles. *PLoS ONE*. 6(1). http://journals.plos.org/plosone/article?id=10.1371/journal.pone.0016198

Chen, H., Zheng, X., Chen, Y., et al. (2014). Influence of copper nanoparticles on the physical-chemical properties of activated sludge. *PLoS ONE*. 9(3). http://www.plosone.org/article/fetchObject.action?uri=info:doi/10.1371/journal.pone.0092871&representation=PDF

Chen, M. and von Mikecz, A. (2005). Formation of nucleoplasmic protein aggregates impairs nuclear function in response to SiO_2 nanoparticles. *Experimental Cell Ressearch*. pg. 51-62. http://www.ncbi.nlm.nih.gov/pubmed/15777787

- "Since SiO(2) nanoparticles trigger a subnuclear pathology resembling the one occurring in expanded polyglutamine neurodegenerative disorders, we suggest that integrity of the functional architecture of the cell nucleus should be used as a read out for cytotoxicity and considered in the development of safe nanotechnology."

Chen, Y., Hung, Y., Liau, I. and Huang, G. (2009). Assessment of the *in vivo* toxicity of gold nanoparticles. *Nanoscale Research Letters*. 4(8). pg. 858-64. http://www.ncbi.nlm.nih.gov/pmc/articles/PMC2894102/pdf/1556-276X-4-858.pdf

Chen, Z., Meng, H.l, Xing, G., et al. (2006). Acute toxicological effects of copper nanoparticles in vivo. *Toxicological Letters*. 163(2). pg. 109-20. http://www.sciencedirect.com/science/article/pii/S0378427405003176

- "Results indicate a gender dependent feature of nanotoxicity. Several factors such as huge specific surface area, ultrahigh reactivity, exceeding consumption of H^+, etc. that likely cause the grave nanotoxicity observed in vivo are discussed."

Cherednichenko, G., Zhang, R., Bannister, R., et al. (2012). Triclosan impairs excitation–contraction coupling and Ca^{2+} dynamics in striated muscle. *PNAS*. http://www.pnas.org/content/109/35/14158.full.pdf

- "Here, we report that TCS impairs ECC of both cardiac and skeletal muscle in vitro and in vivo."

Cheung, W. W. L. et al. (2013). Shrinking of fishes exacerbates impacts of global ocean changes on marine ecosystems. *Nat. Clim. Change*. 3. pg. 254-8.

Chiazze, L. and Ference, L. (1981). Mortality among PVC-fabricating employees. *Environmental Health Perspective*. 41. pg. 137-43. http://www.ncbi.nlm.nih.gov/pmc/articles/PMC1568854/

Chlorine Chemistry Council. (2014). *Backyard trash burning: The wrong answer*. American Chemistry Council. http://www.dioxinfacts.org/sources_trends/trash_burning2.pdf

Cho, Renee. (2012). *State of the planet: What happens to all that plastic?* http://blogs.ei.columbia.edu/2012/01/31/what-happens-to-all-that-plastic/

Cho, S., Maysinger, D., Jain, M., et al. (2007). Long-term exposure to CdTe quantum dots causes functional impairments in live cells. *Langmuir*. 23. pg. 1974-80. http://www.ncbi.nlm.nih.gov/pubmed/17279683

- "CdTe QDs induce cell death via mechanisms involving both Cd2+ and ROS accompanied by lysosomal enlargement and intracellular redistribution."

Choi, A., Levy, J., Ian, J., et al. (2006). Does living near a superfund site contribute to higher polychlorinated biphenyl (PCB) exposure? *Environmental Health Perspectives*. 114(7). pg. 1092-8.

- "We assessed determinants of cord serum polychlorinated biphenyl (PCB) levels among 720 infants born between 1993 and 1998 to mothers living near a PCB-contaminated Superfund site in Massachusetts, measuring the sum of 51 PCB congeners (capital sigmaPCB) and ascertaining maternal address, diet, sociodemographics, and exposure risk factors."

- "The geometric mean of capital sigmaPCB levels was 0.40 (range, 0.068-18.14) ng/g serum. Maternal age and birthplace were the strongest predictors of capital sigmaPCB levels. Maternal consumption of organ meat and local dairy products was associated with higher and smoking and previous lactation with lower capital sigmaPCB levels. Infants born later in the study had lower capital sigmaPCB levels, likely due to temporal declines in exposure and site remediation in 1994-1995."

- "No association was found between capital sigmaPCB levels and residential distance from the Superfund site. Similar results were found with light and heavy PCBs and PCB-118. Previously reported demographic (age) and other (lactation, smoking, diet) correlates of PCB exposure, as well as local factors (consumption of local dairy products and Superfund site dredging) but not residential proximity

146

to the site, were important determinants of cord serum PCB levels in the study community."

Choi, S., Oh, J. and Choy, J. (2009). Toxicological effects of inorganic nanoparticles on human lung cancer A549 cells. *Journal of Inorganic Biochemistry*. 103(3). pg. 463-71. http://www.ncbi.nlm.nih.gov/pubmed/19181388

Choong, A., Teo, S., Leow, J., et. al. (2006). A preliminary ecotoxicity study of pharmaceuticals in the marine environment. *Journal of toxicology and environmental health*. 69(1). http://www.tandfonline.com/doi/abs/10.1080/15287390600751371?url_ver=Z39.88-2003&rfr_id=ori:rid:crossref.org&rfr_dat=cr_pub%3dpubmed#.VQ2EkY7F-Jc

Choy, C. and Drazen, J. (2013). Plastic for dinner? Observations of frequent debris ingestion by pelagic predatory fishes from the central North Pacific. *Marine Ecology Progress Series*. 485. pg. 155-63. http://www.int-res.com/articles/meps_oa/m485p155.pdf

- "Species with the highest incidences of debris ingestion are thought to be primarily mesopelagic and unlikely to come into contact with surface waters containing known debris fields.

- "These observations…suggest that more attention should be given to marine debris in subsurface waters as well as to poorly understood organismal and food web implications."

Christaki, U., Dolan, J., Pelegri, S. and Rassoulzadegan, F. (1998). Consumption of picoplankton-size particles by marine ciliates: Effects of physiological state of the ciliate and particle quality. *Limnology and Oceanography*. 43. pg. 458-64. http://www.obs-vlfr.fr/~dolan/html/PFD/1998/ChristLO98.pdf

Christensen, Thomas H., Kjeldsen, Peter, Bjerg, Poul L., et al. (2000). Review: Biogeochemistry of landfill leachate plumes. *Applied Geochemistry*. 16. pg. 659-718. http://geoweb.tamu.edu/Faculty/Herbert/geol420/docs/BGCLandfillLeachate.pdf

- "Leachate from landfills contains a wide range of contaminants: dissolved organic matter, inorganic cations and anions, heavy metals, and xenobiotic organic compounds. Where leachate enters the groundwater, significant changes in water quality are observed and complicated biogeochemical patterns develop in the leachate pollution plume."

- "Diverse microbial communities have been identified in leachate plumes and are believed to be responsible for the redox processes. Dissolved organic C… apparently acts as substrate for the microbial redox processes."

Christensen, V., Coll, M., Piroddi, C., et al. (2014). A century of fish biomass decline in the ocean. *Marine Ecology Progress Series*. 512. pg. 155-66. http://www.int-res.com/articles/theme/m512p155.pdf

- "We performed a global assessment of how fish biomass has changed over the…period from 1880 to 2007."

- "The biomass of predatory fish in the world oceans has declined by two-thirds over the last 100 yr. This decline is accelerating, with 54% occurring in the last 40 yr."

- "In the future ocean,… the composition of fish assemblages will be very different from current ones, with small prey fish dominating. Our results show that the trophic structure of marine ecosystems has changed at a global scale, in a manner consistent with fishing down marine food webs."

Christensen, V., Guenette, S., Heymans, J.J., et al. (2003). Hundred-year decline of North Atlantic predatory fishes. *Fish and Fisheries*. 4. pg. 1-24. http://onlinelibrary.wiley.com/doi/10.1046/j.1467-2979.2003.00103.x/abstract

Chua, E., Shimeta, J., Nugegoda, D., et al. (2014). Assimilation of polybrominated diphenyl ethers from microplastics by the marine amphipod, *Allorchestes compressa*. *Environmental Science and Technology*. 48(14). pg. 8127-34. http://pubs.acs.org/doi/abs/10.1021/es405717z

- "MPPs in the environment …can transfer PBDEs into a marine organism. Therefore, MPPs pose a risk of contaminating aquatic food chains with the potential for increasing public exposure through dietary sources."

Cincinelli, A., Stortini, A., Perugini, M., et al. (2001). Organic pollutants in sea-surface microlayer and aerosol in the coastal environment of Leghorn (Tyrrhenian Sea). *Marine Chemistry*. 76. pg. 77-98. http://arca.unive.it/bitstream/10278/30810/1/Cincinelli1_Stortini.pdf

Citizens Budget Commission. (2012). *Taxes in, garbage out: The need for better solid waste disposal policies in New York City*. Citizens Budget Commission. New York, NY. http://www.cbcny.org/sites/default/files/REPORT_SolidWaste_053312012.pdf

- "This year, New York City will spend over $2 billion in tax dollars to throw out its garbage. More than $300 million of the bill represents the cost of disposing of the garbage – usually in out-of-state landfills. About three-quarters of city garbage goes to landfills, with 98 percent of that shipped to Ohio, Pennsylvania, South Carolina and Virginia."

- "Of the non-recyclable waste managed by the City, three million tons are sent to distant landfills, and private sector garbage almost doubles that figure."

- "The waste that New York City sends to landfills generates about 679,000 metric tons of greenhouse gases per year."

- "This report makes the case for a significant change in New York City's solid waste disposal practices, a shift from heavy reliance on long-distance exporting to landfills to greater reliance on use of local waste-to-energy facilities."

- New York city refuse disposal destinations by state, 2010: Connecticut and New Jersey, 2%; South Carolina, 8%; Pennsylvania, 48%; Ohio, 11%; Virginia, 31%.

Claessens, M., Meester, S., Landuyt, L., et al. (2011). Occurrence and distribution of microplastics in marine sediments along the Belgian coast. *Marine Pollution Bulletin*. 62. pg. 2199-204. http://www.no-sea-and-earth-pollution.org/ALBATROSS-ARTICLES_files/marine-pollution-bulletin-2011.pdf

- "Particles were found in large numbers in all samples, showing the wide distribution of microplastics in Belgian coastal waters."

- "Microplastics (plastic particulates 61 mm) may become widely distributed in the marine environment through hydrodynamic processes and ocean currents."

- "Statistical analysis showed that the average microplastic concentration of the harbour sediments (166.7 ± 92.1 particles kg^{-1} dry sediment) was significantly higher than the concentration found for both the BCS (97.2 ± 18.6 particles kg^{-1} dry sediment; Mann–Whitney U test, $p = 0.007$) and the beach sediments (92.8 ± 37.2 particles kg^{-1} dry sediment.)"

- "Spatial variation in microplastic concentrations was observed on a relatively small scale. A clear relationship between local human activities and microplastic concentrations was absent."

- "Freshwater rivers are a potentially important source of microplastics."

- "The temporal trends which were observed in this study imply that the concentrations of microplastics in sediments are increasing."

Claessens, M., Van Auwenberghe, L., Vandegehuchte, M. B. and Janssen, C. R. (2013). New techniques for the detection of microplastics in sediments and field collected organisms. *Marine Pollution Bulletin*. 70. pg. 227-33. http://www.sciencedirect.com/science/article/pii/S0025326X13001495

- "The method developed for sediments involves a volume reduction of the sample by elutriation, followed by density separation using a high density NaI solution...The use of these two techniques will result in a more complete assessment of marine microplastic concentrations."

Clark, S., Steele, K., Spicher, J., et al. (2008). Roofing materials' contributions to storm-water runoff pollution. *Journal of Irrigation and Drainage Engineering*. http://www.harvesth2o.com/adobe_files/RoofingMaterialsContributionToPollution.pdf

Cleary, J. J. and Stebbing, A. R. D. (1987). Organotin in the surface microlayer and subsurface waters of Southeast England. *Marine Pollution Bulletin*. 18. pg. 238-46. http://www.sciencedirect.com/science/article/pii/0141113691900438

Colabuono, F., Taniguchi, S. and Montone, R. (2010). Polychlorinated biphenyls and organochloride pesticides in plastics ingested by seabirds. *Marine Pollution Bulletin*. 60. pg. 630-4. http://www.ncbi.nlm.nih.gov/pubmed/20189196

- " PCBs were detected in plastic pellets (491 ng g(-1)) and plastic fragments (243-418 ng g(-1)). Among the OCPs, p,p'-DDE had the highest concentrations, ranging from 68.0 to 99.0 ng g(-1)."

- "Plastics are an important source carrying persistent organic pollutants in the marine environment."

Colborn, Theo and Clement, C. (1992). *Chemically-induced alterations in sexual and functional development: The wildlife/human connection*. Princeton Scientific, Princeton, NJ.

Colborn, Theo, vom Saal, F. and Soto, A. (1993). Developmental effects of endocrine-disrupting chemicals in wildlife and humans. *Environmental Health Perspectives*. 101. pg. 378-84. http://www.ncbi.nlm.nih.gov/pmc/articles/PMC1519860/pdf/envhper00375-0020.pdf

- "Large numbers and large quantities of endocrine-disrupting chemicals have been released into the environment since World War II...transgenerational exposure can result from the exposure of the mother to a chemical at any time throughout her life before producing offspring due to persistence of endocrine-disrupting chemicals in body fat, which is mobilized during...pregnancy and lactation."

Colborn, Theo, Dumanoski, D. and Myers, J. (1997). Widespread pollutants with endocrine-disrupting effects. In: *Our Stolen Future*. Plume, NY. http://www.ourstolenfuture.org/basics/chemlist.htm

150

- "The central point of the *Our Stolen Future* is that some man-made chemicals interfere with the body's own hormones. These compounds find their way into our bodies through a variety of pathways. They build up over time, often over years."

- "Some of these chemicals alter sexual development. Some undermine intelligence and behavior. Others make our bodies less resistant to disease. Sometimes the effects don't appear until a child reaches puberty or afterward, even though the exposure took place in the womb."

- "We are working from a data base of over 4,000 scientific publications. Over 100 scientists have participated directly in deliberations that have produced a series of consensus statements about the nature of the problem."

- Among the first and most important publications on endocrine disrupting chemicals.

- A now antiquarian listing of hundreds of the most important EDCs.

Colborn, Theo. (2004). Neurodevelopment and endocrine disruption. *Environmental Health Perspectives*. 112(9). pg. 944-9. http://www.ncbi.nlm.nih.gov/pmc/articles/PMC1247186/pdf/ehp0112-000944.pdf

- "The exquisite sensitivity of the embryo and fetus to thyroid disturbance…provide evidence of human in utero exposure to contaminants that can interfere with the thyroid."

- "The evidence that certain hormones operate at parts per trillion and parts per billion and equivalent exposure to endocrine-active chemicals is equivalent or higher reveals the extreme vulnerability of development to chemical perturbation."

Colborn, Theo and Carroll, Lynn. (2007). Pesticides, sexual development, reproduction, and fertility: Current perspective and future direction. *Human and Ecological Risk Assessment*. 13. pg. 1078-1110. http://www.beyondpesticides.org/documents/Colborn%20Multigenerational%20Effects.pdf

- "Improvements in chemical analytical technology and non-invasive sampling protocols have made it easier to detect pesticides and their metabolites at very low concentrations in human tissues."

- "Monitoring has revealed that pesticides penetrate both maternal and paternal reproductive tissues and organs, thus providing a pathway for initiating harm to their offspring."

- "Pesticide residues are found on more than 70% of conventionally grown fresh produce, whether produced domestically or imported."

- "The overwhelming complexity that exists because of the vast mixtures of pesticides and industrial chemicals in use, the many exposure pathways, the multiple mechanisms of action of each active ingredient, the timing and life stage of the tissue exposed, the long-term delayed expression of many exposures, and the amazing lack of knowledge about the intricacies of human development on the part of the biomedical, scientific community all contribute to the inability to make specific causal links."

Cole, A., Steffen, A., Eckley, C., et al. (2014). A survey of mercury in air and precipitation across Canada: Patterns and trends. *Atmosphere*. 5(3). pg. 635-68. http://www.mdpi.com/2073-4433/5/3/635/htm

Cole, M., Lindeque, P., Halsband, C. and Galloway, T. (2011). Microplastics as contaminants in the marine environment: A review. *Marine Pollution Bulletin*. 62. pg. 2588-97. http://www.sciencedirect.com/science/article/pii/S0025326X11005133

- "Ingestion of microplastics has been demonstrated in a range of marine organisms, a process which may facilitate the transfer of chemical additives or hydrophobic waterborne pollutants to biota."

Cole, M., Lindeque, P., Fileman, E., et al. (2013). Microplastic ingestion by zooplankton. *Environmental Science and Technology*. 47. pg. 6646-55. http://www.researchgate.net/profile/Tamara_Galloway2/publication/236926420_Micro plastic_ingestion_by_zooplankton/links/0046353712b144e30f000000.pdf

- "Ingestion of microplastics by marine biota, including mussels, worms, fish, and seabirds, has been widely reported."

- "Thirteen zooplankton taxa had the capacity to ingest 1.7–30.6 μm polystyrene beads, with uptake varying by taxa, life-stage and bead-size."

- "Post-ingestion, copepods egested faecal pellets laden with microplastics."

- "Exposure of the copepod Centropages typicus to natural assemblages of algae with and without microplastics showed that 7.3 μm microplastics (>4000 mL−1) significantly decreased algal feeding."

- "Very small microplastics (0.4–3.8 μm) became lodged between the filamental hairs and setae of the antennules, furca, and the swimming legs."

- "Microplastics are still considered to be an under-researched fraction of marine litter, with no consistent data relating to plastic detritus <333 μm in diameter."

- "Better knowledge of the extent of microplastic contamination of oceans waters is now a research imperative."

Cole, M., Webb, H., Lindeque, P., et. Al. (2014). Isolation of microplastics in biota-rich seawater samples and marine organisms. *Scientific Reports*. 4(4528). pg. 1-8.

Collette, B., Carpenter, K., Polidoro, B. et al. (2011). High value and long life—double jeopardy for tunas and billfishes. *Science*. 333. pg. 291-2. http://fishdb.sinica.edu.tw/pdf/896.pdf

- "We present here the first standardized data on the global distribution, abundance, population trends, and impact of major threats for all known species of scombrids [tunas, bonitos, mackerels, and Spanish mackerels] and billfishes [swordfish and marlins]."

- "As these large-bodied scombrids and billfishes are at the top of the pelagic food web, population reduction of these predators may have significant effects on the upper trophic levels of the epipelagic ecosystem and lead to cascading effects on lower trophic levels."

- "The IUCN Red List Criteria provide a transparent, standardized, peer-reviewed means of global conservation status assessment."

Colton, J., Knapp, F. and Burns, B. (1974). Plastic particles in surface waters of the Northwestern Atlantic. *Science*. 185. pg. 491-7. http://5gyres.org/media/Plastic_particles_in_surface_water_of_the_northwestern_Atlantic.pdf

- "The widespread distribution of polystyrene spherules and polyethylene disks in rivers, estuaries, and the open ocean suggests that improper waste-water disposal is common practice in the plastics industry."

Collignon, A., Hecq, J., Glagani, F., et al. (2012). Neustonic microplastic and zooplankton in the North Western Mediterranean Sea. *Marine Pollution Bulletin*. 64(4). pg. 861-4 . http://www.expeditionmed.eu/fr/wp-content/uploads/2012/03/MED-Publication-Scient-2012-.pdf

Colvin, V. (2003). The potential environmental impact of engineered nanomaterials. *Nature Biotechnology*. 21. pg. 1166-70.
http://www.ncbi.nlm.nih.gov/pubmed/14520401

Commission for Environmental Cooperation. (2006). *North American regional action plan for Lindane and other hexachlorocyclohexane (HCH) isomers.*
http://www.cec.org/files/PDF/POLLUTANTS/LindaneNARAP-Nov06_en.pdf

Connolly, Sarah. (2009). The role of reactive oxygen species in nanotoxicity. *Biological Applications*. The 2009 NNIN REU Research Accomplishments.
http://www.nnin.org/sites/default/files/files/2009reura/2009NNINreuConnelly.pdf

Constantini, L., Gilberti, R. and Knecht, D. (2011). The phagocytosis and toxicity of amorphous silica. *PLoS ONE*. 6(2).
http://www.plosone.org/article/fetchObject.action?uri=info:doi/10.1371/journal.pone.0014647&representation=PDF

Cook, B., Ault, T. and Smerdon, J. (2015). Unprecedented 21st century drought risk in the American Southwest and Central Plains. *Science Advances*. 1(1).
http://advances.sciencemag.org/content/1/1/e1400082

Cooper, David and Corcoran, Patricia. (2010). Effects of mechanical and chemical processes on the degradation of plastic beach debris on the island of Kauai, Hawaii. *Marine Pollution Bulletin*. 60. pg. 650-4.
http://www.researchgate.net/publication/41165503_Effects_of_mechanical_and_chemical_processes_on_the_degradation_of_plastic_beach_debris_on_the_island_of_Kauai_Hawaii/links/0046352320bbe1040f000000.pdf

- "Plastic surfaces contain fractures, horizontal notches, flakes, pits, grooves, and vermiculate textures. The mechanically produced textures provide ideal loci for chemical weathering to occur which further weakens the polymer surface leading to embrittlement."

- "Degradation of cm-size plastics results in microscopic particles that remain in Earth's environment indefinitely."

Coors, A., Jones, P., Giesy, J. and Ratte, H. (2003). Removal of estrogenic activity from municipal waste landfill leachate assessed with a bioassay based on reporter gene expression. *Environmental Science and Technology*. 37. pg. 3430-4.
http://www.ncbi.nlm.nih.gov/pubmed/12966991

- "Bisphenol A was responsible for the majority of estrogenic activity in the raw and treated leachate."

COPC (Contaminants of Potential Concern) Committee. (2003). *World Trade Center*

154

indoor environment assessment: Selecting contaminants of potential concern and setting health-based benchmarks. World Trade Center Indoor Air Task Force Working Group. http://epa.gov/WTC/reports/contaminants_of_concern_benchmark_study.pdf

Corcoran, P., Biesinger, M. and Grifi, M. (2009). Plastics and beaches: A degrading relationship. *Marine Pollution Bulletin*. 58. pg. 80-4. http://www.surfacesciencewestern.com/pdf/mpb09_biesinger.pdf

- "Our preliminary results show that beaches represent excellent depositional settings for the diminution of plastic debris."

- "Unfortunately, plastics do not degrade rapidly through mineralization, and may remain in microscopic form indefinitely."

Corredor, C., Hou, W., Klein, S., et al. (2013). Disruption of model cell membranes by carbon nanotubes. *Carbon*. 60. pg. 67-75. http://www.sciencedirect.com/science/article/pii/S0008622313002765

- "Carbon nanotubes (CNTs) have one of the highest production volumes among carbonaceous engineered nanoparticles (ENPs) worldwide."

- "Electrophysiological measurements show that MWCNTs in a concentration range of 1.6–12 ppm disrupt lipid membranes by inducing significant transmembrane current fluxes, which suggest that MWCNTs insert and traverse the lipid bilayer membrane, forming transmembrane carbon nanotubes channels that allow the transport of ions."

Corsolini, S., Romeo, T., Ademollo, N., et al. (2002). POPs in key species of marine Antarctic ecosystem. *Microchemistry*. 43. pg. 187-93. http://www.ccpo.odu.edu/~klinck/Reprints/PDF/corsoliniMicroChem02.pdf

- "PCB concentrations were higher than HCB and p,p9-DDE by two orders of magnitude."

Costa, L. G. and Giordano, G. (2007). Developmental neurotoxicity of polybrominated diphenyl ether (PBDE) flame retardants. *NeuroToxicology*. 28. pg. 1047-67. http://www.ncbi.nlm.nih.gov/pmc/articles/PMC2118052/pdf/nihms34875.pdf

Costa, M., Ivar do Sul, J., Siilva-Cavalcanti, J., et al. (2009). On the importance of size of plastic fragments and pellets on the strandline: A snapshot of a Brazilian beach. *Environmental Monitoring Assessment*. http://www.algalita.org/wp-content/uploads/2014/05/on_the_importance_of_size_of_plastic_fragments_and_pellets_on_the_strandline.pdf

- "Plastic items of all types eventually undergo some form of degradation and subsequent fragmentation leading to the formation of small fragments that have suffered thermal, photochemical (sun), chemical (salting and burial in sand rich in organic matter), or physical (wind, waves, and sand abrasion) degradation."

- "Despite their distinct sources, at some stage in the fragmentation process, virgin plastic pellets and plastic fragments become similar in size and behave similarly in the environment."

Covaci, A., Gerecke, A. C., Law, R. J., et al. (2006). Hexabromocyclododecanes (HBCDs) in the environment and humans: A review. *Environ Sci Technol*. 40. pg. 3679-88.

Cózar, Andres, Echevarría, E., González-Gordillo, J., et al. (2014). Plastic debris in the open ocean. *Proceedings of the National Academy of Sciences*. 111(28). http://www.pnas.org/content/111/28/10239

Cózar, Andres, Sanz-Martin, Marina, Marti, Elisa, et al. (2015). Plastic accumulation in the Mediterranean Sea. *PLoS ONE*. 10(4). http://journals.plos.org/plosone/article?id=10.1371/journal.pone.0121762

- "The average density of plastic… [is] comparable to the accumulation zones described for the five subtropical ocean gyres."

Cozzarelli, I. M., Bohlke, J. K., Masoner, J., et al. (2011). Biogeochemical evolution of a landfill leachate plume, Norman, Oklahoma. *Ground Water*. 49(5). pg. 663-87. http://www.ncbi.nlm.nih.gov/pubmed/21314684

- "In the United States alone, approximately 135 million tons of municipal solid waste (MSW) was deposited in landfills in 2008, making landfills the most common method of MSW disposal."

- "Landfills alone accounted for approximately 23% of total U.S. anthropogenic methane (CH_4) emissions to the atmosphere in 2007."

Cross, J. N., Hardy, J. T., Hose, J. E., et al. (1987). Contaminant concentrations and toxicity of sea-surface microlayer near Los Angeles, California. *Marine Environmental Research*. 23. pg. 307-23.

Cui, D., Tian, F., Ozkan, C., et al. (2005). Effect of single wall carbon nanotubes on human on human HEK293 cells. *Toxicological Letters*. 155. pg. 73-85. http://www.ncbi.nlm.nih.gov/pubmed/15585362

- "SWCNTs can inhibit HEK293 cells growth by inducing cell apoptosis and decreasing cellular adhesion ability."

Curtis, Rick. (1998).OA guide to water purification. In: *The backpacker's field manual*. Random House, NY.

Daly, Matthew and Lindsey, Bruce. (1996). *Occurrence and concentrations of volatile organic compounds in shallow ground water in the lower Susquehanna River basin, Pennsylvania and Maryland*. U.S. Geological Survey. http://pa.water.usgs.gov/reports/wrir_96-4141/report.html

Daly, G. and Wania, F. (2005). Organic contaminants in mountains. *Environmental Science and Technology*. 39(2). pg. 385-98. http://snobear.colorado.edu/Markw/Ecuador/WatershedBio/POC/daly.pdf

- "High rates of snow deposition in mid- and high-latitude mountains may lead to a large contaminant release during snowmelt."

- "Future efforts should further focus on the bioaccumulation and potential effects of contaminants in the upper trophic levels of alpine food chains, on measuring more water-soluble, persistent organic contaminants, and on studying how climate change may affect contaminant dynamics in mountain settings."

Damstra, T. (2002). Potential effects of certain persistent organic pollutants and endocrine disrupting chemicals on the health of children. *J. Toxicol. Clin. Toxicol*. 40. pg. 457-65. http://www.ncbi.nlm.nih.gov/pubmed/12216998

Danoun, Rashad. (2007). *Desalination plants: Potential impacts of brine discharge on marine life*. The Ocean Technology Group, University of Sydney, Australia. http://ses.library.usyd.edu.au/bitstream/2123/1897/1/Desalination%20Plants.pdf

Darbre, P. (2006). Environmental oestrogens, cosmetics and breast cancer. *Best Practices Research Clinical Endocrinology and Metabolism*. 20(1). pg. 121-43. http://www.ncbi.nlm.nih.gov/pubmed/16522524

- "A range of organochlorine pesticides and polychlorinated biphenyls possess oestrogen-mimicking properties and have been measured in human breast adipose tissue and in human milk."

- "The breast is also exposed to a range of oestrogenic chemicals applied as cosmetics…allowing a more direct dermal absorption route for breast exposure to oestrogenic chemicals."

Darnerud, P. O. (2003). Toxic effects of brominated flame retardants in man and in wildlife. *Environment International*. 29(6). pg. 841-53. http://www.ncbi.nlm.nih.gov/pubmed/12850100

Das, Jayanta and Felty, Quentin. (2014). PCB153-induced overexpression of ID3 contributes to the development of microvascular lesions. *PLoS ONE*. 9(8). http://www.plosone.org/article/fetchObject.action?uri=info:doi/10.1371/journal.pone.0104159&representation=PDF

Daughton, C. and Ternes, T. (1999). Pharmaceuticals and personal care products in the environment: Agents of subtle change? *Environmental Health Perspectives*. 107. pg. 907-938. http://www.epa.gov/esd/bios/daughton/errata.pdf

- "During the last three decades, the impact of chemical pollution has focused almost exclusively on the conventional "priority" pollutants."

- "Another diverse group of bioactive chemicals receiving comparatively little attention as potential environmental pollutants includes the pharmaceuticals and active ingredients in personal care products."

- "These compounds and their bioactive metabolites can be continually introduced to the aquatic environment as complex mixtures via a number of routes …organisms are captive to continual life-cycle, multigenerational exposure."

Dave, G. and Aspegren, D. (2010). Comparative acute toxicity of 52 leachates from synthetic textiles made of plastic fibres to Daphnia magna. *Ecotoxicology and Environmental Safety*. 73(7). http://www.sciencedirect.com/science/article/pii/S0147651310001338

- "The environmental aspects of textiles are very complex and include production, processing, transport, usage, and recycling. Textiles are made from a variety of materials and can contain a large number of chemicals. Chemicals are used during production of fibres, for preservation and colouring and they are released during normal wear and during washing…Eco-labeled products were evenly distributed on a toxicity scale, which means that eco-labelling in its present form does not necessarily protect users or the environment from exposure to toxic chemicals."

Davenport, Coral. (October 28, 2015). A close-up look at Greenland, melting away. *The New York Times*. pg. A1, A10.

Davidson, T. (2012). Boring crustaceans damage polystyrene floats under docks polluting marine waters with microplastic. *Marine Pollution Bulletin*. 64(9). pg. 1821-8. http://www.ncbi.nlm.nih.gov/pubmed/22763283

- "Boring isopods damage expanded polystyrene floats under docks and, in the process, expel copious numbers of microplastic particles."

- " Floats from aquaculture facilities and docks were heavily damaged by thousands of isopods and their burrows…[in] multiple sites in Asia, Australia, Panama, and the USA."

- "One isopod creates thousands of microplastic particles when excavating a burrow; colonies can expel millions of particles."

DeBusk, K., Hunt, W., Osmond, D. and Cope, G. (2009). *Water quality on rooftop runoff*. College of Agriculture & Life Sciences. http://www.bae.ncsu.edu/stormwater/PublicationFiles/RooftopRunoff2009.pdf

Dedman, Craig J. (2014). *Investigating microplastic ingestion by zooplankton*. Master thesis. https://ore.exeter.ac.uk/repository/bitstream/handle/10871/17179/DedmanC_TPC.pdf?sequence=2

- "Size is the first and most obvious factor that will determine a particle's bioavailability, as a small size of an item increases its availability to a number of organisms across a larger range of trophic levels."

- "The concentration of microplastics sized <100 μm within the natural environment is currently unknown."

- This study of the bioavailability, ingestion, and adherence of microplastic by zooplankton utilized 10 to 20 um sized microplastics.

DEFRA. (2011). *Environmental impact: Biodiversity indicator DE5: Farmland bird populations*. Department for Environment Food and Rural Affairs. http://archive.defra.gov.uk/evidence/statistics/foodfarm/enviro/observatory/indicators/documents/DE5.pdf

De Guise, S., Shaw, S., Barclay, J., et. al. (2001). Consensus statement: Atlantic coast contaminants workshop 2000. *Environmental Health Perspectives*. 109. pg. 1301-2. http://www.ncbi.nlm.nih.gov/pmc/articles/PMC1240514/pdf/ehp0109-001301.pdf

De Jong, W. H., Hagens, W. I., Krystek, P., et al. (2008). Particle size-dependent organ distribution of gold nanoparticles after intravenous administration. *Biomaterials*. 29. pg. 1912-9.

Depledge, M., Galgani, F., Panti, C., et al. (2013). Plastic litter in the sea. *Marine Environmental Research*. 92. pg. 279-81. http://onemoregeneration.org/wp-content/uploads/2012/07/Plastic-litter-in-the-sea.pdf

- "A horizon scan of global conservation issues recently identified microplastics as one of the main global emerging environmental threats."

- "Few, if any, practical measures have been put in place to manage the situation."

Desai, M., Labhasetwar, V., Amidon, G. et al. (1996). Gastrointestinal uptake of biodegradable microparticles: Effects of particle size. *Pharmaceutical Research*. 13(12). pg. 1838-45. http://deepblue.lib.umich.edu/bitstream/handle/2027.42/41451/11095_2004_Article_30 6778.pdf?sequence=1

- "In general, the efficiency of uptake of 100 nm size particles by the intestinal tissue was 15-250 fold higher compared to larger size microparticles."

- "This has important implications in designing of nanoparticle-based oral drug delivery systems, such as an oral vaccine system."

Derfus, A., Chan, W. and Bhatia, S. (2004). Probing the cytotoxicity of semiconductor quantum dots. *Nanotechnology Letters*. 4. pg. 11-8. http://lmrt.mit.edu/sites/default/files/Derfus2004_NanoLett.pdf

- "With their bright, photostable fluorescence, semiconductor quantum dots show promise as alternatives to organic dyes for biological labeling."

- "CdSe QDs were found to induce cell death due to their inherent chemical composition."

- "We found that CdSe-core QDs were indeed acutely toxic under certain conditions…cytotoxicity correlates with the liberation of free Cd2+ ions due to deterioration of the CdSe lattice. When appropriately coated, CdSe-core QDs can be rendered nontoxic and used to track cell migration and reorganization in vitro."

- "For these QDs, an organic capping layer must bear the burden of preventing surface oxidation, and therefore cytotoxicity."

Derraik, J. (2002). The pollution of the marine environment by plastic debris: A review. *Marine Pollution Bulletin*. 44. pg. 842-52. http://www.marineconnection.org/docs/Derraik_J_2002_plastic_po.pdf

De Witte, B., Devriese, L., Bekaert, K., et al. (2014). Quality assessment of the blue mussel (Mytilus edulis): Comparison between commercial and wild types. *Marine Pollution Bulletin*. 85. pg. 146-55. http://www.sciencedirect.com/science/article/pii/S0025326X14003671

- "The number of total microplastics varied from 2.6 to 5.1 fibers/10 g of mussel. A higher prevalence of orange fibers at quaysides is related to fisheries activities.

Chemical contamination of polycyclic aromatic hydrocarbons and polychlorobiphenyls could be related to industrial activities and water turbidity."

Dhasmana, A., Jamal, Q., Mir, S., et al. (2014). Titanium dioxide nanoparticles as guardian against environmental carcinogent Benzo[alpha]Pyrene. *PLoS ONE.* http://www.plosone.org/article/fetchObject.action?uri=info:doi/10.1371/journal.pone.0107068&representation=PDF

Diamanti-Kandarkis, E., Bourguignon, J., Giudice, L., et al. (2009). Endocrine disrupting chemicals: An Endocrine Society scientific statement. *Endocrine Review.* 30(4). pg. 293-342.

Diaz, R. J. and Rosenberg, R. (2008). Spreading dead zones and consequences for marine ecosystems. *Science.* 321(5891). pg. 926-9.

Dirinck, E., Jorens, P., Covaci, A., et. al. (2011). Obesity and persistent organic pollutants: Possible obesogenic effect of organochlorine pesticides and polychlorinated biphenyls. *Obesity.* 19(4). pg. 709-14. http://onlinelibrary.wiley.com/doi/10.1038/oby.2010.133/epdf

- "Persistent organic pollutants (POPs) are endocrine-disrupting chemicals associated with the development of the metabolic syndrome and type 2 diabetes."

- " The diabetogenic effect of low-dose exposures to POPs might be more complicated than a simple obesogenic effect."

Dishaw, L., Powers, C., Ryde, I., et al. (2011). Is the PentaBDE replacement, Tris (1,3-dichloro-2-propyl) Phosphate (TDCPP), a developmental neurotoxicant? Studies in PC12 Cells. *Toxicological Applied Pharmacology.* 256(3). pg. 281-9. http://www.ncbi.nlm.nih.gov/pmc/articles/PMC3089808/pdf/nihms266652.pdf

- "Organophosphate flame retardants (OPFRs) are used as replacements for the commercial PentaBDE mixture that was phased out in 2004. OPFRs are ubiquitous in the environment and detected at high concentrations in residential dust, suggesting widespread human exposure."

- "Different OPFRs show divergent effects on neurodifferentiation, suggesting the participation of multiple mechanisms of toxicity… OPFRs may affect neurodevelopment with similar or greater potency compared to known and suspected neurotoxicants."

Di Ventra, M., Evoy, S. and Heflin, J. (2004). *Introduction to nanoscale science and technology.* Springer, Belin, Heidelberg, Germany.

Dodson, R., Nishioka, M., Sandley, L., et. al. (2012). Endocrine disruptors and asthma-associated chemicals in consumer products. *Environmental Health perspectives*. 120(7). pg. 935-43.
http://www.ncbi.nlm.nih.gov/pmc/articles/PMC3404651/pdf/ehp.1104052.pdf

- "Chemicals contained in consumer products are ubiquitous in human tissues, sometimes at high concentrations."

- "Common products contain complex mixtures of EDCs and asthma-related compounds…a consumer who used the alternative surface cleaner, tub and tile cleaner, laundry detergent, bar soap, shampoo and conditioner, facial cleanser and lotion, and toothpaste (a plausible array of product types for an individual) would potentially be exposed to at least 19 compounds: two parabens, three phthalates, MEA, DEA, five alkylphenols, and seven fragrances."

- "Many detected chemicals were not listed on product labels."

- "Consumers should be able to avoid some target chemicals—synthetic fragrances, BPA, and regulated active ingredients—using purchasing criteria. More complete product labeling would enable consumers to avoid the rest of the target chemicals."

Domingo, J. L., Falcó, G., Llobet, J. M., et al. (2003). Polycyclic naphthalenes in foods: estimated dietary intake by the population of Catalonia, Spain. *Environmental Science and Technology*. 37. pg. 2332-5.

Donaldson, K. and Stone, V. (2003a). Current hypotheses on the mechanisms of toxicity of ultrafine particles. *Annali dell'Istituto Superiore di Sanita*. 39(3). pg. 405-10.
https://www.nasa.gov/centers/johnson/pdf/486006main_Donaldson.pdf

- "Particle as small as ultrafine particles have a very large surface area and particle number per unit mass."

- "To obtain 10 $\mu g/m^3$ of 2 μm diameter particles you only need 1.2 particles per ml of air and the total surface area of particles is 24 $\mu m^2/ml$; the same airborne mass concentration of 20 nm particles requires 2.4 million particles with a surface area of 3,016 $\mu m^2/ml$. The lung is likely to respond quite differently to 2.4 million particles with their concomitant huge surface area than to a relatively small number of larger particle."

- "Ultrafine particles are present in large numbers in ambient air, with outside background levels in the range 5000-10,000 particles per ml rising during pollution episodes to 3,000,0000 particles/ml."

- "Vacuum cleaning has been reported to generate more than 8,000,000 particles/cubic foot."

Donaldson, K., Stone, V., Gilmour, P., et al. (2003b). Ultrafine particles: Mechanisms of lung injury. *Philosophical Transactions of the Royal Society of London*. 358. pg. 2741-9. http://rsta.royalsocietypublishing.org/content/roypta/358/1775/2741.full.pdf

- "Many ultrafine particles comprised classically of low–toxicity, low–solubility materials such as carbon black and titanium dioxide have been found to have greater toxicity than larger, respirable particles made of the same material."

- "Using latex particles in three sizes — 64, 202 and 535 nm — revealed that the smallest particles (64 nm) were profoundly inflammogenic but that the 202 and 535 nm particles had much less activity, suggesting that the cut–off for ultrafine toxicity lies somewhere between 64 and 202 nm."

- "Oxidative stress is a likely process by which the ultrafines have their effects."

Donaldson, K., Tran, L., Jimenez, L., et al. (2005) Combustion-derived nanoparticles: a review of their toxicology following inhalation exposure part. *Fibre Toxicology*. 2(10). http://www.ncbi.nlm.nih.gov/pubmed/16242040

- "Combustion-derived nanoparticles (CDNP)…originate from a number of sources…diesel soot, welding fume, carbon black and coal fly ash."

- "Key CDNP-associated properties of large surface area and the presence of metals and organics all have the potential to produce oxidative stress. CDNP may also exert genotoxic effects, depending on their composition."

Donaldson, K. and Poland, C. (2009). Nanotoxicology: New insights into nanotubes. *Nature Nanotechnology*. 4(11). pg. 708-10. http://www.nature.com/nnano/journal/v4/n11/full/nnano.2009.327.html

Donohue, M., Boland, R., Sramek, C. and Antonelis, G. (2001). Derelict fishing gear in the Northwestern Hawaiian Islands: Diving surveys and debris removed in 1999 confirm threat to coral reef ecosystems. *Marine Pollution Bulletin*. 42. pg. 1301-12. http://www.pifsc.noaa.gov/cred/derelict_fishing_gear.php

Dow Industrial. (2014). Material property data: Dow saran wrap 3 plastic film. *Matweb: Material Property Data*. http://www.matweb.com/search/datasheet.aspx?matguid=57c29e222a7749d58267c18e9e18b637&ckck=1

Downs, C., Kramarsky-Winter, E., Fauth, J., et al. (2013). Toxicological effects of the sunscreen UV filter, benzophenone-2, on planulae and in vitro cells of the coral,

Stylophora pistillata. Ecotoxicology. 23(2). pg. 175-91.
http://link.springer.com/article/10.1007/s10646-013-1161-y

Downs, C., Kramarsky-Winter, E., Segal, R., et al. (2015). Toxicopathological effects
of the sunscreen UV filter, oxybenzone (benzophenone-3), on coral planulae and
cultured primary cells and its environmental contamination in Hawaii and the U. S.
Virgin Islands. *Arch. Environ. Contam. Toxicol.*
http://link.springer.com/article/10.1007%2Fs00244-015-0227-7

Doyle, M., Watson, W., Bowlin, N. and Sheavley, S. (2011). Plastic particles in coastal
pelagic ecosystems of the Northeast Pacific ocean. *Marine Environmental Research.*
71. pg. 41-52. http://www.pmel.noaa.gov/foci/publications/2011/doyl0755.pdf

- "Nets with 0.505 mm mesh were used to collect surface samples during all
  cruises."

- "The ubiquity of such particles in the survey areas and predominance of sizes
  <2.5 mm implies persistence in these pelagic ecosystems as a result of
  continuous breakdown from larger plastic debris fragments."

- "It is likely that the particles encountered in this study continue to fragment and
  degrade to smaller and smaller particles (<0.5 mm)."

- "Ongoing plankton sampling programs in large marine ecosystems should be
  considered as sources of data for continued assessment of the abundance,
  distribution and potential impact of plastic debris particles in productive pelagic
  environments."

Dresselhaus, M. and Dresselhaus, P. (2001). *Carbon nanotubes: Synthesis, structure,
properties, and applications.* Springer, Berlin. Heidelberg, Germany.

Duan, J., Yu, Y., Shi, H., et al. (2013a). Toxic effects of silica nanoparticles on
zebrafish embryos and larvae. *PLoS ONE.* 8(9).
http://www.plosone.org/article/fetchObject.action?uri=info:doi/10.1371/journal.pone.00
74606&representation=PDF

- "Our data indicated that SiNPs caused embryonic developmental toxicity,
  resulted in persistent effects on larval behavior."

Duan, J., Yu, Yongbu, Li, Yang, et al. (2013b). Toxic effect of silica nanoparticles on
endothelial cells through DNA damage response via Chk1-dependent G2/M checkpoint.
PLoS ONE.
http://www.plosone.org/article/fetchObject.action?uri=info:doi/10.1371/journal.pone.00
62087&representation=PDF

Dubaish, Fatehi and Liebezeit, Gerd. (2013). Suspended microplastics and black carbon particles in the Jade System, Southern North Sea. *Water, Air and Soil Pollution*. 224. pg. 1352.
https://www.google.com/url?sa=t&rct=j&q=&esrc=s&source=web&cd=1&ved=0CB4QFjAA&url=http%3A%2F%2Fwww.researchgate.net%2Fprofile%2FFatehi_Dubaish%2Fpublication%2F257673562_Suspended_Microplastics_and_Black_Carbon_Particles_in_the_Jade_System_Southern_North_Sea%2Flinks%2F00463536961851d847000000.pdf&ei=m7IFVduuJoqgNrrRgqgI&usg=AFQjCNHBJy5Gt1SC9kTVA9Ff_CI1X3ifKg&sig2=F2-OMV5WE4xnjViXQNw9Hg&cad=rja

- "On average, 64 ± 194 granular particles. 88 ± 82 fibres and 30 ± 41 BC particles per litre were recorded...all freshwater sources discharged microplastics."

- "The large numbers encountered might give rise to a potentially high impact on filter feeding organisms such as blue mussels or oysters. In addition zooplankton and fish feeding indiscriminantly on particulate matter may also take up suspended microplastic particles."

Dubilier, N., McFall-Ngai, M., and Zhao, L. (2015). Create a global microbiome effort. *Nature*. 526. Pg. 631-634. http://www.nature.com/news/microbiology-create-a-global-microbiome-effort-1.18636

- "...to be successful, microbiome research will require a coordinated effort across the international community of biologists, chemists, geologists, mathematicians, physicists, computer scientists and clinical experts...science is only just realizing the full importance of the microbial world...currently only 35 bacterial and archaeal phyla are recognized...sequencing efforts in the past few years have pushed the number closer to 1,000."

- "Microbes in the oceans produce 50% of the oxygen we breathe, and—through photosynthesis—remove roughly the same proportion of carbon dioxide from the atmosphere."

- "Billions of tones of human-made toxic chemicals have overwhelmed the degrading and recycling capacity of microbiomes."

Dulvy, N., Fowler, S., Musick, J. et al. (2014). Extinction risk and conservation of the world's sharks and rays. *eLife*.

- "The rapid expansion of human activities threatens ocean-wide biodiversity loss. Numerous marine animal populations have declined."

- "Large-bodied, shallow-water species are at greatest risk and five out of the seven most threatened families are rays."

- "Population depletion… is particularly prevalent in the Indo-Pacific Biodiversity Triangle and Mediterranean Sea."

Dulvy, N., Sadovy, Y. and Reynolds, J. D. (2003). Extinction vulnerability in marine populations. *Fish and Fisheries*. 4. pg. 25-64. http://onlinelibrary.wiley.com/doi/10.1046/j.1467-2979.2003.00105.x/abstract

Dye, C., Schlabach, M., Green, J., et al. (2007) *Bronopol, resorcinol, m-cresol and triclosan in the Nordic environment*. Nordic Council of Ministers, Copenhagen, TemaNord: 585. http://nordicscreening.org/index.php?module=Pagesetter&type=file&func=get&tid=5&fid=reportfile&pid=2

Eckely, N. (2000). *From regional to global assessment: Learning from persistent organic pollutants: ENRP Discussion paper 2000-23*. Kennedy school of Government, Harvard University. http://www.hks.harvard.edu/gea/pubs/2000-23.pdf

Ecocycle. (2011). *Microplastics in compost*. Woods End Laboratories and Ecocycle. http://www.ecocycle.org/files/pdfs/microplastics_in_compost_summary.pdf

- "In order to eliminate microplastics from finished compost, all plastic-coated paper products should be excluded from the composting process."

Edyvane, K., Dalgetty, A., Hone, P., et al. (2004). Long-term marine litter monitoring in the remote Great Australian Bight, South Australia. *Marine Pollution Bulletin*. 48. pg. 1060-75. http://www.ncbi.nlm.nih.gov/pubmed/15172812

Ejlertsson, E., Karlsson, A., Lagerkvist, A., et al. (2003). Effects of co-disposal of wastes containing organic pollutants with municipal solid waste: A landfill simulation study. *Advances in Environmental Research*. 7. 949-60. http://naulibrary.org/dglibrary/admin/book_directory/Environmental_management/4872.pdf

- "Landfills receive solid waste from municipalities and industries and often sludge from sewage treatment plants. The landfilled waste will undergo transformations through chemical and biological conversion and degradation leading to a development of four typical phases during the ageing of a landfill."

- "An initial oxic phase (1) characterized by aerobic degradation, which is followed by an acid fermentation phase (2), where volatile fatty acids (VFAs) and different alcohols may reach molar concentrations."

- "Next is a methane formation phase (3), which is the longest degradation phase of the microbially active landfill. Finally, an oxic phase (4) occurs when the substrates, which may be converted to methane and carbon dioxide (biogas), have been depleted."

- "In general, the main part of the active life cycle of a landfill takes place when the waste is under anoxic conditions. Therefore, the microbiological degradation of organic materials is due mainly to anaerobic bacteria and their dynamic interaction."

- "The overall aim of the present investigation has been to study the effects of co-disposal of wastes containing OPs with MSW."

- "The results obtained in the study showed that OPs released from the waste had effects on the microbial processes and that the methanogenic consortia degrading MSW were hampered, probably due to the presence of Freons."

Eloe, E., Fadrosh, D., Novotny, M., et al. (2011). Going deeper: Metagenome of a hadopelagic microbial community. *PLoS ONE*. 6(5). http://bartlettlab.ucsd.edu/Publications_files/Eloe2011.pdf

Endo, S., Takizawa, R., Okuda, K., et al. (2001) Concentration of polychlorinated biphenyls (PCBs) in beached resin pellets: Variability among individual particles and regional differences. *Environmental Science and Technology*. 35. pg. 318-24. http://www.ncbi.nlm.nih.gov/pubmed/11347604

Endo, S., Yuyama, M. and Takada, H. (2013). Desorption kinetics of hydrophobic organic contaminants from marine plastic pellets. *Marine Pollution Bulletin*. 74(1). pg. 125-131. http://www.ncbi.nlm.nih.gov/pubmed/23906473

- "This study investigated the desorption behavior of polychlorinated biphenyls (PCBs) from marine plastic pellets."

- "After 128 d of the experiment, the…major fractions (90-99%) of highly chlorinated congeners remained in the pellets."

- "The desorption half-lives are estimated to 14d to 210 years for CB 8 to CB 209."

Engler, Richard. (2012). The complex interaction between marine debris and toxic chemicals in the ocean. *Environmental Science and Technology*. 46. pg. 12302-15. http://water.epa.gov/type/oceb/marinedebris/upload/The-Complex-Interaction-between-MD-and-Toxic-Chemicals-in-the-Ocean.pdf

- "In 2008, plastics represented about 12% of the municipal solid waste (MSW) generated in the United States, or about 30 million tons, of which over 94% (by weight) was discarded."

- "It is not currently possible to accurately calculate how much of the toxic chemical load present in human diet is attributable to this mechanism."

Environmental Integrity Project. (2012). *Air quality profile of Curtis Bay, Brooklyn, and Hawkins Point, Maryland*. Environmental Integrity Project, Washington, DC. http://www.environmentalintegrity.org/news_reports/documents/FINALBAYBROOKREPORT_003.pdf

- "Each year, from 2005 to 2009, the Curtis Bay zip code was among the top ten zip codes in the country for highest quantity of toxic air pollutants released by stationary (nonmobile) facilities."

- "In 2007 and 2008, Curtis Bay ranked first in the entire country for quantity of these releases, with 20.6 and 21.6 million pounds released respectively each year. In 2009, it ranked second in the nation after the quantity decreased to 13.8 million pounds."

- "In 2010, due to pollution control technology upgrades at two coal-fired power plants, this number decreased to 2.2 million pounds, dropping the Curtis Bay zip code to 74th in the nation out of 8,949c zip codes reporting toxic emissions."

- This report contains an extensive description of PM 2.5 standards and EPA information sources.

Environmental Working Group. (2007). *Bisphenol A: Toxic plastics chemical in canned food*. EWG. http://www.ewg.org/research/bisphenol

- "Independent laboratory tests found a toxic food-can lining ingredient associated with birth defects of the male and female reproductive systems in over half of 97 cans of name-brand fruit, vegetables, soda, and other commonly eaten canned goods."

- "The study… targeted the chemical bisphenol A (BPA), a plastic and resin ingredient used to line metal food and drink cans. There are no government safety standards limiting the amount of BPA in canned food."

- "The nation's system of public health protections from industrial chemicals like BPA are embodied in the Toxic Substances Control Act, a law passed in 1976 that is the only major environmental or public health statute that has never been updated."

- See Volume 3 of the Phenomenology of Biocatastrophe publication series for a more detailed listing of some EWG research findings.

Environmental Working Group. (2013). *Dirty dozen list of endocrine disruptors*. EWG. http://static.ewg.org/pdf/kab_dirty_dozen_endocrine_disruptors.pdf

- The EWG lists the following as the most important endocrine disrupting hormone altering chemicals: BPA, dioxin, atrazine, phthalates, perchlorate, fire retardants, lead, arsenic, mercury, perfluorinated chemicals, organophosphate pesticides, and glycol ethers.

Eriksen, M., Mason, S., Wilson, S., et al. (2013). Microplastic pollution in the surface waters of the Laurentian Great Lakes. *Marine Pollution Bulletin*. 77(1-2). pg. 177-82. http://www.marcuseriksen.com/wp-content/uploads/2013/10/Microplastic-pollution-in-the-surface-waters-of-the-Laurentian-Great-Lakes.pdf

- "Neuston samples were collected…using a 333 um mesh manta trawl."

- "Average abundance was approximately 43,000 microplastic particles/km^2 , station 20, downstream from two major cities, contained over 466,000 particles/km^2…SEM analysis determined nearly 20% of particles less than 1 mm, which were initially identified as microplastic by visual observation, were aluminum silicate from coal ash."

Eriksen, M., Lebreton, L., Carson, H., et al. (2014). Plastic pollution in the world's oceans: More than 5 trillion plastic pieces weighing 250,000 tons afloat at sea. *PLoS ONE*. 9(12). http://www.plosone.org/article/info%3Adoi%2F10.1371%2Fjournal.pone.0111913

- "When comparing between four size classes, two microplastic <4.75mm and meso- and macroplastic >4.75mm, a tremendous loss of microplastics is observed from the sea surface compared to expected rates of fragmentation, suggesting there are mechanisms at play that remove <4.75mm plastic particles from the ocean surface."

Erikson, Britt. (2002). Analyzing the ignored environmental contaminants. *Environmental Science and Technology*. 141A-5A. http://www.rachel.org/files/document/Analyzing_the_Ignored_Environmental_Contaminan.pdf

Eriksson, C., Burton, H. (2003). Origins and biological accumulation of plastic particles in fur seals from Macquarie Island. *Ambio*. 32(6). pg. 380-4. http://www.ncbi.nlm.nih.gov/pubmed/14627365

Eriksson, P., Fischer, C. and Frederiksson, A. (2006). Polybrominated diphenyl ethers, a group of brominated flame retardants, can interact with polychlorinated biphenyls in enhancing developmental neurobehavioral defects. *Toxicological Science.* 94(2). pg. 302-309. http://toxsci.oxfordjournals.org/content/94/2/302.full.pdf+html

- "Polybrominated diphenyl ethers (PBDEs) and polychlorinated biphenyls (PCBs) can interact and enhance developmental neurobehavioral defects when the exposure occurs during a critical stage of neonatal brain development."

Eriksson, C., Burton, H., Fitch, S., et al. (2013). Daily accumulation rates of marine debris on sub-Antarctic island beaches. *Marine Pollution Bulletin.* 66. pg. 199-208. http://www.sciencedirect.com/science/article/pii/S0025326X12004262

- "The daily accumulation rate of plastic debris on Macquarie Island was an order of magnitude higher than that estimated from monthly surveys during the same 4 months in the previous 5 years. This finding suggests that estimates of the oceans' plastic loading are an order of magnitude too low."

Espí, E., Salmerón, A., Fontecha, A., et al. (2006). Plastic films for agricultural applications. *Journal of Plastic Film and Sheeting.* 22(85). http://campus.extension.org/file.php/424/Supplemental_Reading/02%20Structures/Plastic%20Films%20for%20Agricultural%20Applications.pdf

- "The most important agricultural applications of plastic films are greenhouse, walk-in tunnel and low tunnel covers, and mulching. The raw materials are usually low density polyethylene and ethylene-vinyl acetate or ethylene-butyl acrylate copolymers for the covers and linear low density polyethylene for mulching."

Estes, J., Terborgh, J., Brashares, J. et al. (2011). Trophic downgrading of planet Earth. *Science.* 333(6040). pg. 301-6. http://www.cof.orst.edu/leopold/papers/Estes_etal2011.pdf

European Environment Agency. (2012). *The impacts of endocrine disrupters on wildlife, people, and their environments—The Weybridge +15 (1996-2011) report.* European Environment Agency, Copenhagen, Denmark. http://www.eea.europa.eu/publications/the-impacts-of-endocrine-disrupters/download

European Bioplastics. (2013). *Bioplastics facts and figures.* European bioplastics, Berlin, Germany. http://en.european-bioplastics.org/wp-content/uploads/2013/publications/EuBP_FactsFigures_bioplastics_2013.pdf

European Union. (2011). Commission Regulation (EU) No 1259/2011 of 2 December 2011 amending Regulation (EC) No 1881/2006 as regards maximum levels for dioxins, dioxin-like PCBs and non dioxin-like PCBs in foodstuffs. *Official Journal of the European Union.* L320. pg.18-23.

Evenset, A., Christensen, G. N. and Kallenborn, R. (2005). Selected chlorobornanes, polychlorinated naphthalenes and brominated flame retardants in Bjørnøya (Bear Island) freshwater biota. *Environmental Pollution*. 136. pg. 419-30.

Falandysz, J., Strandberg, L., Bergqvist, P.-A., et al. (1997). Spatial distribution and bioaccumulation of polychlorinated naphthalenes (PCNs) in mussel and fish from the Gulf of Gdafisk, Baltic Sea. *Science of the Total Environment*. 203. pg. 93-104.

Famiglietti, J., Lo, S., Ho, L, et. al. (2011). Satellites measure recent rates of groundwater depletion in California's Central Valley. *Geophysical Research Letters*. 38. pg. L0343. http://onlinelibrary.wiley.com/doi/10.1029/2010GL046442/epdf

- "Our results show that the Central Valley lost 20.4 ± 3.9 mm/yr of groundwater duringthe 78-month period, or 20.3 km^3 in volume. Continued groundwater depletion at this rate may well be unsustainable,with potentially dire consequences for the economic and food security of the United States."

- "Nearly 2 billion people rely on groundwater as a primary source of drinking water and for irrigated agriculture. However, in many regions of the world, groundwater resources are under stress due to a number of factors, including salinization, contamination and rapid depletion."

- "When coupled with the pressures of changing climate and population growth, the stresses on groundwater supplies will only increase in the decades to come…no comprehensive framework for monitoring the world's groundwater resources currently exists."

Fång, J., Nyberg, E., Bignert, A. and Bergman, A. (2013). Temporal trends of polychlorinated dibenzo-p-dioxins and dibenzofurans and dioxin-like polychlorinated biphenyls in mothers' milk from Sweden, 1972-2011. *Environment International*. 60. pg. 224-31. http://www.ncbi.nlm.nih.gov/pubmed/24080458

- "The rate of which $\sum$PCDDs, $\sum$PCDFs $\sum$DL-PCBs and the $\sum$TEQ are decreasing (on pg/g fat WHO-TEQ2005) is steeper in the last decade compared to the 40 year period, 1972–2011. The declines for PCDDs, PCDFs, DL-PCBs and $\sum$TEQs are 10%, 7.3%, 12% and 10% per year, last decade, compared to 6.1%, 6.1%, 6.9%% and 6.5% per year, 1972–2011."

Fant, M., Nyman, M., Helle, E. and Rudback, E. (2001). Mercury, cadmium, lead and selenium in ringed seals (*Phoca hispida*) from the Baltic Sea from Svalbard. *Environmental Pollution*. 111. pg. 493-501. http://www.sciencedirect.com/science/article/pii/S0269749100000786

Farrell, P. and Nelson, K. (2013). Trophic level transfer of microplastic: *Mytilus edulis* to *Carcinus maenas*. *Environmental Pollution*. 117. pg. 1-3. http://plasticsoupfoundation.org/wp-content/uploads/2013/03/Trophic-level-transfer-of-microplastic-Mytilus-edulis-L.-to-Carcinus-maenas-L.-2.pdf

- "Mussels (*Mytilus edulis*) were exposed to 0.5 um fluorescent polystyrene microspheres, then fed to crabs (*Carcinus maenas*)."
- "This study is the first to show 'natural' trophic transfer of microplastic, and its translocation to haemolymph and tissues of a crab. This has implications for the health of marine organisms, the wider food web and humans."

Federici, G., Shaw, B., and Handy, R. (2007). Toxicity of titanium dioxide nanoparticles to rainbow trout (*Oncorhynchus mykiss*): Gill injury, oxidative stress, and other physiological effects. *Aquatic Toxicology*. 84. pg. 415-30. http://www.ncbi.nlm.nih.gov/pubmed/17727975

- "Sub-lethal toxicity involves oxidative stress, organ pathologies, and the induction of anti-oxidant defences, such as glutathione."

Fenchel, T. (1982a). Ecology of heterotrophic microflagellates II: Bioenergetics and growth. *Marine Ecology Progress Series*. 8. pg. 225-31. http://www.int-res.com/articles/meps/8/m008p225.pdf

Fenchel, T. (1982b). Ecology of heterotrophic microflagellates IV: Bioenergetics and growth. *Marine Ecology Progress Series*. 9. pg. 35-42. http://www.int-res.com/articles/meps/9/m009p035.pdf

Fendall, L. and Sewell, M. (2009). Contributing to marine pollution by washing your face: Microplastics in facial cleansers. *Marine Pollution Bulletin*. 58. pg. 1225-8. https://facultystaff.richmond.edu/~sabrash/110/Chem%20110%20Spring%202014%20Articles/Contributing%20to%20Marine%20Pollution%20By%20Washing%20Your%20Face.pdf

Fernanda, P., Mário B., Monica, C., et al. (2011). Plastic debris ingestion by marine catfish: An unexpected fisheries impact. *Marine Pollution Bull*. 62(5). pg. 1098-102. http://www.caseinlet.org/uploads/Plastic_debris_ingestion_by_marine_catfish_An_unexpected_fisheries_impact_1_.pdf

Fernández, Pilar and Grimalt, Joan. (2003). On the global distribution of persistent organic pollutants. *Environmental Analysis*. 57(9). http://www.cid.csic.es/homes/pilarf/pdfs/39_FernandezCHIMIAPAPER.pdf

- "The global distribution of persistent organic pollutants (POPs) has become one of the main environmental problems in the last decade…their semivolatile character and high environmental half-lives result in long-range atmospheric transport and global planetary distribution."

- "Most POPs are organochlorine compounds (OCs) used for diverse applications, both industrial and agricultural. Their intentional use has been restricted or banned in many developed countries, but in some cases, e.g. 4,4'-DDT, their application in developing countries is allowed or unregulated."

- "Other POPs like dioxins, furans or polycyclic aromatic hydrocarbons (PAH) are mainly produced in combustion processes, which complicates the control of their emissions to the environment."

- "Transport and deposition of polycyclic aromatic hydrocarbons are mainly linked to atmospheric particles. Consequently, wet and dry deposition are the main removal processes of these compounds from the atmosphere."

- "Global Distillation Effect theory as proposed for semivolatile persistent pollutants, indicating that this transport mechanism not only involves transfer from low to high latitudes, but also preferential accumulation of the less volatile compounds in high altitude regions of mid-latitude areas."

Ferrey, Mark. (2011). *Wastewater treatment plant endocrine disrupting chemical monitoring study*. Minnesota Pollution Control Agency. http://www.pca.state.mn.us/index.php/view-document.html?gid=15610

- "EACs are hormonally active at very low, non-toxic concentrations, and may alter normal physiological functions in organisms that are exposed to them. An unknown number of the more than 87,000 chemicals that are manufactured worldwide may possess endocrine active properties…Endocrine active chemicals, or EACs, do not usually exhibit acute toxicity at the levels normally found in the environment, but instead can alter the normal functioning and growth of the exposed organism at very low concentrations."

- "EACs, including pharmaceuticals, are widespread at low concentrations in Minnesota's rivers and lakes."

- "Pharmaceuticals, triclosan, nonylphenol, nonylphenol ethoxylates, octylphenol, octylphenol ethoxylates, bisphenol A, phytoestrogens, and steroid hormones were detected in wastewater, surface water and sediment samples."

- "The pharmaceuticals carbamazepine (an anti-seizure medication used to treat attention deficit disorders) and sulfamethoxazole (a common antibiotic) were detected in 96 percent of effluent water samples."

- "Other pharmaceuticals found in water samples included trimethoprim (an antibiotic), fluoxetine (an antidepressant found in 80 percent of the effluent samples and 17 percent of downstream water samples); bupropion (an anti-depressant, detected in 80 percent of effluent samples and in 46 percent of downstream water samples); hydroxybupropion (a degradation product of bupropion), and diphenhydramine (an antihistamine) found in greater than 50 percent of the wastewater effluent samples."

- "Bisphenol A, nonylphenol, nonylphenol ethoxylates, octylphenol ethoxylates, and musk fragrances were found in WWTP effluent. However, these chemicals were detected in water and sediment upstream of WWTPs, sometimes at higher concentrations than downstream of the plants, indicating that there were other upstream sources of these chemicals."

- "The effluents from most of the WWTPs that were tested were estrogenic."

- "It is unclear to what degree other sources of contaminants – including private septic systems, agricultural feedlots, row cropping, biosolids application, and urban runoff – affect surface water."

Fiedler, H. (1998). Polychlorinated biphenyls (PCBs): Uses and environmental releases. *Proceedings of the Subregional Awareness Raising Workshop on Persistent Organic Pollutants at Abu Dhabi, United Arab Emirates.* http://www.chem.unep.ch/pops/pops_inc/proceedings/bangkok/fiedler1.html

- "PCBs are produced by chlorination of biphenyl; its commercial production started about 60 years ago. The total amount produced world-wide is estimated at 1.5 million tons. Depending on the degree of chlorination of the PCBs, their physico-chemical properties, like inflammability or electric conductivity, brought about a wide field of application. Thus, PCBs have been used as electric fluids in transformers and capacitors, as pesticide extenders, adhesives, dedusting agents, cutting oils, flame retardants, heat transfer fluids, hydraulic lubricants, sealants, paints, and in carbonless copy paper."

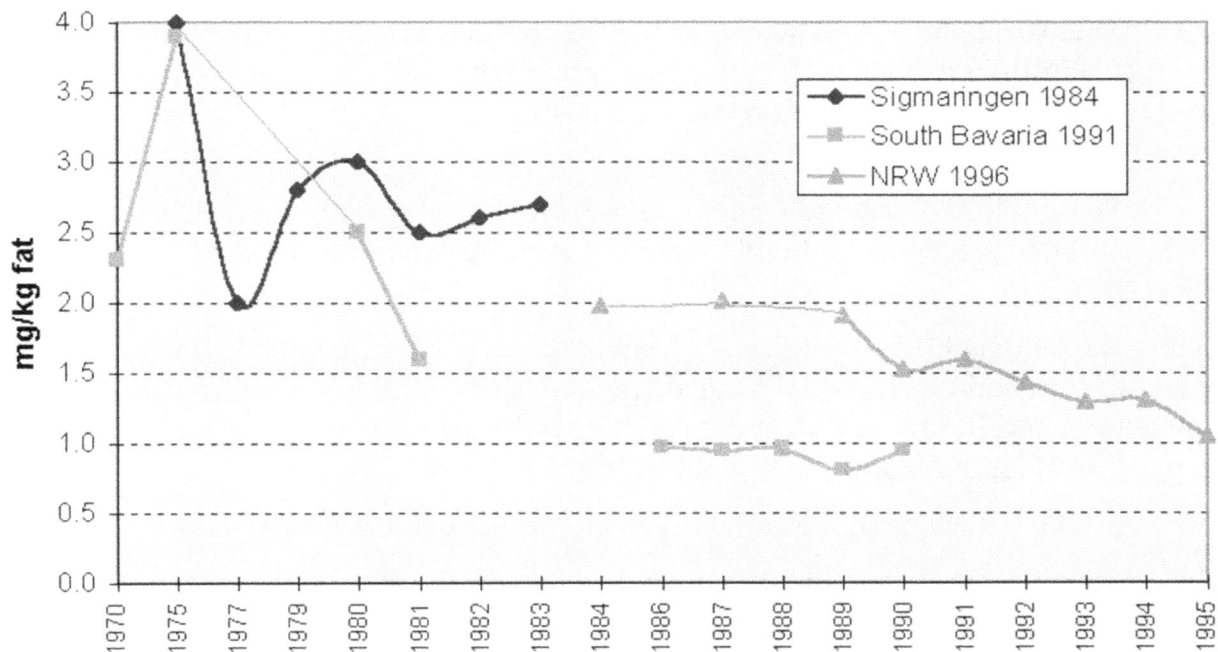

PCB Time Trends in Mother's Milk - Germany

Legend:
- Sigmaringen 1984
- South Bavaria 1991
- NRW 1996

Y-axis: mg/kg fat (0.0 to 4.0)

X-axis years: 1970, 1975, 1977, 1979, 1980, 1981, 1982, 1983, 1984, 1986, 1987, 1988, 1989, 1990, 1991, 1992, 1993, 1994, 1995

Fiedler, H., Hutzinger, O., Welsch-Pausch, K. and Schmiedinger, A. (2000). *Evaluation of the occurrence of PCDD/PCDF and POPs in wastes and their potential to enter the foodchain*. European Commission, DG Environment.
http://ec.europa.eu/environment/archives/dioxin/pdf/001_ubt_final.pdf

Fischer, Hans and Chan, Warren. (2007). Nanotoxicity: the growing need for in vivo study. *Current opinion in Biotechnology*. 18. pg. 565-71.
http://www.ncbi.nlm.nih.gov/pubmed/18160274

- "Nanotoxicology refers to the study of the interactions of nanostructures with biological systems with an emphasis on elucidating the relationship between the physical and chemical properties (e.g. size, shape, surface chemistry, composition, and aggregation) of nanostructures with induction of toxic biological responses."

Fischer, H., Liu, L., Pang, K. and Chan, W. (2006). Pharmacokinetics of nanoscale quantum dots: In vivo distribution, sequestration, and clearance in the rat. *Advanced Functional Materials*. 16(10). pg. 1299-305.
http://onlinelibrary.wiley.com/doi/10.1002/adfm.200500529/abstract

Fisheries and Agriculture Organization (FAO). (2014). *The state of world fisheries and aquaculture 2014*. Food and Aquaculture Organization of the United Nations, Rome, Italy. http://www.fao.org/3/a-i3720e.pdf

- "The world's marine fisheries expanded continuously to a production peak of 86.4 million tonnes in 1996 but have since exhibited a general declining trend. Global recorded production was 82.6 million tonnes in 2011 and 79.7 million tonnes in 2012."

- "The fraction of assessed stocks fished within biologically sustainable levels has exhibited a decreasing trend, declining from 90 percent in 1974 to 71.2 percent in 2011."

Fisner, M., Taniguchi, S., Moreira, F., Márcia, B. and Turra, A. (2013). Polycyclic aromatic hydrocarbons (PAHs) in plastic pellets: Variability in the concentration and composition at different sediment depths in a sandy beach. *Marine Pollution Bulletin.* 70. pg. 219-26. http://resodema.org/publications/publication13.pdf

- "Plastic pellets have the ability to adsorb organic pollutants such as PAHs."

- "The total PAHs varied, with the highest concentrations in the surface layer."

Flegal, K., Carroll, M., Ogden, C. and Curtin, L. (2010). Prevalence and trends in obesity among US adults, 1999-2008. *JAMA.* 303. pg. 235-41. https://www.health.ny.gov/prevention/obesity/statistics_and_impact/docs/obesity_trends_in_adults_1999-2008.pdf

- "In 2007-2008, the age-adjusted prevalence of obesity was 33.8% (95% confidence interval [CI], 31.6%-36.0%) overall, 32.2% (95% CI, 29.5%-35.0%) among men, and 35.5% (95% CI, 33.2%-37.7%) among women. The corresponding prevalence estimates for overweight and obesity combined (BMI 25) were 68.0% (95% CI, 66.3%-69.8%), 72.3% (95% CI, 70.4%-74.1%), and 64.1% (95% CI, 61.3%- 66.9%). Obesity prevalence varied by age group and by racial and ethnic group for both men and women."

Flint, S., Markle, T., Thompson, S. and Wallace, E. (2012). Bisphenol A exposure, effects and policy: A wildlife perspective. *Journal of Environmental Management.* 104. pg. 19-34. http://www.consbio.umn.edu/download/Flint_et_al_2012_BPA.pdf

- "Although it degrades quickly, it is pseudo-persistent in the environment because of continual inputs."

- "Many of the thousands of anthropogenic chemicals currently released into the environment are endocrine-disrupting compounds."

- "Some metabolites of BPA are more estrogenic than the original compound and environmental characteristics may alter degradation rates or biological impacts."

- It contains a very extensive bibliography on Bisphenol A.

Focazio, M., Kolpin, D. and Buxton, H. (2003). Pharmaceuticals, hormones, personal-care products, and other organic wastewater contaminants in water resources: Recent research activities of the U.S. Geological Survey's Toxic Substances Hydrology Program. *GeoHealth News*. 2(1). http://energy.er.usgs.gov/images/medical_geology_images/vol2_no1_web.pdf

Foekema, E., de Gruijter, C., Mergia, M., et al. (2013). Plastic in North Sea fish. *Environmental Science and Technology*. 47(15). pg. 8818-24. http://pubs.acs.org/doi/abs/10.1021/es400931b

- "1203 individual fish of seven common North Sea species were investigated… Plastic particles were found in 2.6% of the examined fish."

- "Only particles larger than 0.2 mm, being the diameter of the sieve used, were considered for the data analyses."

- "In addition, small fibers were initially detected in most of the samples."

- No mention of the methods of measuring micro or nanoplastics < 2 mm.

Fogarty, Lisa. (2007). *Bacteria and emerging chemical contaminants in the St. Clair River/Lake St. Clair Basin, Michigan*. USGS. http://pubs.usgs.gov/of/2007/1083/pdf/OFR2007-1083.pdf

Food and Drug Administration. (2014). *Pesticide monitoring program 2011 pesticide report*. U.S. Food and Drug Administration, Washington, DC. http://www.fda.gov/downloads/Food/FoodborneIllnessContaminants/Pesticides/UCM382443.pdf

Franeker, J., Blaize, C., Danielsen, J., et al. (2011). Monitoring plastic ingestion by the northern fulmar *Fulmarus glacialis* in the North Sea. *Environmental Pollution*. 159(10). pg. 2609-15. http://www.ncbi.nlm.nih.gov/pubmed/21737191

Frank, Adam. (January 18, 2015). Is climate disaster inevitable? *The New York Times*. pg. 6-7. http://www.nytimes.com/2015/01/18/opinion/sunday/is-a-climate-disaster-inevitable.html?_r=0

- "But the basic physics of energy, heat and work known as thermodynamics tell us that waste, or what we physicists call entropy, must be generated and dumped back into the environment in the process."

- "Combustion always produces chemical byproducts, and those byproducts can't just disappear."

- "When it comes to building world-girdling civilizations, there are no planetary free lunches."

Frederiksen, M., Vorkamp, K., Thomsen, M. and Knudsen, L. (2009). Human internal and external exposure to PBDEs—a review of levels and sources. *International Journal of Hygienic Environmental Health*. 212. pg. 109-35. http://www.ncbi.nlm.nih.gov/pubmed/18554980

- "Food items like fish from high trophic levels or lipid-rich oils have been found to contain relatively high concentrations of PBDEs, thus presenting an important exposure pathway to humans."

- Dust and air are thus not the only exposure pathways for PBDEs. This report summarizes data from 98 surveys of PBDE levels and sources in over 30 countries collected during the previous 20 years (milk, serum and placenta). It has an extensive bibliography.

Free, C., Jensen, O., Mason, S., et al. (2014). High-levels of microplastic pollution in a large, remote, mountain lake. *Marine Pollution Bulletin*. https://marine.rutgers.edu/pubs/private/Free_etal_MPB_in_press.pdf

Frenc, Sammy, Ben-Moshe, tal, Dror, Ishai, et al. (2013). Effect of metal oxide nanoparticles on microbial community structure and function in two different soil types. *PLoS ONE*. 8(12). http://journals.plos.org/plosone/article?id=10.1371/journal.pone.0084441

- "Because the soil bacterial community is a major service provider for the ecosystem and humankind, it is critical to study the effects of ENP exposure on soil bacteria."

- "ENPs are potentially harmful to soil environments."

Frias, J. P. G. L., Sobral, P. and Ferreira, A. M. (2010). Organic pollutants in microplastics from two beaches of the Portuguese coast. *Marine Pollution Bulletin*. 60. pg. 1988-92.

Friedman, C., Burgess, R., Perron, M., et al. (2009). Comparing polychaete and polyethylene uptake to assess sediment resuspension effects on PCB bioavailability. *Environmental Science and Technology*. 43(8). pg. 2865-70. http://cfpub.epa.gov/si/si_public_record_Report.cfm?dirEntryId=192163&CFID=19689118&CFTOKEN=78296479&jsessionid=cc30c746fc1a15e0607031172634656f1455

Fries, E. and Zarfl, C. (2012). Sorption of polycyclic aromatic hydrocarbons (PAHs) to low and high density polyethylene (PE). *Environmental Science Pollution Research international*. 19. pg. 1296-1304.
http://www.researchgate.net/publication/51798298_Sorption_of_polycyclic_aromatic_hydrocarbons_(PAHs)_to_low_and_high_density_polyethylene_(PE)

- "According to their high sorption capacity polyethylene (PE) passive samplers are often used for the analysis of polycyclic aromatic hydrocarbons (PAHs) in the aquatic environment."

- "Sorption equilibrium seemed to be driven by parameters other than, or in addition to, organic carbon. For both plastic types, diffusion coefficients decreased while the molecular weight of the PAHs increased."

Fries, E., Dekiff, J., Willmeyer, J., et al. (2013). Identification of polymer types and additives in marine microplastic particles using pyrolysis-GC/MS and scanning electron microscopy. *Environmental Science Processes*. 1949(15).
http://pubs.rsc.org/En/content/articlepdf/2013/em/c3em00214d

- "Any assessment of plastic contamination in the marine environment requires knowledge of the polymer type and the...organic plastic additives (OPAs)."

- "When polymer–TiO2 composites are degraded in the marine environment, TiO2-NPs are probably released...marine microplastics may act as a TiO2-NP source."

Fromme, H., Kuchler, T., Otto, T., et al. (2002). Occurrence of phthalates and bisphenol A and F in the environment. *Water Research*. 36. pg. 1429-38.
http://www.ncbi.nlm.nih.gov/pubmed/11996333

Fu, P., Xia, Q., Hwang, H., et al. (2014). Mechanisms of nanotoxicity: Generation of reactive oxygen species. *Journal of Food and Drug Analysis*. 22(1). pg. 64-75.
http://www.jfda-online.com/article/S1021-9498(14)00006-4/pdf

- "An important mechanism of nanotoxicity is the generation of reactive oxygen species (ROS). Overproduction of ROS can induce oxidative stress, resulting in cells failing to maintain normal physiological redox-regulated functions. This in turn leads to DNA damage, unregulated cell signaling, change in cell motility, cytotoxicity, apoptosis, and cancer initiation."

- "There are critical determinants that can affect the generation of ROS. These critical determinants, discussed briefly here, include: size, shape, particle surface, surface positive charges, surface-containing groups, particle dissolution, metal ion release from nanometals and nanometal oxides, UV light activation, aggregation, mode of interaction with cells, inflammation, and pH of the medium."

- "ROS have a short half-life, in the range of about 109 seconds, and thus, without derivatization, are difficult to be isolated directly for structural identification."

- "Understanding the mechanism of nanomaterial-induced toxicity is the first defense for hazard prevention."

- "As a result of their reactivity and ability to damage biological targets, hydroxyl radicals can serve as an ideal representative ROS for investigation."

- "The hydroxyl radical is the most powerful ROS in causing biological damage."

- "Some endogenous and dietary antioxidants are effective in scavenging hydroxyl radicals. Dietary antioxidants may play an important role in limiting oxidative damage and reducing the risk of numerous chronic diseases related to advancing age."

Fujioka, K, Hanada, S., Inoue, Y., et al. (2014). Effects of silica and titanium oxide particles on a human natural stem cell line: Morphology, mitochondrial activity and gene expression of differentiation markers. *International Journal of Molecular Science*. 15. pg. 11,742-59. http://www.mdpi.com/1422-0067/15/7/11742/pdf

- "Several in vivo studies suggest that nanoparticles (smaller than 100 nm) have the ability to reach the brain tissue."

- "To evaluate its effects on neural stem cells, we assayed a human neural stem cell (hNSCs) line exposed in vitro to three types of silica particles (30 nm, 70 nm, and <44um)."

- "hNSCs aggregated and exhibited abnormal morphology when exposed to the particles…Notably, 30-nm silica particles exhibited acute membrane permeability at concentrations ≥ 62.5 µg/mL in 24 h…indicate the potential toxicity of accumulated particles for long-term usage or continuous exposure."

- "There is a concern that some characteristics of nanomaterials, such as their tube- or fiber-like structures with rigid properties or certain sizes, might cause toxicity similar to that of asbestos."

Fujimasa, I. (1996). *Micromachines: A new era in mechanical engineering*. Oxford University Press, Oxford.

Fürst, Peter. (2006). Dioxins, polychlorinated biphenyls and other organohalogen compounds in human milk. Levels, correlations, trends and exposure through breastfeeding. *Molecular Nutrition and Food Research*. 50(10). pg. 922-33.

Gadd, G. M. (1993). Microbial formation and transformation of organometallic and organometalloid compounds. *FEMS Microbiology Reviews*. 11(4). pg. 297-316. http://www.sciencedirect.com/science/article/pii/016864459390003R

- "Microbial formation and transformation of organometallic and organometalloid compounds comprise significant components of biogeochemical cycles for the metals mercury, lead and tin and the metalloids arsenic, selenium, tellurium and germanium."

- "The major microbial methylating agents are methylcobalamin (CH_3CoB_{12}), involved in the methylation of mercury, tin and lead, and *S*-adenosylmethionine (SAM), involved in the methylation of arsenic and selenium."

- "Microorganisms can accumulate organometal(loid)s, a phenomenon relevant to toxicant transfer to higher organisms."

Gagne, F., Auclair, J., Turcotte, P., et al. (2003). Ecotoxicity of CdTe quantum dots to freshwater mussels: Impacts on immune system, oxidative stress and genotoxicity. *Aquatic Toxicology*. 86. pg. 333-40. http://www.ncbi.nlm.nih.gov/pubmed/18160110

Galy-Lacaux, C., Laouali, D., Descroix, L., et al. (2009). Long term precipitation chemistry and wet deposition in a remote dry savanna site in Africa (Niger). *Atmospheric Chemistry Physics*. 9. pg. 1579-95. http://www.atmos-chem-phys.net/9/1579/2009/acp-9-1579-2009.pdf

Galgani, F., Leaute, J., Moguedet, P., et al. (2000). Litter on the sea floor along European coasts. *Marine Pollution Bulletin*. 40(6). pg. 516-27. http://resodema.org/publications/publication6.pdf

Galil, G., Golik, A. and Turkayk, M. (1995). Litter at the bottom of the sea: A sea bed survey in the Eastern Mediterranean. *Marine Pollution Bulletin*. 30(1). pg. 22-4. http://www.sciencedirect.com/science/article/pii/0025326X9400103G

Gao, X., Ma, J., Chen, Y. and Wang, H. (2014). *OR18-3: Evaluation of the Rapid Effects of Bisphenol S (BPS) on the Heart: Impact on Arrhythmogenesis and Cardiac Calcium Handling*. Endocrine Society's 96th Annual Meeting and Expo, June 21–24, 2014 – Chicago. http://press.endocrine.org/doi/abs/10.1210/endo-meetings.2014.ED.3.OR18-3

- "Bisphenol A (BPA) is an environmental endocrine disrupting chemical that has been widely used in the production of plastic consumer goods. Because of BPA's potential adverse impact on human health, another member of the bisphenol family, bisphenol S (BPS), is increasingly becoming an alternative to BPA in production of plastic bottles (often labeled "BPA-free") and thermal paper…the potential health impact of BPS exposure is poorly defined…our results suggest that exposure to BPS, a substitute for the well-studied BPA, may have acute cardiac toxicity similar to that of BPA."

Gardner, T. A., Cote, I. M., Gill, J. A. et al. (2003). Long-term region-wide declines in Caribbean corals. *Science*. 301. pg. 958-60. http://www.sciencemag.org/content/301/5635/958.full.pdf

Gassel, M., Harwani, S., Park, J. and Jahn, A. (2013). Detection of nonylphenol and persistent organic pollutants in fish from the North Pacific Central Gyre. *Marine Pollution Bulletin*. 64. pg. 2374-9. http://www.ncbi.nlm.nih.gov/pubmed/23746941

- "Research on food web contamination from chemicals in plastic is limited."

- " Ingestion of synthetic debris occurred in ~10% of the sample population. PCBs and DDTs were 352±240 (mean±SD) and 1425±1118 ng/g lw, respectively. PBDEs were 9.08±10.6 ng/g lw, with BDEs-47, 99, and 209 representing 90% of PBDEs. Nonylphenol (NP) was detected in one-third of the yellowtail…[at] 67±72.3 ng/g ww."

Gatoo, M., Naseem, S., Arfat, M., et al. (2014). Physicochemical properties of nanomaterials: Implication in associated toxic manifestations. *Biomed Research International*. 498420. http://downloads.hindawi.com/journals/bmri/2014/498420.pdf

- "Physicochemical characteristics of nanoparticles and engineered nanomaterials including size, shape, chemical composition, physiochemical stability, crystal structure, surface area, surface energy, and surface roughness generally influence the toxic manifestations of these nanomaterials. This compels the research fraternity to evaluate the role of these properties in determining associated toxicity issues."

Gatti, A., Tossini, D. and Gambarelli, A. (2004). *Investigations of trace elements in bread through environmental scanning electron microscope and energy dispersive system*. 2nd International IUPAC Symposium, Brussels.

Gatti, A. and Montanari, S. (2008). Nanopollution: The invisible fog of future wars. *Futurist*. 42(3). pg. 32. http://connection.ebscohost.com/c/articles/31535416/nanopollution-invisible-fog-future-wars

- "The solution to these medical mysteries may be found in a new word: nanopathology, the study of diseases caused by micrometric and nanometric particles. Dust at the nanoscale (smaller than one-thousandth of a millimeter) can elude physiological barriers and easily enter the bloodstream, where it is very likely to reach all internal organs and tissues."

- "From what we have observed in our laboratory, nano-scale particles can interact with proteins, bacteria, and viruses to potentially cause not-yet-studied phenomena and develop new diseases with unusual symptoms."

Gaw, S., Thomas, K. and Hutchison, T. (2013). Sources, impacts and trends of pharmaceuticals in the marine and coastal environment. *Philosophical Transactions of the Royal Society*. 369. http://rstb.royalsocietypublishing.org/content/royptb/369/1656/20130572.full.pdf

Gaylor, M., Harvey, E. and Hale, R. (2012). House crickets can accumulate polybrominated diphenyl ethers (PBDEs) directly from polyurethane foam common in consumer products. *Chemosphere*. 86. pg. 500-5. http://www.ncbi.nlm.nih.gov/pubmed/22071374

- "Polybrominated diphenyl ether (PBDE) flame retardants are added at percent levels to many polymers and textiles abundant in human spaces and vehicles, wherein they have been long assumed to be tightly sequestered. However, the mgkg(-1) burdens recently detected in indoor dust testify to substantial releases. The bulk of released PBDEs remain in the terrestrial environment."

- "Insects/arthropods, such as crickets, are the most abundant invertebrate organisms and facilitate the trophic transfer of contaminants by breaking down complex organic matter (including discarded polymers) and serving as food for other organisms."

Geiser, M., Rothen-Rutlshauser, B., Knapp, N., et al. (2005). Ultrafine particles cross cellular membranes by nonphagocytic mechanisms in lungs and in cultured cells. *Environmental Health Perspectives*. 113(11). pg. 1555-60. http://www.ncbi.nlm.nih.gov/pmc/articles/PMC1310918/

- "High concentrations of airborne particles have been associated with increased pulmonary and cardiovascular mortality, with indications of a specific toxicologic role for ultrafine particles (UFPs; particles < 0.1 μm)."

- "Within hours after the respiratory system is exposed to UFPs, the UFPs may appear in many compartments of the body, including the liver, heart, and nervous system."

- "The fast-growing nanotechnology industry generates new UFPs daily, which may become aerosolized at some stage and may present additional health risks."

- "A relatively large surface area per unit mass facilitates adsorption of various organic compounds from the ambient air and enhances interaction with biological molecules within the organism."

- "Because everyone on earth inevitably inhales thousands to millions of UFPs with each breath, it is important to assess health risks by UFP air pollution."

- "To date, the mechanisms by which UFPs penetrate boundary membranes and the distribution of UFPs within tissue compartments of their primary and secondary target organs are largely unknown."

- "The toxic potential of UFPs is greatly enhanced by their free location and movement within cells, which promote interactions with intracellular proteins and organelles and even the nuclear DNA."

- "Potential health implications of our findings are related not only to ambient UFPs but also to engineered "nanoscaled particles," which may be released into our environment during their production, transport, and aging, or during waste disposal."

GESAMP. (1995). *The sea-surface microlayer and its role in global change*. Report No. 59. GESAMP, The Joint Group of Experts on the Scientific Aspects of Marine Environmental Protection. London.
http://www.gesamp.org/data/gesamp/files/media/Publications/Reports_and_studies_59/gallery_1358/object_1388_large.pdf

- "The sea surface of the ocean comprises a series of sublayers. These include a thin surface nanolayer ($< \sim 1\mu m$) containing high densities of particles and microorganisms; and the surface millilayer ($< \sim 10mm$) inhabited by small animals and the eggs and larvae of fish and invertebrates. The sea-surface microlayer is operationally defined in this report as the uppermost $\sim 1000\mu m$ (1mm) of the ocean surface. It, together with an overlying atmospheric layer of thickness 50-500μm, constitutes the boundary layer between the ocean and atmosphere."

- "Natural surface-active substances (surfactants) are often enriched in the sea surface compared to subsurface water. These include amino acids, proteins, fatty acids, lipids, phenols, and a variety of other organic compounds. The biota of the underlying water column are the primary source of such naturally-derived organic materials. Plankton produce dissolved compounds as part of their metabolic processes. Air bubbles, rising through the water column scavenge such chemicals and bring them to the surface. In addition, as plankton die and disintegrate some particles and many of the breakdown products (e.g., oils, fats and proteins) are either buoyant or are actively transported to the surface."

- "There is increasing evidence for the importance of surface films in the transfer of mass, heat and momentum across the air-sea interface… Material accumulated in the sea-surface microlayer is ejected into the atmosphere in an enriched form as part of the sea-salt aerosol produced by bursting bubbles. This provides a mechanism for the selective transfer of materials to terrestrial environments. Documented examples of such aerosol transport from sea-surface microlayers include bacteria, viruses, 'red tide' dinoflagellates and artificial radionuclides."

- "As might be anticipated, lipophilic organic compounds of anthropogenic origin introduced by way of atmospheric transport or aqueous and particulate runoff should be enriched in the sea-surface microlayer… High concentrations of toxic chemicals are also often found in the surface microlayer compared to the subsurface bulkwater in coastal environments."

- "Anthropogenic contributions include point sources related to industrial processes, agricultural runoff, and spills of petroleum products (catastrophic and chronic). In addition, municipal wastewater discharges are frequently highly enriched in surfactants which enter coastal seas and sediments."

- "The single largest source of surfactants is thought to be production by autochthonous marine organisms, principally phytoplankton, which exude natural surfactants as metabolic byproducts."

- "Plankton in the water column produce an abundance of particulate and dissolved organic material, some of which is transported to the surface either passively by floatation or actively by bubble transport."

- "The quantities and types of anthropogenic chemicals entering the earth's atmosphere continues to grow. Many of these chemicals, some of which are highly toxic, are now globally distributed in the atmosphere and deposit to the sea surface even in remote areas…Contaminants tend to concentrate in the microlayer due to the same processes that concentrate the organic materials and organisms."

- "Permanent inhabitants of the surface layer, such as bacterioneuston, hytoneuston, zooneuston and ichthyoneuston, often reach much higher densities than similar organisms found in subsurface waters. Temporary inhabitants of the neuston, particularly the eggs and larvae of a great number of fish and invertebrate species, utilize the surface during a portion of their embryonic and larval development… Piconeuston (<2 μm), like picoplankton are being increasingly recognized as important links in the recycling of organic matter in aquatic ecosystems… The enrichment of bacteria in the surface microlayer results, at least in part, from the greater degree of hydrophobicity of bacterioneuston compared to bacterioplankton and, thus, their adhesion to organically-enriched surface microlayers. Bacteria also adhere to the air-water interface of bubbles and may be injected into the atmosphere as part of sea-salt aerosol."

- This contains a comprehensive, if antiquarian (pre-1995) bibliography on the sea surface microlayers.

- There is no mention of microplastics in the microlayer, nor of PNP transport via microzooplankton discharge and aerosol transport.

GESAMP. (2008). *GESAMP Reports and Studies 76: Assessment and communication of environmental risks in coastal aquaculture*. GESAMP, The Joint Group of Experts on the Scientific Aspects of Marine Environmental Protection. http://www.fao.org/3/a-i0035e/i0035e00.pdf

- "Most aquaculture production of fish, crustaceans and molluscs continues to occur in freshwater environments (56.6% by quantity and 50.1% by value). Mariculture contributes 36.0% of production quantity and 33.6% of the total value."

- "Aquaculture continues to grow more rapidly than all other animal food-producing sectors, with an average annual growth rate for the world of 8.8 percent per year since 1970."

- "Although brackish-water production represented only 7.4% of production quantity in 2004, it contributed 16.3% of the total value, reflecting the prominence of high-value crustaceans and finfish."

- "The most common causes of environmental concern from coastal aquaculture are nutrient release, habitat change and loss, effects on wild fish and shellfish populations, chemical pollution, and secondary effects on other production systems. Many of the interactions with the environment are subtle and cumulative, they can be highly dispersed in space and time, and often the magnitude and probability of environmental changes can be unclear."

GESAMP. (2010). *Report of the thirty-seventh session of GESAMP, Bangkok, 14-19 February 2010*. GESAMP, The Joint Group of Experts on the Scientific Aspects of Marine Environmental Protection.

http://www.gesamp.org/data/gesamp/files/media/Publications/Reports_and_studies_81/gallery_1376/object_1530_large.pdf

GESAMP. (2012). *Sources, fate, and effects of microplastics in the marine environment: A global assessment.* GESAMP's Report of the Inception Meeting. GESAMP, The Joint Group of Experts on the Scientific Aspects of Marine Environmental Protection. http://www.marinelittersolutions.com/cust/documentrequest.aspx?DocID=53872

- "GESAMP was created in 1969 to provide a number of UN Agencies, having marine and maritime interests and responsibilities, with access to authoritative and independent scientific advice on a wide range of issues."

- "The potential influence of microplastic particles on the transfer of pollutants, from seawater to biota, was raised as an emerging issue within GESAMP."

- "Whether microplastics are having a significant ecological impact is perhaps the most important question WG40 should address."

- "Available time-series do not show convincing trends in microplastic concentrations, implying we are missing important pathways (e.g. sinking particles)."

- This publication contains no discussions of the significance of the devolution of microplastics into plastic nanoparticles.

Ghosh, S., Pal, S. and Ray, S. (2013). Study of microbes having potentiality for biodegradation of plastics. *Environ. Sci. Pollut. Res.* DOI 10:1007/s11356-013-1707-x.

Gilan, I., Hadar, Y. and Sivan, A. (2004). Colonization, biofilm formation and biodegradation of polyethylene by a strain of R. Ruber. *Applied Microbiology and Biotechnology.* 65. pg. 97-104. http://link.springer.com/article/10.1007%2Fs00253-004-1584-8#page-1

Gilfillan, L., Ohman, M., Doyle, M. and Watson, W. (2009). Occurrence of plastic micro-debris in the southern California current system.*California Cooperative Oceanic and Fisheries Investigations Report.* pg. 123-33. https://swfsc.noaa.gov/publications/CR/2009/2009Gilf.pdf

- "Plastic micro-debris is widespread in the California Current system off the southern California coast."

- "Our results also suggest that not only is plastic micro-debris widely distributed, it has been present in the northeast Pacific Ocean water column for at least 25 years."

- "Further investigation is needed of the occurrence, distribution, and fate of plastic micro-particles in the California Current system."

Gillis, Justin and Richtel, Matt. (April 6, 2015). Beneath California crops groundwater crisis grows. *The New York Times*. pg. A1, A10. http://www.nytimes.com/2015/04/06/science/beneath-california-crops-groundwater-crisis-grows.html?_r=0

Ginebreda, A., Muñoz, I., de Alda, M., et al. (2009). Environmental risk assessment of pharmaceuticals in rivers: Relationships between hazard indexes and aquatic macroinvertebrate diversity indexes in the Llobregat River (NE Spain). *Environment International*. http://www.clipmedia.net/galera/icra/rdp/031209_estudi_llobregat/environment_international.pdf

Glassmeyer, S., Furlong, E., Kolpin, D., et al. (2005). Transport of chemical and microbial compounds from known wastewater discharges: Potential for use as indicators of human fecal contamination. *Environmental Science and Technology*. 39(14). pg. 5157-69. http://digitalcommons.unl.edu/cgi/viewcontent.cgi?article=1066&context=usgsstaffpub

- "The concentrations of the majority of the chemicals present in the samples…were either nonexistent or at trace levels in the upstream samples, had their maximum concentrations in the WWTP effluent samples, and then declined in the two downstream samples."

Gleick, Peter H. (2014). Water, drought, climate change, and conflict in Syria. *American meteorological society AMS online journal*. http://journals.ametsoc.org/doi/abs/10.1175/WCAS-D-13-00059.1

Goddard, W., Brenner, D., Lyshevski, S. and Iafrate, G. (2012). *Handbook of nanoscience, engineering, and technology, 3rd Edition*. CRC Press. Boca Raton, FL.

Gogotsi, Y. (2006). *Nanomaterials handbook*. CRC Press, Taylor & Francis Group. Boca Raton, FL.

- "Currently, nanomaterials play a role in numerous industries, e.g., (1) carbon black particles (about 30 nm in size) make rubber tires wear-resistant; (2) nano phosphors are used in LCDs and CRTs to display colors."

- "The nanomaterials area alone is so broad that it is virtually impossible to cover all materials in a single volume…The handbook consists of 27 chapters written by leading researchers from academia, national laboratories, and industry, and covers the latest material developments in America, Asia, Europe, and Australia."

Goldberg, E. D. (1997). Plasticizing the seafloor: An overview. *Environ. Technol.* 18. pg. 195-202.

Goldsmith, Gregg. (2010). *Environmental impacts of military range use.* AEPI and USAWC Civilian Research Project, Army Environmental Policy Institute. http://www.aepi.army.mil/awcfellows/docs/Environmental_Impacts_of_Military_Rang e_Use.pdf

Goldstein, M., Rosenberg, M. and Cheng, L. (2012). Increased oceanic microplastic debris enhances oviposition in an endemic pelagic insect. *Biology Letters.* http://rsbl.royalsocietypublishing.org/content/roybiolett/early/2012/04/26/rsbl.2012.029 8.full.pdf

- "The dynamics of hard-substrate-associated organisms may be important to understanding the ecological impacts of oceanic microplastic pollution."

Goldstein, M., Titmus, A. and Ford, M. (2013). Scales of spatial heterogeneity of plastic marine debris in the Northeast Pacific Ocean. *PLoS ONE.* 8(11). http://www.plosone.org/article/fetchObject.action?uri=info:doi/10.1371/journal.pone.00 80020&representation=PDF

- "Most microplastic was found on the sea surface, with the highest densities detected in low-wind conditions…high variability of surface microplastic will make future changes in abundance difficult to detect without substantial sampling effort."

Gomara, B., Herrero, L., Ramos, J., et al. (2007). Distribution of polybrominated diphenyl ethers in human umbilical cord serum, paternal serum, maternal serum, placentas, and breast milk from Madrid population, Spain. *Environmental Science and Technology.* 41. pg. 6961-8. http://www.ncbi.nlm.nih.gov/pubmed/17993135

- "Total PBDE concentrations…are in the same range as those recently reported by other European and Asian studies and lower than those conducted in the U.S.A."

Gonzalez, L., Lison, D. and Kirsch-Volders, M. (2008). Genotoxicity of engineered nanomaterials: A critical review. *Nanotoxicology.* 2. pg. 252-73. http://informahealthcare.com/doi/abs/10.1080/17435390802464986

Gopee, N., Roberts, D., Webb, P., et al. (2007). Migration of intradermally injected quantum dots to sentinel organs in mice. *Toxicological Sciences.* 98(1). pg. 249-57. http://toxsci.oxfordjournals.org/content/98/1/249.full.pdf+html

- "Topical exposure to nanoscale materials is likely from a variety of sources including sunscreens and cosmetics."

- "Sentinel organs are effective locations for monitoring transdermal penetration of nanoscale materials into animals."

Gore, Andrea. (2007). *Endocrine-disrupting chemicals: From basic research to clinical practice (Contemporary Endocrinology)*. Human Press. Totowa, NJ.

Gorelick, D., Iwanowicz, L., Hung, A., et al. (2014). Transgenic zebrafish reveal tissue-specific differences in estrogen signaling in response to environmental water samples. *Environmental Health Perspectives*. 122(4). http://ehp.niehs.nih.gov/1307329/

- "We observed selective patterns of Estrogen Receptor activation in transgenic fish exposed to river water samples from the Mid-Atlantic United States, with several samples preferentially activating receptors in embryonic and larval heart valves. We discovered that tissue specificity in ER activation was due to differences in the expression of ER subtypes."

Goryka, M. (2009). *Environmental risks of microplastics*. Vrije University, Amsterdam, Netherlands. http://www.cleanup-sa.co.za/Images/Environmental_Risks_Microplastics.pdf

Gouin, T., Roche, N., Lohmann, R. and Hodges, G. (2011). A thermodynamic approach for assessing the environmental exposure of chemicals absorbed to microplastic. *Environmental Science and Technology*. 45(4). pg. 1466-72. http://pubs.acs.org/doi/abs/10.1021/es1032025

- "In this study we assess…the physicochemical properties of chemicals that are most likely absorbed by microplastic and therefore ingested by biota."

Gouin, T., Thomas, G. O., Chaemfa, C., Harner, T., Macay, D. and Jones, K. C. (2006). Concentrations of decabromodiphenyl ether in air from southern Ontario: Implications for particle-bound transport. *Chemosphere*. 64. pg.256-61.

Graham, E. and Thompson, J. (2009). Deposit- and suspension-feeding sea cucumbers (Echinodermata) ingest plastic fragments. *Journal of Experimental Marine Biology and Ecology*. 368(1). pg. 22-9. http://www.cabdirect.org/abstracts/20093325773.html;jsessionid=497844DF8C4103D7FA588D1B501DF798

- "Four species of deposit-feeding and suspension-feeding sea cucumbers (Echinodermata, Holothuroidea)…ingest significantly more plastic fragments than predicted given the ratio of plastic to sand grains in the sediment."

Gramatica, P., Pozzi, S., Consonni, V., et al. (2002). Classification of environmental pollutants for global mobility potential. *SAR QSAR Environ. Res*. 13. pg. 205-17

Grandjean, P. and Landrigan, P. J. (2006). Developmental neurotoxicity of industrial chemicals. *Lancet*. 368. pg. 2167-78. http://www.env-health.org/IMG/pdf/06tl9094page.pdf

- "Neurodevelopmental disorders such as autism, attention deficit disorder, mental retardation, and cerebral palsy are common, costly, and can cause lifelong disability. Their causes are mostly unknown."

- "A few industrial chemicals (eg, lead, methylmercury, polychlorinated biphenyls [PCBs], arsenic, and toluene) are recognised causes of neurodevelopmental disorders and subclinical brain dysfunction."

- "Another 200 chemicals are known to cause clinical neurotoxic effects in adults…many additional chemicals have been shown to be neurotoxic in laboratory models."

- "One in every six children has a developmental disability…[including] certain neurodevelopmental disorders—autism and attention deficit and hyperactivity disorder."

Gratton, S., Ropp, P., Pohlhaus, P., et al. (2008). The effect of particle design on cellular internalization pathways. *PNAS*. 105(33). http://www.pnas.org/content/105/33/11613.full.pdf+html

Gregory, Murray. (1983). Virgin plastic granules on some beaches of eastern Canada and Bermuda. *Marine Environmental research*. 10. pg. 73-92. http://www.researchgate.net/publication/222509002_Virgin_plastic_granules_on_some_beaches_of_Eastern_Canada_and_Bermuda

Gregory, Murray. (1996). Plastic "scrubbers" in hand cleansers: A further (and minor) source for marine pollution identified. *Marine Pollution Bulletin*. 32(12). pg. 867-71. http://www.sciencedirect.com/science/article/pii/S0025326X96000471

Gregory, Murray, and Ryan, P. (1997). Pelagic plastics and other seaborne persistent synthetic debris: A review of Southern Hemisphere perspectives. In: *Marine Debris – Sources, Impacts and Solutions*. Coe, James M. and Rogers, Donald, eds. Springer - Verlag, New York. pg. 49-66. http://link.springer.com/chapter/10.1007/978-1-4613-8486-1_6

Gregory, Murray. (2009). Environmental implications of plastic debris in marine settings—entanglement, ingestion, smothering, hangers-on, hitch-hiking and alien invasions. *Philosophical Transactions of the Royal Society of London Biological*

Sciences. 364(1526). pg. 2013-25.
http://www.ncbi.nlm.nih.gov/pmc/articles/PMC2873013/

- "Over the past five or six decades, contamination and pollution of the world's enclosed seas, coastal waters and the wider open oceans by plastics and other synthetic, non-biodegradable materials (generally known as 'marine debris') has been an ever-increasing phenomenon. The sources of these polluting materials are both land- and marine-based, their origins may be local or distant, and the environmental consequences are many and varied. The more widely recognized problems are typically associated with entanglement, ingestion, suffocation and general debilitation, and are often related to stranding events and public perception."

Griffit, R., Luo, J., Gao, J., et al. (2008). Effects of particle composition and species on toxicity of metallic nanomaterials in aquatic organisms. *Environmental Toxicological Chemistry.* 27. pg. 1972-8. http://www.ncbi.nlm.nih.gov/pubmed/18690762

- "Our results indicate that nanosilver and nanocopper cause toxicity in all organisms tested, with 48-h median lethal concentrations as low as 40 and 60 microg/L, respectively, in Daphnia pulex adults, whereas titanium dioxide did not cause toxicity in any of the tests."

Grindler, N., Allsworth, J., Macones, G., Kannan, K., Roehl, K. and Cooper, A. (2015). Persistent organic pollutants and early menopause in U.S. women. *PLoS ONE.* January 28.
http://www.plosone.org/article/fetchObject.action?uri=info:doi/10.1371/journal.pone.01
16057&representation=PDF

- "This analysis examined 111 EDCs and focused on known reproductive toxicants or chemicals with half-lives >1 year."

- "EDC-exposed women were up to 6 times more likely to be menopausal than non-exposed women."

- "We identified 15 EDCs that warrant closer evaluation because of their persistence and potential detrimental effects on ovarian function."

- "Exposure to these compounds has been linked to reduced sperm quality, earlier age of puberty, declines in fecundity, and increased rates of pregnancy complications."

Groß, R., Bunke, D., Gensch, C., et al. (2008). *Study on hazardous substances in electrical and electronic equipment, not regulated by the RoHS directive.* Öko-Institut

for Applied Ecology.
http://ec.europa.eu/environment/waste/weee/pdf/hazardous_substances_report.pdf

- A detailed analysis of 14 substances in EEE not already restricted for use (ROHS directive 2007)

- Contains a complete inventory of the hundreds of substances already restricted from use in EEE.

- Also contains an extensive bibliography.

Grün, F. and Blumberg, B. (2006). Environmental obesogens: Organotins and endocrine disruption via nuclear receptor signaling. *Endocrinology.* 147(6 Suppl). pg. S50–5. http://www.ncbi.nlm.nih.gov/pubmed/16690801

- "The environmental obesogen hypothesis predicts that inappropriate receptor activation by organotins will lead directly to adipocyte differentiation and a predisposition to obesity and/or will sensitize exposed individuals to obesity and related metabolic disorders under the influence of the typical high-calorie, high-fat Western diet."

Grün, F. and Blumberg. (2007). Perturbed nuclear receptor signaling by environmental obesogens as emerging factors in the obesity crisis. *Rev. Endocrine Metabolism Disorders.* 8. pg. 161-71. http://blumberg-serv.bio.uci.edu/reprints/grun-riem-final.pdf

- Included is a discussion of bisphenol A, diethylstilbestrol, tributyltin chloride, and triphenyltin chloride as endocrine disrupting chemicals (EDCs).

- Also contains an extensive bibliography.

Grün, F. and Blumberg, B. (2009). Endocrine disrupters as obesogens. *Molecular Cellular Endocrinology.* 304(1-2). pg. 19-29. http://www.ncbi.nlm.nih.gov/pmc/articles/PMC2713042/pdf/nihms120122.pdf

- "Evidence points to endocrine disrupting chemicals that interfere with the body's adipose tissue biology, endocrine hormone systems or central hypothalamic-pituitary-adrenal axis as suspects in derailing the homeostatic mechanisms important to weight control."

- "The obesogen hypothesis proposes that perturbations in metabolic signaling that result from exposure to novel environmental influences are superimposed on these trends of energy intake and expenditure. These obesogens include, but are not limited to endocrine disrupting compounds that alter fat cell differentiation or function and that initiate or exacerbate misregulation of homeostatic controls."

Guérin, T., Sirot, V., Volatier, J. L., et al. (2007). Organotin levels in seafood and its implications for health risk in high-seafood consumers. *Sci. Total Environ.* 388(1-3). pg. 66-77.

Guvenius, D., Aronsson, A., Ekman-Ordeberg, G., et al. (2003). Human prenatal and postnatal exposure to polybrominated diphenyl ethers, polychlorinated biphenyls, polychlorobiphenylols, and pentachlorphenol. *Environmental Health Perspectives*. 111. pg. 1235-41.

Guzman, K., Taylor, M. and Banfield, J. (2006). Environmental risks of nanotechnology: National Nanotechnology Initiative Funding, 2000-2004. *Environmental Science and Technology*. 40. pg. 1401-7. http://pubs.acs.org/doi/pdf/10.1021/es0515708

- "Though literature exits on the exposure, transport, and toxicity of incidental nanoparticles, little work has been published on the environmental risks of engineered nanoparticles."

- Includes a listing of NNI funding and an extensive bibliography.

Haack, S., Metge, D., Fogarty, L., Meyer, M., Barber, L., Harvey, R., LeBlanc, D. and Kolpin, D. (2012). Effects on groundwater microbial communities of an engineered 30 day in situ exposure to the antibiotic sulfamethoxazole. *Environmental Science and Techology*. 46(14). pg. 7478-86. http://www.academia.edu/6751434/Effects_on_Groundwater_Microbial_Communities_ of_an_Engineered_30-_Day_In_Situ_Exposure_to_the_Antibiotic_Sulfamethoxazole

Hagens, W., Oomen, A., de Jong, W., Cassee, F. and Sips, A. (2007). What do we (need to) know about the kinetic properties of nanoparticles in the body? Regulatory Toxicological Pharmacology. 49(3). pg. 217-29. http://www.ncbi.nlm.nih.gov/pubmed/17868963

Hagstrom, A., Azam, F., Andersson, A., Wikner, J. and Rassoulzadegan, F. (1988). Microbial loop in an oligotrophic pelagic marine ecosystem: Possible roles of cyanobacteria and nanoflagellates in the organic fluxes. *Marine Ecology Progress*. 49. pg. 171-178. http://www.int-res.com/articles/meps/49/m049p171.pdf

Hahn, M. (2002). Biomarkers and bioassays for detecting dioxin-like compounds in the marine environment. *Science of the Total Environment*. 289(1-3). pg. 49-69. http://www.ncbi.nlm.nih.gov/pubmed/12049406

Hale, R. C., La Guardia, M. J., Harvey, E. P., et al. (2001). Polybrominated diphenyl ether flame retardants in Virginia freshwater fishes (USA). *Environ Sci Technol*. 35 pg. 4585-91.

Hale R. C., LaGuardia M. J. and Harvey E. P. (2002). Potential role of fire retardant-treated polyurethane foam as a source of brominated diphenyl ethers to the US environment. *Chemosphere*. 46. pg. 729-35.
http://www.researchgate.net/publication/11370635_Potential_role_of_fire_retardant-treated_polyurethane_foam_as_a_source_of_brominated_diphenyl_ethers_to_the_US_environment

- "While consumption of BDE formulations has recently decreased in some European nations, due to potential adverse health and environmental effects, overall global demand for these products has increased, reaching 67,125 metric tons in 1999."

Hall, N. M., Berry, K. L. E., Rintoul, L. and Hoogenboom, M. O. (2015). Microplastic ingestion by scleractinian corals. *Marine Biology*. 162(3). pg. 725-32.
http://link.springer.com/article/10.1007%2Fs00227-015-2619-7

- "Experimental feeding trials revealed that corals mistake microplastics for prey and can consume up to ~50 µg plastic $cm^{-2}h^{-1}$, rates similar to their consumption of plankton and *Artemia* nauplii."
- "Ingested microplastics were found wrapped in mesenterial tissue within the coral gut cavity, suggesting that ingestion of high concentrations of microplastic debris could potentially impair the health of corals."

Halpern, B. S. et al. (2008). A global map of human impact on marine ecosystems. *Science*. 319. pg. 948-52. https://www.nceas.ucsb.edu/globalmarine

Ham, S., Kim, Y. and Lee, D. (2008). Leaching characteristics of PCDDs/DFs and dioxin-like PCBs from landfills containing municipal solid waste and incineration residues. *Chemosphere*. 70. pg. 1685-93.
http://www.hia21.eu/dwnld/20131216_Leaching%20characteristics%20of%20PCDDs DFs%20and%20dioxin-like%20PCBs.pdf

- "The leaching concentration of PCDDs/DFs decreased with increasing ratio of non-biodegradable wastes, especially incineration residues, but increased with increasing ratio of biodegradable wastes."

Hämer, J., Gutow, L., Kohler, A. and Saborowski, R. (2014). Fate of microplastics in the marine isopod *Idotea emarginata*. *Environmental Science and Technology*. 48(22). pg. 13451-8.
http://www.researchgate.net/profile/Reinhard_Saborowski/publication/266625593_The _fate_of_microplastics_in_the_marine_isopod_Idotea_emarginata/links/545b9a2a0cf24 9070a7a7751.pdf

- "At least 32 marine invertebrate species including pelagic (e.g., copepods and euphausids) and benthic representatives (e.g., mussels, lobsters, and polychaetes) have been reported to ingest microplastics. Similar to larger animals, small invertebrates may suffer from clogging of digestive organs, reduced appetite, and incorporation of microplastics into body tissue."

Hammer, J., Kraak, M. and Parsons, J. (2012). Plastics in the marine environment: The dark side of a modern gift. In: *Reviews of Environmental Contamination and Toxicology, vol. 220.* Springer, New York, NY.

Handy, R. and Shaw, B. (2007). Toxic effects of nanoparticles and nanomaterials: Implications for public health, risk assessment and the public perception of nanotechnology. *Health Risk and Society.* 9. pg. 125-44. http://www.tandfonline.com/doi/abs/10.1080/13698570701306807?journalCode=chrs20

Handy, R., Owen, R. and Vlsami-Jones, E. (2008). The ecotoxicology of nanoparticles and nanomaterials: Current status, knowledge gaps, challenges, and future needs. *Ecotoxicology.* 17. pg. 315-325. http://www.ncbi.nlm.nih.gov/pubmed/18408994

Hansen, S., Larsen, B., Olsen, S. and Baun, A. (2007). Categorization framework to aid hazard identification of nanomaterials. *Nanotoxicology.* 1(3). pg. 243-50. http://www.researchgate.net/publication/232047148_Categorization_framework_to_aid_hazard_identification_of_nanomaterials/links/0912f50d8964239fd1000000

Hansen, L.G. (2002). Persistent organic pollutants in food supplies. *Journal of Epidemiological Community Health.* 56. pg. 820-821. http://jech.bmj.com/content/56/11/820.full.pdf+html

- "The Stockholm Convention is having an accelerating effect on the global decline of POP manufacture and use."

- "Reservoirs from previous misuses are much larger than current manufacture and efforts should be directed at containing these reservoirs."

Hardman, Ron. (2006). A toxicologic review of quantum dots: Toxicity depends on physicochemical and environmental factors. *Environmental Health Perspectives.* 114. pg. 165-72. http://www.ncbi.nlm.nih.gov/pmc/articles/PMC1367826/

- "With the nanotechnology economy estimated to be valued at $1 trillion by 2012…the vastness and novelty of the nanotechnology frontier leave many areas unexplored…such as the potential adverse human health effects resulting from exposure to novel nanomaterials."

- "QD size, charge, concentration, outer coating bioactivity (capping material and functional groups), and oxidative, photolytic, and mechanical stability have each been implicated as determining factors in QD toxicity."

- "Each QD type will need to be characterized individually as to its potential toxicity."

Hardy, John and Cleary, John. (1992). Surface microlayer contamination and toxicity in the German Bight. *Inter-Research*. 91. pg. 203-10. http://www.int-res.com/articles/meps/91/m091p203.pdf

- "Large quantities of anthropogenic pollutants enter the North Sea by atmospheric deposition, riverine input and ocean dumping."

- "Results indicate that...Echnioderm larvae, molluscan bivalve larvae and copepod toxicity tests all represent feasible options for offshore pollution assessment."

Harrad, S. and Diamond, M. (2006). New directions: Exposure to polybrominated diphenyl ethers (PBDEs) and polychlorinated biphenyls (PCBs): Current and future scenarios. *Atmospheric Environment*. 40. pg. 1187-8. http://www.researchgate.net/publication/248343746_New_Directions_Exposure_to_polybrominated_diphenyl_ethers_%28PBDEs%29_and_polychlorinated_biphenyls_%28PCBs%29_Current_and_future_scenarios

Harris, Gardiner. (May 31, 2015). Holding your breath in India. *The New York Times*. pg. 1, 6-7. http://www.nytimes.com/2015/05/31/opinion/sunday/holding-your-breath-in-india.html?_r=0

- "India, in fact, has 13 of the world's 25 most polluted cities, while Lanzhou is the only Chinese city among the worst 50; Beijing ranks 79th."

- "Nearly half of Delhi's children have permanent lung damage."

- "In Beijing, PM2.5 levels that exceed 500 make international headlines; here, levels twice that high are largely ignored."

Harrison, E., Oakes, S., Hysell, M. and Hay, A. (2006). Organic chemicals in sewage sludges. *Science of the Total Environment*. 367. pg. 481-97. http://cwmi.css.cornell.edu/Sludge/organicchemicals.pdf

- "Sewage sludges are residues resulting from the treatment of wastewater released from various sources including homes, industries, medical facilities, street runoff and businesses."

- Sewage sludges "are widely used as soil amendments. They also, however, contain contaminants including metals, pathogens, and organic pollutants."

- "Data were found for 516 organic compounds which were grouped into 15 classes."

- "Eighty-three percent of the 516 chemicals were not on the EPA established list of priority pollutants and 80% were not on the EPA's list of target compounds. Thus analyses targeting these lists will detect only a small fraction of the organic chemicals in sludges."

- "Monitoring sludges for priority pollutants will not capture the vast majority of chemicals that may be present."

Harrison, J., Sapp, M., Schratzberger, M. and Osborn, A. (2011). Interactions between microorganisms and marine microplastics: A call for research. *Marine Technology Society Journal*. 45(2). pg. 12-20. http://www.researchgate.net/publication/230759508_Interactions_between_microorganisms_and_marine_microplastics_a_call_for_research

- "Microplastics (≤5-mm fragments) are rapidly emerging pollutants in marine ecosystems that may transport potentially toxic chemicals into macrobial food webs."

- "This commentary…identifies the lack of microbial research into microplastic contamination as a significant knowledge gap."

- "Microorganisms (bacteria, archaea, and picoeukaryotes) in coastal sediments represent a key category of life with reference to understanding and mitigating the potential adverse effects of microplastics due to their role as drivers of the global functioning of the marine biosphere and as putative mediators of the biodegradation of plastic-associated additives, contaminants, or even the plastics themselves."

- There is no mention of the breakdown of microplastics into nanoplastics and their propensity to sorb anthropogenic ecotoxins.

Harrison, J., Ojeda, J. and Romero-Gonzalez, M. (2012). The applicability of reflectance micro-Fourier-transform infrared spectroscopy for the detection of synthetic microplastics in marine sediments. *Science of the Total Environment*. 416(0). pg. 455-63. http://www.ncbi.nlm.nih.gov/pubmed/22221871

- "Synthetic microplastics (≤5-mm fragments) are globally distributed contaminants within coastal sediments that may transport organic pollutants and additives into food webs."

- "Molecular mapping… was explored in the presence and absence of 150-μm PE fragments…our results emphasize the urgency of developing efficient and reproducible techniques to separate microplastics from sediments."

- No mention is made of the urgency of measuring nanoplastics <150 nm.

Harrison, J., Schratzberger, M., Sapp, M. and Osborn, A. (2014). Rapid bacterial colonization of low-density polyethylene microplastics in coastal sediment microcosms. *BMC Microbiology*. 14(232). http://download.springer.com/static/pdf/710/art%253A10.1186%252Fs12866-014-0232-4.pdf?auth66=1423317321_dce45574946deb614b54b4cc2b42134c&ext=.pdf

Hart, Michael. (1991). Particle captures and the method of suspension feeding by echinoderm larvae. *Biology Bulletin*. 180(1). pg. 12-27. http://biostor.org/cache/pdf/62/49/54/6249548d9f4d32b660b1ffe720a768da.pdf

- "All feeding echinoderm larvae employ the same mechanisms to concentrate food particles from suspension."

Hasaneen, Mohammed Nagib. (2012). *Herbicides: Properties, synthesis and control of weeds*. Intech. http://cdn.intechweb.org/pdfs/25624.pdf

Hashim, Ahmed and Hajjaj, Muneer. (2005). Impact of desalination plants fluid effluents on the integrity of seawater, with the Arabian Gulf in perspective. *Science*. 182(1-3). pg. 373-93. http://www.sciencedirect.com/science/article/pii/S0011916405004534

Hassan, A., Rylander, C., Brustad, M. and Sandanger, T. (2013). Persistent organic pollutants in meat, liver, tallow and bone marrow from semi-domesticated reindeer (*Rangifer tarandus tarandus* L.) in Northern Norway. *Acta Veterinaria Scandinavica*. 55(57). http://www.ncbi.nlm.nih.gov/pmc/articles/PMC3751901/pdf/1751-0147-55-57.pdf

Hatch, E., Nelson, J., Qureshi, M., et. al. (2008). Association of urinary phthalate metabolite concentrations with body mass index and waist circumference: A cross-sectional study of NHANES data, 1999-2002. *Environmental Health*. 7. pg. 27. http://www.ncbi.nlm.nih.gov/pubmed/18522739

- "This exploratory, cross-sectional analysis revealed a number of interesting associations with different phthalate metabolites and obesity outcomes, including notable differences by gender and age subgroups."

- "Further research on the potential for phthalates to act as obesogens is warranted."

Hauk, T., Anderson, R. and Fischer, H. *In vivo* quantum dot toxicity assessment. *Small*. 4. pg. 26-49. http://www.ncbi.nlm.nih.gov/pubmed/19743433

Hauser, R., Duty, S., Godfrey-Bailey, L. and Calafat, A. M. (2004). Medications as a source of human exposure to phthalates. *Environ. Health Perspect*. 112. pg. 751-3.

Hauser, R. and Calafat, A.M. (2005a). Phthalates and human health. *Occupational Environental Medicine*. 62. pg. 806-18. http://www.ncbi.nlm.nih.gov/pmc/articles/PMC1740925/pdf/v062p00806.pdf

Hauser, R., Williams, P. Altshul, L. and Calafat, A. (2005b). Evidence of interaction between polychlorinated biphenyls and phthalates in relation to human sperm motility. *Environmental Health Perspectives*. 113. pg. 425-30. http://www.ncbi.nlm.nih.gov/pmc/articles/PMC1278482/pdf/ehp0113-000425.pdf

- "We found evidence of interactions between some PCBs and phthalates in relation to alterations in human sperm motility."

Hauser, R., Meeke, J., Singh, N., et al. (2007). DNA damage in human sperm is related to urinary levels of phthalate monoester and oxidative metabolites. *Human Reproduction*. 22. pg. 688-95. http://humrep.oxfordjournals.org/content/22/3/688.full.pdf+html

- "Among 379 men from an infertility clinic, urinary concentrations of phthalate metabolites were measured."

- "The urinary levels of phthalate metabolites among these men were similar to those reported for the US general population, suggesting that exposure to some phthalates may affect the population distribution of sperm DNA damage."

Hayes, T., Case, P., Chui, S., et. al. (2006). Pesticide mixtures, endocrine disruption, and amphibian declines: Are we underestimating the impact? *Health Perspectives*. 144(S1). pg. 40-50. https://www.savethefrogs.com/threats/pesticides/atrazine/images/publications/Hayes-2006-Pesticides%20Immunosupression%20and%20delayed%20metamorphosis.pdf

Health & Consumer Protection Directorate-General. (2000). *Opinion of the Scientific Committee on Animal Nutrition on the dioxin contamination of feedingstuffs and their contribution to the contamination of food of animal origin*. European Commission. http://ec.europa.eu/food/committees/scientific/out55_en.pdf

- "For humans, it is well known that more than 90% of the daily dioxin intake comes from food. In the same way it can be assumed that the animal dioxin body burden derives mainly from feeding. Therefore feedingstuffs, and in some cases soil, are of special concern as potential sources of dioxins."

- "Considering the sources of the contamination of feed materials by dioxins, PCBs and dioxin-like PCBs, the following points must be stressed: The ubiquitous environmental distribution of these compounds causes a background contamination affecting all terrestrial plants directly grazed (pastures) or used as raw materials for animal feeding as well as the aquatic feed chain. This is also true for the soil that might contaminate feed materials, or which can be directly ingested by the animals."

van der Heijden, M. G. A., Bardgett, R. D. and van Straalen, N. M. (2008). The unseen majority: Soil microbes as drivers of plant diversity and productivity in terrestrial ecosystems. *Ecology Letters*. 11(3). pg. 296-310. http://onlinelibrary.wiley.com/doi/10.1111/j.1461-0248.2007.01139.x/pdf

- "Microbes are the unseen majority in soil and comprise a large portion of life's genetic diversity."

- "The impact of soil microbes on ecosystem processes is still poorly understood."

- "Mycorrhizal fungi and nitrogen-fixing bacteria are responsible for c. 5–20% (grassland and savannah) to 80% (temperate and boreal forests) of all nitrogen, and up to 75% of phosphorus, that is acquired by plants annually."

- "Conservative estimates suggest that c. 20,000 plant species are completely dependent on microbial symbionts for growth and survival."

- "It has been estimated that one gram of soil contains as many as 10^{10}–10^{11} bacteria, 6,000–50,000 bacterial species, and up to 200 m fungal hyphae."

- Soil microbes as a vector for environmental chemicals in terrestrial ecosystems is not discussed.

Helm, P., Gewurtz, S., Whittle, D., et al. (2008). Occurrence and biomagnifications of polychlorinated naphthalenes and non- and mono-ortho PCBs in Lake Ontario sediments and biota. *Environmental Science and Technology*. 42(4). pg. 1024-31. http://www.ncbi.nlm.nih.gov/pubmed/18351067

Henry, M., Beguin, M., Requier, F., et al. (2012). A common pesticide decreases foraging success and survival in honey bees. *Science Xpress*. http://sciences.blogs.liberation.fr/files/abeilles-pesti-2.pdf

- "Non-lethal exposure of honey bees to thiamethoxam (neonicotinoid systemic pesticide) causes high mortality due to homing failure at levels that could put a colony at risk of collapse."

Herbert, B. M. J., Halsall, C. J., Villa, S., et al. (2005). Polychlorinated naphthalenes in air and snow in the Norwegian Arctic: a local source or an Eastern Arctic phenomenon? *Science of the Total Environment*. 342. pg. 145-60.

Herbst, A. and Bern, H. (1981). *Developmental effects of diethylstilbestrol (DES) in pregnancy.* Thieme-Stratton, New York, NY.

Hershey, Ida and Wolf, Nicole. (2014). *The dangers of backyard trash burning.* Oklahoma State University Division of Agricultural Sciences and Natural Resources. http://pods.dasnr.okstate.edu/docushare/dsweb/Get/Document-7930/AGEC-1027web.pdf

Heskett, M., Takada, H., Yamashita, R., et al. (2012). Measurement of persistent organic pollutants (POPs) in plastic resin pellets from remote islands: toward establishment of background concentrations for International Pellet Watch. *Marine Pollution Bulletin*. 64. pg. 445-8. http://www.ncbi.nlm.nih.gov/pubmed/22137935

- "Concentrations of PCBs (sum of 13 congeners) in the pellets were 0.1-9.9 ng/g-pellet. These were 1-3 orders of magnitude smaller than those observed in pellets from industrialized coastal shores."

Hett, Annabelle. (2004). *Nanotechnology: Small matter, many unknowns*. Swiss Reinsurance Company. Zurich, Switzerland. http://www.nanowerk.com/nanotechnology/reports/reportpdf/report93.pdf

- "What happens when products manufactured using nanotechnology end up on waste disposal sites and their particles are released into the environment?"

- "Any material reduced to the size of nanoparticles can suddenly behave much differently than it did before."

- "Nanotechnology manufactured products have been retailed for some time now without any particular labeling by regulators and have thus not been recognized by consumers for what they are."

- "Today's systematic use and manipulation of individual nanoparticles was only realized through the invention of special tools, such as the scanning tunnel microscope (STM) and the atomic force microscope (AFM), and their improved versions in the 1980s."

- "In contrast to larger microparticles, nanoparticles have almost unrestricted access to the human body. The possibility of absorption through the skin is currently under discussion, while the entry of certain nanoparticles into the bloodstream by inhalation via the lungs is considered a certainty… Even obstacles as hard to overcome as the blood-brain barrier seem to present no major problem to some nanoparticles."

- This same observation describes the surface to area ratio of plastic nanoparticles (PNP) < 100 nm.

- "Via the water cycle, nanoparticles could spread rapidly all over the globe, possibly also promoting the transport of pollutants."

- Worst-case scenarios: In view of the prevailing uncertainties, worst-case scenarios easily come to the fore. What would happen if certain nanoparticles did exert a harmful influence on the environment? Would it be possible to withdraw them from circulation? Would there be any way of removing nanoparticles from the water, earth or air?

Hidalgo-Ruz, Valeria and Thiel, Martin. (2013). Distribution and abundance of small plastic debris on beaches in the SE Pacific (Chile): A study supported by a citizen science project. *Marine Environmental Research*. Advance copy. pg. 1-7. http://www.cientificosdelabasura.cl/docs/publicaciones/2_hidalgoruz_thiel.pdf

- "The average abundance obtained was 27 small plastic pieces per m^2 for the continental coast of Chile, but the samples from Easter Island had extraordinarily higher abundances (>800 items per m^2)."

- "The widespread distribution and abundance of small plastic debris on Chilean beaches underscores the need to extend plastic debris research to ecological aspects of the problem and to improve waste management."

Hilborn, R. et al. (2003). State of the world's fisheries. *Annu. Rev. Environ. Resour*. 28. pg. 359-99.

Hillie, T. and Hlophe, M. (2007). *Nanotechnology and the challenge of clean water*. National Centre for Nano-structured Materials. http://www.nature.com/nnano/journal/v2/n11/pdf/nnano.2007.350.pdf

Hinojosa, I. and Thiel, M. (2009). Floating marine debris in fjords, gulfs and channels of southern Chile, *Marine Pollution Bulletin*. 58. pg. 341-50. http://www.ncbi.nlm.nih.gov/pubmed/19124136

- "Our results indicate that sea-based activities (mussel farming and salmon aquaculture) are responsible for most FMD in the study area."

Hirai, H., Takada, H., Ogata, Y., et al. (2011). Organic micropollutants in marine plastic debris from the open ocean and remote and urban beaches. *Marine Pollution Bulletin*. 62(8). pg. 1683-92. http://www.ncbi.nlm.nih.gov/pubmed/21719036

- "Hydrophobic organic compounds such as PCBs and PAHs were sorbed from seawater to the plastic fragments."

Hites, R. A. (2004). Polybrominated diphenyl ethers in the environment and in people: A meta-analysis of concentrations. *Environmental Science and Technology*. 38. pg. 945-56. http://www.researchgate.net/publication/8680195_Polybrominated_diphenyl_ethers_in _the_environment_and_in_people_a_meta-analysis_of_concentrations

- "The environment and people from North America are very much more contaminated with PBDEs as compared to Europe and that these PBDE levels have doubled every 4-6 yr."

- "Herring gull eggs from the Great Lakes region now have PBDE concentrations of ±7000 ng/g lipid, and these levels have doubled every approximately 3 yr."

Hites, R., Foran, J., Carpenter, D., et al. (2004). Global assessment of organic contaminants in farmed salmon. *Science*. 303(5655). pg. 226-9. http://www.iberica2000.org/documents/SubmittedScienceHites.pdf

- "Analysis of more than two metric tons of salmon from around the globe reveals that concentrations of organochlorine contaminants, including polychlorinated biphenyls, dioxins, toxaphene, and dieldrin, are significantly higher in farmed salmon than in wild, significantly higher in European farmed salmon as compared to those from North and South America, and occur at levels that may present a human health risk if farmed salmon are consumed more than once or twice per month."

- "The annual global production of farmed salmon has increased by a factor of 40 during the past two decades."

- "Having analyzed over 2 metric tons of farmed and wild salmon from around the world for organochlorine contaminants, we show that concentrations of these contaminants are significantly higher in farmed salmon than in wild."

- "European-raised salmon have significantly greater contaminant loads than those raised in North and South America."

- "Risk analysis indicates that consumption of farmed Atlantic salmon may pose health risks that detract from the beneficial effects of fish consumption."

Hladik, Michele L. and Calhoun, Daniel. (2012). Analysis of the herbicide diuron, three diuron degradates, and six neonicotinoid insecticides in water—method details and application to two Georgia streams. U.S. Geological Survey Scientific Investigations Report 2012–5206. USGS. http://pubs.usgs.gov/sir/2012/5206/pdf/sir20125206.pdf

Hladik, Michele L., Kolpin, Dana W. and Kuivila, Kathryn M. (2014). Widespread occurrence of neonicotinoid insecticides in streams in a high corn and soybean producing region, USA. *Environmental Pollution*. 193. pg. 189-96. http://www.sciencedirect.com/science/article/pii/S0269749114002802

Hoegh-Guldberg, O. et al. (2007). Coral reefs under rapid climate change and ocean acidification. *Science*. 318. pg. 1737-42. https://www.sciencemag.org/content/318/5857/1737.abstract

Hoellein, T., Rojas, M., Pink, A., et al. (2014). Anthropogenic litter in urban freshwater ecosystems: Distributions and microbial interactions. *PLoS ONE*. 9(6). http://www.plosone.org/article/fetchObject.action?uri=info:doi/10.1371/journal.pone.0098485&representation=PDF

- "We suggest the term anthropogenic litter [AL] is most useful because 1) AL differentiates the material from natural litter or debris accumulations (e.g. leaf litter or woody debris), 2) AL describes the material independent of its collection site, and 3) the term could promote an expanded perspective on the spatial dynamics and entire 'life cycle' of AL to unify terrestrial, freshwater, and marine ecosystem research on the topic."

Hoet, P., Bruske-Hohlfeld, I., Salata, O. (2004). Nanoparticles – known and unknown health risks. *Journal of Nanobiotechnology*. 2(12). http://www.jnanobiotechnology.com/content/pdf/1477-3155-2-12.pdf

- "The pharmaco-kinetic behaviour of different types of nanoparticles requires detailed investigation and a database of health risks associated with different nanoparticles (e.g. target organs, tissue or cells) should be created."

- "Nanoparticles designed for drug delivery or as food components need special attention."

Hofmann, G. E. et al. (2010). The effect of ocean acidification on calcifying organisms in marine ecosystems: An organism-to-ecosystem perspective. *Annu. Rev. Ecol. Evol. Syst.* 41. pg. 127-47.

Hogberg, J. (2008). Phthalate diesters and their metabolites in human breastmilk, blood or serum, and urine as biomarkers of exposure in vulnerable populations. *Environmental Health Perspectives*. 116. pg. 334-9. http://www.ncbi.nlm.nih.gov/pubmed/17090632

Hoh, E. and Hites, R. (2005). Brominated flame retardants in the atmosphere of the East-Central United States. *Environmental Science and Technology*. 39. pg. 7794-804. http://pubs.acs.org/doi/abs/10.1021/es050718k

Holbrook, R. D., Murphy, K. E., Morrow, J. B., et al. (2008). Trophic transfer of nanoparticles in a simplified invertebrate food web. *Nature Nanotechnology*. 3. pg. 352-5. http://www.ncbi.nlm.nih.gov/pubmed/18654546

Holgate, S. (2010). Exposure, uptake, distribution and toxicity of nanomaterials in humans. *Journal of Biomedical Nanotechnology*. 6(1). pg. 1-19. http://www.ncbi.nlm.nih.gov/pubmed/20499827

Hollman, P., Bouwmeester, H. and Peters, R. (2013). *Microplastics in the aquatic food chain*. RIKILT Wageningen UR (University & Research centre) RIKILT report. 2013(003). http://edepot.wur.nl/260490

- "Prolonged exposure to UV light and physical abrasion cause the plastic items to fragment, despite the durability of the polymers. Especially on shorelines, photo degradation will make plastic brittle and abrasion through wave action will enhance fragmentation. Fragmentation is of concern because these smaller fragments have the potential to be ingested by a much wider array of organisms, and additionally, they are difficult to remove from the environment. Plastic particles with a diameter smaller than 5 mm are generally designated as microplastics. Although it has not been quantified as of yet, it is quite likely that also nanoscale particles are produced during weathering of plastic debris."

- "Microplastics can adsorb all kinds of toxic contaminants that are already present in sea water, river water and in sediments. Contaminants able to adsorb to these particles include polychlorinated biphenyls (PCBs), polyaromatic hydrocarbons (PAHs), organochlorine pesticides (e.g. DDT, HCH) together with other persistent organic pollutants. These contaminants generally are hydrophobic and therefore have a high affinity for microplastics that is orders of magnitude higher than that for water. In addition, the small particle size (high surface to volume ratio) of microplastics strongly increases the amount adsorbed per gram plastic. As a consequence, microplastics efficiently extract and concentrate contaminants."

- "Toxicity of nanoplastic particles:

- They have been shown to be able to translocate through cellular membranes
- They may even pass through the blood-brain barrier, and potentially also the placenta.
- They may reach and penetrate all organs
- Data on toxicity are largely incomplete
- Toxicity will very likely be highly dependent on their physic-chemical properties, but how?
- Data on nanoparticles are almost exclusively limited to metal and metal oxide particles"

- "Bioavailability of additives and contaminants in plastic particles is not known."

- "Research on the potential health risks of food microplastics is needed. The following is proposed: Development of a quantitative method to measure microplastics in foods. The method should be able to distinguish between microparticles (0.1-5µm) and nanoparticles (< 0.1 µm). The polymer composition of the particles should be determined. Determination of selected additives and adhering contaminants should be considered."

- "A recent UNEP/WHO report on the state of the science of endocrine disrupting chemicals expressed concerns over endocrine disrupters because of the high incidence and increasing trends of many endocrine-related disorders in humans and of endocrine-related effects in wildlife populations."

- This report contains the following two tables:

Plastics production in the USA in 2005

	% total production
Low Density Polyethylene (LDPE)	22.3
High Density Polyethylene (HDPE)	20.4
Polypropylene (PP)	13.8
Polyethylene Terephtalate (PET)	9.9
Polystyrene (PS)	9.0
Polyvinyl Chloride (PVC)	5.7
Other	19.0
Polystyrene (PS)	
Polycarbonate (PC)	
Polyester (PES)	
Polyamides (PA)	

Leachable chemicals from plastics

	Function
Phtalates	Plasticizer
monomethyl phtalate (MMP)	
dimethyl phtalate (DMP)	
diethylhexyl phtalate (DEHP)	
butylbenzyl phtalate (BBzP)	
monobutyl phtalate (MBP)	
dibutyl phtalate (DBP)	
Alkylphenols	Plastizer/stabilizer
trisnonylphenol phosphites (TNP)	
nonylphenol (NP)	
octylphenol (OP)	
Bisphenol A (BPA)	Monomer/additive
Organotin compounds	Stabilizer
mono- en dialkyltin carboxylates	
tin mercaptans	
tin sulphides	
Polybrominated diphenyl ethers (PBDEs)	Flame retardant
tetrabromobisphenol A (TBBPA)	

Holmes, L., Turner, A. and Thompson, R. (2012). Adsorption of trace metals to plastic resin pellets in the marine environment. *Environmental Pollution*. 160(1). pg. 42-8. http://www.ncbi.nlm.nih.gov/pubmed/22035924

- "Results suggest that plastics may represent an important vehicle for the transport of metals in the marine environment."

Holtcamp, Wendee. (2012). Obesogens: An environmental link to obesity. *Environmental Health Perspectives*. 120(2). pg. A63-8. http://www.ncbi.nlm.nih.gov/pmc/articles/PMC3279464/pdf/ehp.120-a62.pdf

Hood, E. (2004). Nanotechnology: Looking as we leap. *Environmental Health Perspectives*. 112(13). pg. A741-9. http://www.ncbi.nlm.nih.gov/pmc/articles/PMC1247535/pdf/ehp0112-a00740.pdf

- "Nanotechnology is poised to become a major factor in the world's economy and part of our everyday lives in the near future. The science of the very small is going to be very big, very soon."

- "The ability to create unusual nanostructures such as bundles, sheets, and tubes holds promise for new and powerful drug delivery systems, electronic circuits, catalysts, and light-harvesting materials."

Hoor, James and Islam, Rafiq. (2010). *Understanding soil microbes and nutrient recycling*. The Ohio State University. http://ohioline.osu.edu/sag-fact/pdf/0016.pdf

Hoshino, A., Fujioka, K., Oku, T., et al. (2004). Physicochemical properties and cellular toxicity of nanocrystal quantum dots depend on their surface modification. *Nanotechnology Letters*. 4. pg. 2163-70. http://www.oalib.com/references/13897320

Howard, C. (2004). Small particles, big problems. *International Laboratory News*. 34. pg. 28-9.

Howard, P.H. and Muir, D.C.G. (2010). Identifying new persistent and bioaccumulative organics among chemicals in commerce. *Environmental Science and Technology*. 44. pg. 2277-85. http://www.ncbi.nlm.nih.gov/pubmed/21740030

- "The goal of this study was to identify commercial pharmaceuticals that might be persistent and bioaccumulative (P&B) and that were not being considered in current wastewater and aquatic environmental measurement programs."

- "Of the 275 drugs detected in the environment, 92 were rated as potentially bioaccumulative, 121 were rated as potentially persistent, and 99 were HPV pharmaceuticals."

Howdeshell, K., Hotchkiss, A., Thayer, K., et al. (1999). Exposure to bisphenol A advances puberty. *Nature*. 401. pg. 763-4. http://www.nature.com/nature/journal/v401/n6755/full/401763a0.html

- "Low environmental doses of EEDCs may…increase genital abnormality in boys and earlier sexual maturation in girls."

Howdeshell, K., Wilson, V., Furr, J., et al. (2008). A mixture of five phthalate esters inhibits fetal testicular testosterone production in the Sprague Dawley rat in a cumulative dose additive manner. *Toxicological Science*. 105. pg. 153-65. http://www.researchgate.net/publication/5441935_A_mixture_of_five_phthalate_esters_inhibits_fetal_testicular_testosterone_production_in_the_sprague-dawley_rat_in_a_cumulative_dose-additive_manner/links/004635294d0f7d7f3e000000

Hoyt, V. and Mason, E. (2008). Nanotechnology, emerging health issues. *Journal of Chemical Health and Safety*. http://www.researchgate.net/publication/257701286_Nanotechnology_Emerging_health_issues

Hsiao, I. and Huang, Y. (2011). Effects of various physicochemical characteristics on the toxicities of ZnO and TiO$_2$ nanoparticles toward human lung epithelial cells. *Science of the Total Environment*. 409(7). pg. 1219-28. http://www.sciencedirect.com/science/article/pii/S0048969710013641

Huang, X., Teng, X., Chen, D., et al. (2010). The effect of the shape of mesoporous silica nanoparticles on cellular uptake and cell function. *Biomaterials*. 31(3). pg. 438-48. http://isites.harvard.edu/fs/docs/icb.topic725337.files/Homework%204.pdf

- "Nanostructures not only passively interact with cells, but also actively engage and mediate the molecular processes that are essential for regulating cell functions."

- "The interplay between cell responses and particle shape will undoubtedly be an important aspect in investigating numerous areas of interest, including the design of targeting strategies for therapeutic applications and the environmental fate of NPs."

Huang, Y., Wong, C., Zheng, J., et al. (2012). Bisphenol A (BPA) in China: A review of sources, environmental levels, and potential human health impacts. *Environment International*. 42. pg. 91-9. http://www.sciencedirect.com/science/article/pii/S0160412011001206

- "The demand and production capacity of BPA in China have grown rapidly. This trend will lead to much more BPA contamination in the environmental media and in the general population in China."

- Top water concentrations reported in the range of 15-63 ng/L, while river water samples ranged up to 3920 ng/L.

- This review contains an extensive bibliography.

Huckzo, A., Lange, H., Calko, E., et al. (2001). Physiological testing of carbon nanotubes: Are they asbestos-like? *Fullerene Science and Technology*. 9(2). pg. 251-4. http://www.researchgate.net/publication/233036326_PHYSIOLOGICAL_TESTING_OF_CARBON_NANOTUBES_ARE_THEY_ASBESTOS-LIKE

Huggett, Clayton and Levin, Barbara. (1987). *Toxicity of the pyrolysis and combustion products of poly (vinyl chlorides): A literature assessment*. U.S. Department of Commerce, Center for Fire Research. Gaithersburg, MD. http://fire.nist.gov/bfrlpubs/fire87/PDF/f87015.pdf

Humblet, O., Sergeyev, O., Altshul, L., et al. (2011). Temporal trends in serum concentrations of polychlorinated dioxins, furans, and PCBs among adult women living in Chapaevsk, Russia: A longitudinal study from 2000 to 2009. *Environmental Health*. 10(62). http://www.ehjournal.net/content/10/1/62

- "The average total toxic equivalency (TEQ) decreased by 30% (from 36 to 25 pg/g lipid), and the average sum of PCB congeners decreased by 19% (from 291 to 211 ng/g lipid)."

- "A general trend towards decreasing dioxin and PCB levels exists in the general population."

Hund-Rincke, K. and Simon, M. (2006). Ecotoxic effect of photocatalytic active nanoparticles (TiO2) on algae and daphnids, *Environmental Science & Pollution Research*. 13(4). pg. 225-32. http://www.ncbi.nlm.nih.gov/pubmed/16910119

Hussain, N., Jaitley, V. and Florence, A. (2001). Recent advances in the understanding of uptake of microparticulates across the gastrointestinal lymphatics. *Advanced Drug Delivery Reviews*. 50(1-2). pg. 107-42.

Hutchings, Jeffrey A. (2000). Collapse and recovery of marine fishes. *Nature*. 406. pg. 882-5. http://www.nature.com/nature/journal/v406/n6798/full/406882a0.html

Hutchings, Jeffrey and Reynolds, John. (2004). Marine fish population collapses: Consequences for recovery and extinction risk. *BioScience*. 54(4). pg. 297-309. http://johnreynolds.org/wp-content/uploads/2012/07/hutchings-reynolds-2004-bioscience1.pdf

- "Rapid declines threaten the persistence of many marine fish. Data from more than 230 populations reveal a median reduction of 83% in breeding population size from known historic levels. Few populations recover rapidly; most exhibit little or no change in abundance up to 15 years after a collapse."

- "Heightened extinction risks were highlighted recently when a Canadian population of Atlantic cod (*Gadus morhua*) was listed as endangered, on the basis of declines as high as 99.9% over 30 years."

- "Among 56 populations of clupeids (including Atlantic herring, *Clupea harengus*), 73% experienced historic declines of 80% or more. Within the Gadidae (including haddock [*Melanogrammus aeglefinus*] and cod [G. *morhua* and other species]), of the 70 populations for which there are data, more than half declined 80% or more. And among 30 pleuronectid populations (flatfishes, including flounders, soles, and halibuts), 43% exhibited declines of 80% or more."

- Other factors noted in addition to overfishing include life history, habitat alteration, changes to species assemblage, genetic responses, and reduction in population growth.

- The growing presence of environmental chemicals in marine biota is not referenced.

Hutchings, J. A., Minto, C., Ricard, D. et al. (2010).Trends in the abundance of marine fishes. *Can. J. Fish. Aquat. Sci*. 67. pg. 1205-10.

Ikonomou, M. G., Rayne, S. and Addison, R. F. (2002). Exponential increases of the brominated flame retardants, polybrominated diphenyl ethers, in the Canadian arctic from 1981 to 2000. *Environ Sci Technol*. 36. pg. 1886-92.

Imhof, H., Ivleva, N., Schmid, J., et al. (2013). Contamination of beach sediments of a subalpine lake with microplastic particles. *Current Biology*. 23(19). pg. 867-8. http://download.cell.com/current-biology/pdf/PIIS0960982213011081.pdf?intermediate=true

- "Microplastic particles (<5 mm) are either directly introduced via sewage discharge or formed by biofouling and mechanical abrasion, making them more prone to consumption by aquatic organisms."

- "Freshwater ecosystems also act, at least temporarily, as a sink for plastic particles."

- "We show that even in a subalpine lake the amount of plastic particles is reaching similar magnitudes as in marine environments."

International Agency for Research on Cancer (IARC). (2006). *IARC monographs on the evaluation of carcinogenic risks to humans: Preamble*. World Health Organization, Lyon, France.

International Atomic Energy Agency. (2010). *Environmental impact assessment of nuclear desalination*. IAEA. http://www-pub.iaea.org/MTCD/publications/PDF/te_1642_web.pdf

Intertech-Pira. (2006). *The future of nanoplastics*. Intertech-Pira. http://hybridplastics.com/_OLD/_pdf_old/Future%20Nano.pdf

- "The promise of nanotechnology has been much talked about, but the time has come to deliver. Over the last twelve months the number of commercial examples of nanotechnology in plastics has grown exponentially and now industries as diverse as automotive, construction and consumer packaging are racing to embrace the opportunities nanoplastics can offer."

- "Nano-enhanced materials…are lighter, thinner, and longer lasting without any compromise in strength or barrier."

- "Economies of scale will inevitably make nanotechnology cheaper as it reaches full commercialization."

- "Whether you want your material to be stronger, lighter, higher barrier, more UV resistant, more conductive or more heat resistant, nanotechnology seems to have the answers. The use of nanotubes, nanoclays, and nanocomposites in plastics offer a wealth of new and improved material performance for an ever widening range of applications."

IPCS. (2002). *Global assessment of state-of-the-science of endocrine disruptors*. World Health Organization, Geneva, Switzerland. http://www.who.int/ipcs/publications/new_issues/endocrine_disruptors/en/

Ivar do Sul, J., Spengler, A. and Costa, M. (2009). Here, there and everywhere. Small plastic fragments and pellets on beaches of Fernando de Noronha (Equatorial Western Atlantic). *Marine Pollution Bulletin*. 58(8). pg. 1236-8. http://link.springer.com/article/10.1007%2Fs10653-014-9623-6#page-2

- "Drag your plankton net almost anywhere in the world oceans…and you will find them: virgin plastic pellets and small plastic fragments…small plastic fragments are the result of successive degradation processes acting on larger plastic debris in the environment."

- "Now these marine debris can be found almost anywhere in the oceans."

Ivar do Sul, J. and Costa, M. (2014). The present and future of microplastic pollution in the marine environment. *Environmental Pollution*. 185. pg. 352-64. http://ac.els-cdn.com/S0269749113005642/1-s2.0-S0269749113005642-main.pdf?_tid=a1e3b046-70dc-11e4-8e90-00000aab0f27&acdnat=1416505543_de0e683392d5d07a28b0de2ce107e708

- "Microplastics are commonly studied in relation to (1) plankton samples, (2) sandy and muddy sediments, (3) vertebrate and invertebrate ingestion, and (4) chemical pollutant interactions. All of the marine organism groups are at an eminent risk of interacting with microplastics according to the available literature."

- "Plastics are a branch of the oil and gas industry (8% of the oil produced is used in plastics production). Therefore, both sectors must meet to collaborate as soon as possible."

Ivask, A., Kurvet, I., Kasemets, K., et al. (2014). Size-dependent toxicity of silver nanoparticles to bacteria, yeast, algae, crustaceans and mammalian cells *in vitro*. *PLoS ONE*. 9(7).

http://www.plosone.org/article/fetchObject.action?uri=info:doi/10.1371/journal.pone.0102108&representation=PDF

Jabr-Milane, L., Van Vlerken, L., Devalapally, H., et al. (2008). Multi-functional nanocarriers for target delivery of drugs and genes. *Journal of Controlled Release*. 130(2). pg. 121-8. http://www.ncbi.nlm.nih.gov/pubmed/18538887

Jackson, J. B. C. et al. (2001). Historical overfishing and the recent collapse of coastal ecosystems. *Science*. 293. pg. 629-37.

- "Evidence from retrospective records strongly suggests that major structural and functional changes due to overfishing occurred worldwide in coastal marine ecosystems over many centuries."

Jacobs, M., Covaci, A. and Schepens, P. (2002). Investigation of selected persistent organic pollutants in farmed Atlantic salmon (Salmo salar), salmon aquaculture feed, and fish oil components of the feed. *Environmental Science and Technology*. 36(13). pg. 2797-2805. http://www.ncbi.nlm.nih.gov/pubmed/12144249

- "Comparison of the samples for all groups of contaminants, except for HCHs, showed an increase in concentration in the order fish oil < feed < salmon."

Jambeck, Jenna R., Geyer, Roland, Wilcox, Chris, et al. (2015). Plastic waste inputs from land into the ocean. *Science*.13. 347(6223). pg. 768-71.

Janesick, A. and Blumberg, B. (2011). Endocrine disrupting chemicals and the developmental programming of adipogenesis and obesity. *Birth Defects Res. C Embryo Today*. 93(1). pg. 34-50.

Jarosova, B., Blaha, L., Vrana, B., et al. (2013). Changes in concentrations of hydrophilic organic contaminants and of endocrine-disrupting potential downstream of small communities located adjacent to headwaters. *Environment International*. 45. pg. 22-31. http://www.ncbi.nlm.nih.gov/pubmed/22572113

- "Endocrine-disruptive potential and concentrations of polar organic contaminants [including those characterized as pharmaceuticals] were measured in seven headwaters flowing through relatively unpolluted areas of the Czech Republic."

- "Increased exposure potential of estrogenic and dioxin-like compounds… downstream of the towns were demonstrated."

Jelic, A., Gros, M., Ginebreda, A., et. al. (2011). Occurrence, partition and removal of pharmaceuticals in sewage water and sludge during wastewater treatment. *Water Research*. 45(3). pg. 1165-76. http://www.ncbi.nlm.nih.gov/pubmed/21167546

- "72 samples…were analyzed to assess the occurrence and fate of 43 pharmaceutical compounds."

- "21 pharmaceuticals accumulated in sewage sludge…the elimination of most of the substances is incomplete and improvements of the wastewater treatment and subsequent treatments of the produced sludge are required to prevent the introduction of these micro-pollutants in the environment."

Jelic, A., Gros, M., Petrovic, M., et. al. (2012) Occurrence and elimination of pharmaceuticals during conventional wastewater treatment. *Emerging and Priority Pollutants in Rivers*. 19. pg. 1-23. http://www.newbooks-services.de/MediaFiles/Texts/6/9783642257216_Excerpt_001.pdf

- "Pharmaceuticals have an important role in the treatment and prevention of disease in both humans and animals. Since they are designed either to be highly active or interact with receptors in humans and animals or to be toxic for many infectious organisms, they may also have unintended effects on animals and microorganisms in the environment."

- "Wastewater treatment plants have been identified as the main point of their collection and subsequent release into the environment, via both effluent wastewater and sludge."

- "Activated sludge process…treatment has been shown to have limited capability of removing pharmaceuticals from wastewater."

Jenkins, S., Wang, J., Eltoum, I., et al. (2011). Chronic oral exposure to Bisphenol A results in nonmonotonic dose response in mammary carcinogenesis and metastasis in MMTV-erbB2 mice. *Environmental Health Perspectives*. 119(11). pg. 1604-9. http://www.ncbi.nlm.nih.gov/pmc/articles/PMC3226508/

- "Bisphenol A is a synthetic compound used to produce plastics and epoxy resins. BPA can leach from these products in appreciable amounts, resulting in nearly ubiquitous daily exposure to humans."

Jensen, B., Kuznetsova, T., Kvamme, B. and Oterhals, A. (2011). Molecular dynamics study of selective adsorption of PCB on activated carbon. *Fluid Phase Equilibria*. https://bora.uib.no/bitstream/handle/1956/5128/Molecular%20Dynamics%20Study%20of%20Selective%20Adsorption.pdf?sequence=1

- "To remain competitive, North European fish oil and fish meal industry needs a cost-effective way to reduce the PCB levels to comply with requirements set by the European legislation."

Jenssen, B., Sormo, E., Baek, K., et al. (2007). Brominated flame retardants in North-East Atlantic marine ecosystems. *Environmental Health Perspective*. 115(1). http://www.ncbi.nlm.nih.gov/pmc/articles/PMC2174400/pdf/ehp0115s1-000035.pdf

- "Being environmentally persistent compounds with high production volumes, PBDEs and HBCD are among the most abundant BFRs detected in the environment."

- "BFRs are lipophilic and many are resistant to physical and biochemical degradation. Such BFRs are therefore bioaccumulative and may biomagnify in food webs, and are thus classified as persistent organic pollutants (POPs)."

- "Concentrations of PBDEs in marine mammals, birds, and fish have been increasing in recent decades, although the concentrations found in the European wildlife are in general lower than those found in North America."

- "Temporal studies have also documented that levels of HBCD are increasing in seals in North America and in harbor porpoises (*Phocoena phocoena*) in the United Kingdom."

- "Levels of BFRs in Arctic North-East Atlantic coastal ecosystems (Spitsbergen) are generally lower than along the Norwegian coast and much lower than in South- and Central-East Atlantic coastal ecosystems. This reflects the distance from the release sources."

Jeuck, Alexandra and Arndt, Hartmut. (2013). A short guide to common heterotrophic flagellates of freshwater habitats based on the morphology of living organisms. *Protist*. 164. pg. 842-60. http://www.sciencedirect.com/science/article/pii/S1434461013000795

- "Heterotrophic flagellates (HF) are very likely the most abundant eukaryotes on Earth, hundreds of specimens occur in each droplet of water even in groundwater and the deep sea. As the main feeders on bacteria they play an essential role in aquatic and terrestrial food webs…Feeding modes include true filter-feeding (e.g. choanoflagellates), direct interception feeding (e.g. chrysomonads) or raptorial feeding (e.g. most benthic forms)."

- "HF are a very heterogenous group with an enormous size range between 1-450 μm (some authors refer to the species smaller than 15 μm as heterotrophic nanoflagellates 'HNF')."

- Included are 15 charts that provide graphic illustrations and detailed descriptions of the many varieties (hundreds if not thousands) of heterotrophic flagellates (HFs).

- These HF play a key link in the translocation of PNP and their sorbed toxins into both terrestrial aquatic and marine food chains.

- Where have all the plastics gone? Any chance they have been transported by biogeochemical cycles, of which HFs are an essential component, into food webs that are the principal pathways to human exposure?

- The report contains an extensive bibliography.

Ji, Z., Zhang, D., Li, L., et al. (2009). The hepatotoxicity of multi-walled carbon nanotubes in mice. *Nanotechnology*. 20(44). pg. 445101. http://www.ncbi.nlm.nih.gov/pubmed/19801780

Jiang, J., Oberdorster, G. and Biswas, P. (2009). Characterization of size, surface charge, and agglomeration state of nanoparticle dispersions for toxicological studies. *Journal of Nanoparticle Research*. 11(1). pg. 77-89. http://cms.springerprofessional.de/journals/JOU=11051/VOL=2009.11/ISU=1/ART=94 46/BodyRef/PDF/11051_2008_Article_9446.pdf

Jiang, R., Jones, M., Sava, F., et al. (2014). Short-term diesel exhaust inhalation in a controlled human crossover study is associated with changes in DNA methylation of circulating mononuclear cells in asthmatics. *Particle and Fibre Toxicology*. 11(71). http://www.particleandfibretoxicology.com/content/pdf/s12989-014-0071-3.pdf

Jirtle, R. L. and Skinner, M. K. (2007). Environmental Epigenomics and disease susceptibility. *Nat. Rev. Genet.* 8. pg. 253-62.

Johnson, M. and Coull, M. (2014). In the ocean, the most harmful plastic is too small to see. *The Conversation.* http://theconversation.com/in-the-ocean-the-most-harmful-plastic-is-too-small-to-see-35336
   - "In some regions of the central Pacific there is now six times as much plankton-sized plastic are there is plankton."
   - "There are at least 268,000 tonnes of plastic floating around in the oceans, according to new research by a global team of scientists."
   - "The world generates 288m tonnes of plastic worldwide each year, just a little more than the annual vegetable crop, yet using current methods only 0.1% of it is found at sea. The new research illustrates as much as anything, how little we know about the fate of plastic waste in the ocean once we have thrown it 'away'."
   - "We don't yet know precisely how plastic nanoparticles interact with marine fauna but we do know that they can be absorbed at the level of individual cells. And what's worse is they're very efficient carriers of organic molecules such as estradiol, the drug used for birth control."
   - "Nasty endocrine disrupting chemicals can be concentrated a million times more than background levels on the surfaces of plastic particles."

- "Plastic pollution of the marine environment is the Cinderella of global issues, garnering less attention than its ugly sisters climate change, acidification, fisheries, invasive species or food waste but it has links to them all and merits greater attention by the scientific community."

Johnson-Restrepo, B., Kannan, K., Addink, R. and Adams, D. H. (2005). Polybrominated diphenyl ethers and polychlorinated biphenyls in a marine foodweb of coastal Florida. *Environ Sci Technol*. 39 pg. 8243-50.

Johnson-Restrepo, B., Adams, D. H. and Kannan, K. (2008). Tetrabromobisphenol A (TBBPA) and hexabromocyclododecanes (HBCDs) in tissues of humans, dolphins, and sharks from the United States. *Chemosphere*. 70. pg. 1935-44.

Johnson-Restrepo, B. and Kanna, K. (2009). An assessment of sources and pathways of human exposure to polybrominated diphenyl ethers in the United States. *Chemosphere*. 76(4). pg. 542-8. http://www.ncbi.nlm.nih.gov/pubmed/19349061

Jones, G., McCormick, M., Srinivasan, M. and Eagle, J. (2004). Coral decline threatens fish biodiversity in marine reserves. *PNAS*. 101(21). pg. 8251-3. http://www.ncbi.nlm.nih.gov/pmc/articles/PMC419589/pdf/1018251.pdf

Jones, K. C. and de Voogt, P. (1999). Persistent organic pollutants (POPs): state of the science. *Environmental Pollution*. 100. pg. 209-21. https://educnet.enpc.fr/pluginfile.php/2457/mod_resource/content/0/Jones-EP-1999-POPs.pdf

- No mention in this now antiquarian report that POPs are also sorbed by micro and nanoplastics in the marine environment and then ingested by microzooplankton and translocated to and biomagnified in higher trophic levels.

Jonker, M. and Koelmans, A. (2002). Sorption of polycyclic aromatic hydrocarbons and polychlorinated biphenyls to soot and soot-like materials in the aqueous environment. *Mechanistic Considerations, Environmental Science and Technology*. 36. pg. 3725-34. http://www.ncbi.nlm.nih.gov/pubmed/12322744

- "Recent studies have shown that sorption of polycyclic aromatic hydrocarbons (PAHs) in soot-water systems is exceptionally strong. As a consequence, soot may fully control the actual fate of PAHs in the aquatic environment."

Jørgensen, Bo Barker and Boetius, Antje. (2007). Feast and famine: microbial life in the deep-sea bed. *Nature Reviews: Microbiology*. 5(10). pg. 770-81. http://www.ecology.uni-jena.de/ecologymedia/Jorgensen+and+Boetius+nrmicro1745+2.pdf

Joseph, T. and Morrison, M. (2006). *Nanotechnology in agriculture and food*. Nanoforum Report. ftp://ftp.cordis.europa.eu/pub/nanotechnology/docs/nanotechnology_in_agriculture_and_food.pdf

Joskow, R., Barr, D. B., Barr, J., et al. (2006). Exposure to bisphenol A from bis-glycidyl dimethacrylate-based dental sealants. *Journal of the American Dental Association*. 137. pg. 353-62. http://www.adajournal.com/content/137/3/353.full.pdf?related-urls=yes&legid=jada;137/3/353

- "Only about one-third of the BPA produced in the United States is used in epoxy resins, including dental sealants."

- "Placement of clinically relevant amounts of Delton LC sealant resulted in low-level BPA exposure…Sealants should remain a useful part of routine preventive dental practice, especially those that leach negligible amounts of BPA."

Joye, S., Teske, A. and Kostka, J. (2014). Microbial dynamics following the macondo oil well blowout across Gulf of Mexico environments. *Bioscience*. 64(9). http://bioscience.oxfordjournals.org/content/64/9/766.full.pdf+html

Judy, J., Unrine, J. and Bertsch, Paul. (2011). Evidence for biomagnification of gold nanoparticles within a terrestrial food chain. *Environmental Science and Technology*. 45(2). pg. 776-81. http://www.sludgenews.org/resources/documents/judy_nano_in_sludge.pdf

Kach, D. and Ward, J. (2008). The role of marine aggregates in the ingestion of picoplankton-size particles by suspension-feeding mollusks. *Marine Biology*. 153. pg. 797-805. http://www.researchgate.net/publication/225591224_The_role_of_marine_aggregates_in_the_ingestion_of_picoplankton-size_particles_by_suspension-feeding_molluscs

- "Results indicate that aggregates significantly enhance the ingestion of 1.0- and 0.5-lm beads by all species of bivalves."

- "Suspension-feeding molluscs can ingest marine aggregates and their constituent picoplankton-size particles. Aggregates contain living and non-living organic material that can be ingested and potentially used as a food source."

Kaiser, Jocelyn. (2010). The dirt on ocean garbage patches. *Science*. 328. pg. 1506. http://www.zo.utexas.edu/courses/Thoc/OceanGarbagePatches.pdf

Kan, C. A. and Meijer, G. A. L. (2007). The risk of contamination of food with toxic substances present in animal feed. *Animal Feed Science and Technology*. 133. pg. 84-108. http://www.animalfeedscience.com/article/S0377-8401(06)00307-5/abstract

Kane, A. and Hurt, R. (2008). Nanotoxicology: The asbestos analogy revisited. *Nature Nanotechnology*. 3(7). pg. 378-9. http://www.nature.com/nnano/journal/v3/n7/full/nnano.2008.182.html

Karapanagioti, H. and Klontza, I. (2008). Testing phenanthrene distribution properties of virgin plastic pellets and plastic eroded pellets found on Lesvos island beaches (Greece). *Marine Environment Research*. 65(4). pg. 283-90. http://www.ncbi.nlm.nih.gov/pubmed/18164383

- "Plastic pellets have been characterized as toxic pollutant carriers throughout the world oceans and coastal zones. However, their sorptive properties are not yet well understood."

- "Although diffusion into the polymer happens with similar rates for both freshwater and saltwater external solutions, apparent diffusion is dependent on the solution salinity because it results in higher equilibrium distribution coefficients."

Kashiwagi, Takashi. (2007). *Flame retardant mechanism of the nanotubes-based nanocomposites*. NIST GCR 07-912. National Institute of Standards and Technology. http://fire.nist.gov/bfrlpubs/fire07/PDF/f07034.pdf

Kawamura, Y., Koyama, Y., Takeda, Y. and Yamada, T. (1998). Migration of bisphenol A from poly-carbonate products. *Journal of the Food Hygienic Society*. 99. pg. 206-12. http://www.bisphenol-a.org/pdf/migration.pdf

Keiter, S., Rastall, A., Kosmehl, T., et al. (2006). Ecotoxicological assessment of sediment, suspended matter and water samples in the upper Danube River. *Environmental Science Pollution Research*. 13(5). pg. 308-19. http://www.researchgate.net/profile/Steffen_Keiter/publication/6729264_Ecotoxicological_assessment_of_sediment_suspended_matter_and_water_samples_in_the_upper_Danube_River._A_pilot_study_in_search_for_the_causes_for_the_decline_of_fish_catches/links/02bfe510768a427336000000.pdf

Keller, A., Garner, K., Miller, R. and Lenihan, H. (2012). Toxicity of nano-zero valent iron to freshwater and marine organisms. *PLoS ONE*. 7(8). http://www.plosone.org/article/fetchObject.action?uri=info:doi/10.1371/journal.pone.0043983&representation=PDF

Kelley, C., Mohtadi, S., Cane, M., et. Al. (2015). Climate change in the Fertile Crescent and implications of the recent Syrian drought. *Proceedings of the National Academy of Science*. 112(11). Pg. 3241-3246. http://www.pnas.org/content/112/11/3241.abstract

Kelly, B., Ikonomou, M., Blair, J., et al. (2007). Food web-specific biomagnifications of persistent organic pollutants. *Science*. 317. http://www.precaution.org/lib/more_pops.070713.pdf

- "Regulatory authorities identify bioaccumulative substances as hydrophobic, fat-soluble chemicals having high octanol-water partition coefficients (K_{OW}) (≥100,000)…can biomagnify to a high degree in food webs containing air-breathing animals (including humans) because of their high octanol-air partition coefficient (K_{OA}) and corresponding low rate of respiratory elimination to air."

- "These low KOW–high KOA chemicals, representing a third of organic chemicals in commercial use, constitute an unidentified class of potentially bioaccumulative substances that require regulatory assessment to prevent possible ecosystem and human-health consequences."

Kelly, B., Ikonomou, M., Blair, J., et al. (2009). Perfluoroalkyl contaminants in an arctic marine food web: Trophic magnification and wildlife exposure. *Environmental Science and Technology*. 43(11). pg. 4037-43. http://www.ncbi.nlm.nih.gov/pubmed/19569327

Kennish, M.J. (1997). *Practical handbook of estatuarine and marine pollution*. CRC Press, Boca Raton.

Keri, R., Ho, S., Hunt, P., et al. (2007). An evaluation of evidence for the carcinogenic activity of bisphenol A. *Reproductive Toxicology*. 24. pg. 240-52. http://www.ncbi.nlm.nih.gov/pmc/articles/PMC2442886/pdf/nihms55062.pdf

- A comprehensive "assessment of the carcinogenic activity of BPA," but without information on exposure sources and pathways.

Khordagui, H. and Abu-Hilal, Ahmad. (1994). Industrial plastic on the southern beaches of the Arabian Gulf and the western beaches of the Gulf of Oman. *Environmental Pollution*. 84. pg. 325-7. http://www.ncbi.nlm.nih.gov/pubmed/15091703

- "By early 1992 alarming levels of fresh plastic pellets were noticed on the Arabian Gulf beaches of the UAE. Large numbers of 25 kg sacks of white plastic spherules manufactured by (SABIC) in Jubail, Saudi Arabia were washed ashore."

- "When compared to other parts of the world, the beaches of the UAE on the Arabian Gulf are considered to be heavily polluted with industrial plastic."

Kielhorn, J., Melber, C., Wahnschaffe, U., et al. (2000). Vinyl chloride: Still a cause for concern. *Environmental Health Perspectives*. 108(7). http://www.mindfully.org/Plastic/Vinyl-ChlorideJul00.htm

Kim, Soong Ho, Knight, Elysse M., saunders, Eric L. et al. (2012). Rapid doubling of Alzheimer's amyloid-ß40 and 42 levels in brains of mice exposed to a nickel nanoparticle model of air pollution. *F1000 Research*. 1(70). http://www.ncbi.nlm.nih.gov/pmc/articles/PMC3782349/

- "Inhalant exposures should be evaluated for their possible roles in contributing to the environmental risk for common forms of AD [Alzheimer's Disease]."

Kirchner, C., Liedl, T., Kudera, S., et. al. (2005). Cytotoxicity of colloidal CdSe and CdSe/ZnS nanoparticles. *Nanotechnology Letters*. 5(2). pg. 331-8. http://www.nanion.de/images/stories/papers/NanoLetters_Cytotoxicity.pdf

- "Three different pathways [exist] by which nanoparticles introduced into an organism could interfere with its function and finally lead to impairment. Most evident, introduced nanoparticles can be composed of toxic materials."

- "Partial decomposition and release of ions is more likely for nanoparticles due to their enhanced surface-to-volume ratio."

- "There might be an effect caused by the shape of the (inert) particles. It has been reported for example, that carbon nanotubes can impale cells like needles."

Klanjscek, T., Nisbet, R., Priester, J. and Holden, P. (2012). Modeling physiological processes that relate toxicant exposure and bacterial population dynamics. *PLoS ONE*. 7(2). http://journals.plos.org/plosone/article?id=10.1371/journal.pone.0026955

Klepeis, N., Nelson, W., Ott, W., et al. (2001). The national human activity pattern survey (NHAPS): A resource for assessing exposure to environmental pollutants. *Exposure Analytical Environmental Epidemiology*. 22. pg. 231-52. http://www.nature.com/jes/journal/v11/n3/pdf/7500165a.pdf

Klosterhaus, S., Stapleton, M., La Guardia, J., and Greig, D. (2012). Brominated and chlorinated flame retardants in San Francisco Bay sediments and wildlife. *Environment International*. 47. pg. 56-65. http://www.ncbi.nlm.nih.gov/pubmed/22766500

- "In addition to PBDEs, brominated and chlorinated flame retardants (hexabromocyclododecane (HBCD) and Dechlorane Plus (DP)) were detected in Bay sediments and wildlife."

- "Two additional flame retardants, pentabromoethylbenzene (PBEB) and 1,2-bis(2,4,6 tribromophenoxy)ethane (BTBPE) were detected in sediments but with less frequency and at lower concentrations."

- "Compared to other locations, concentrations of PBDEs in Bay wildlife were comparable or higher."

- "Future studies should also examine the fate and distribution of the organophosphate flame retardants in the Bay. Further study of tris(1,3-dichloro-2-propyl) phosphate in particular is needed, considering it is a known replacement for penta-BDE (and has the potential to act as a mutagen, carcinogen, neurotoxin, and endocrine disruptor.)"

Knapp, S. (2014). Microplastics in Maine waters: are they in our food? Research begins to measure amounts in Gulf of Maine. *The Working Waterfront*. 6(45). The Island Institute. http://www.workingwaterfront.com/articles/Microplastics-in-Maine-waters-are-they-in-our-food/16140

Koch, Holger and Calafat, A. (2009). Human body burdens of chemicals used in plastic manufacture. *Philosophical Transactions of the Royal Society*. 364. pg. 2063-78. http://rstb.royalsocietypublishing.org/content/364/1526/2063.full-text.pdf

- "We present an overview on the use of biomonitoring in exposure and risk assessment using phthalates and bisphenol A as examples of chemicals used in the manufacture of plastic goods."

Koelmans, A., Besseling, E., Wegner, A. and Foekema, E. (2013). Plastic as a carrier of POPs to aquatic organisms: A model analysis. *Environmental Science and Technology*. 47(14). pg. 7812-20. http://pubs.acs.org/doi/abs/10.1021/es401169n

- "The model accounts for dilution of exposure concentration by sorption of POPs to plastic (POP 'dilution'), increased bioaccumulation by ingestion of plastic-containing POPs ('carrier'), and decreased bioaccumulation by ingestion of clean plastic ('cleaning')."

- Academic obfuscation of the real-world phenomena of the increasing sorption of ecotoxins as microplastics (<5mm) break down into nanoplastic particles (<500 mm).

Koelmans, A., Besseling, E. and Foekema, E. (2014). Leaching of plastic additives to marine organisms. *Environmental Pollution*. 187. pg. 49-54. http://www.sciencedirect.com/science/article/pii/S0269749113006465

Köhler, Michael and Fritzsche, Wolfgang. (2004). *Nanotechnology*. Wiley. New York, NY.

Köhler, A. (2010). Cellular fate of organic compounds in marine invertebrates. *Comparative Biochemistry and Physiology*. 157. http://www.sciencedirect.com/science/article/pii/S1095643310002047

Kolpin, D., Barbash, J. and Gilliom, R. (1998). Occurrence of pesticides in shallow groundwater of the United States: Initial results from the national water-quality assessment program. *Environmental Science and Technology*. 32. pg. 558-66. http://digitalcommons.unl.edu/cgi/viewcontent.cgi?article=1062&context=usgsstaffpub

- "Pesticide results from the 41 landuse studies conducted during 1993-1995 indicate that pesticides were commonly detected in shallow groundwater."

- "Urban and suburban pesticide use significantly contribute to pesticide occurrence in shallow groundwater."

- "MCLs have been established for only 25 of the 46 pesticide compounds examined, do not cover pesticide degradates, and, at present, do not take into account additive or synergistic effects of combinations of pesticide compounds or potential effects on nearby aquatic ecosystems."

Kolpin, D., Furlong, E., Meyer, M., et al. (2002). Pharmaceuticals, hormones, and other organic wastewater contaminants in U.S. streams, 1999-2000: A national reconnaissance. *Environmental Science and Technology*. 36(6). pg. 1202-11. http://digitalcommons.unl.edu/cgi/viewcontent.cgi?article=1064&context=usgsstaffpub

- "33 of the 95 target OWCs are known or suspected to exhibit at least weak hormonal activity with the potential to disrupt normal endocrine function."

- "The results of this study document that detectable quantities of OWCs occur in U.S. streams at the national scale. This implies that many such compounds survive wastewater treatment and biodegradation."

- "Little is known about the potential interactive effects (synergistic or antagonistic toxicity) that may occur from complex mixtures of these compounds in the environment."

- "Select OWCs may be degrading into new, more persistent compounds that could be transported into the environment instead of (or in addition to) their associated parent compound."

Kolpin, D., Thurman, E., Lee, E., et al. (2006). Urban contributions of glyphosate and its degradate AMPA to streams in the United States. *Science of the Total Environment*.

354(2-3). pg. 191-7.
http://www.sciencedirect.com/science/article/pii/S0048969705000690

- "Glyphosate is the most widely used herbicide in the world, being routinely applied to control weeds in both agricultural and urban settings. Microbial degradation of glyphosate produces aminomethyl phosphonic acid (AMPA)."

- "The results document the apparent contribution of WWTP effluent to stream concentrations of glyphosate and AMPA, with roughly a two-fold increase in their frequencies of detection between stream samples collected upstream and those collected downstream of the WWTPs."

Kolpin, D., Hoerger, C., Meyer, M., et al. (2010). Phytoestrogens and mycotoxins in Iowa streams: An examination of underinvestigated compounds in agricultural basins. *Journal of Environmental Quality*. 39(6). pg. 2089-99.
https://www.agronomy.org/publications/jeq/abstracts/39/6/2089

- "Target compounds were frequently detected in stream samples: atrazine (100%), formononetin (80%), equol (45%), deoxynivalenol (43%), daidzein (32%), biochanin A (23%), zearalenone (13%), and genistein (11%)."

- "Atrazine concentrations commonly exceeded 100 ng L-1 (42/75 measurements)."

- "The ecotoxicological effects from long-term, low-level exposures to phytoestrogens and mycotoxins or complex chemicals mixtures including these compounds that commonly take place in surface water are poorly understood and have yet to be systematically investigated in environmental studies."

Kolosnjaj, J., Szwarc, H. and Moussa, F. (2007). Toxicity studies of carbon nanotubes. *Advances in Experimental Medicine and Biology*. 620. pg. 181-204.
http://link.springer.com/chapter/10.1007%2F978-0-387-76713-0_14

Kolosnjaj-Tabi, J., Szwarc, H. and Moussa, F. (2012). *In vivo toxicity studies of pristine carbon nanotubes: A review, the delivery of nanoparticles*. Intech.
http://cdn.intechopen.com/pdfs-wm/36876.pdf

- "Discovered in 1991, carbon nanotubes have attracted considerable attention in many fields of science and technology because of their unique structural, mechanical, and electronic properties...[including] in the fields of materials science...energy storage and energy conversion devices, sensors, field emission displays and radiation sources, hydrogen storage, nanometer-sized semiconductor devices, probes and interconnects."

- "As the reactivity and the general behavior of CNTs in biological media are not completely understood, assessing the safety of these nanoparticles should also include a careful selection of appropriate experimental methods."

Konikow, Leonard F. (2013). *Groundwater depletion in the United States, 1900-2008*. United States Geological Survey. http://pubs.usgs.gov/sir/2013/5079/SIR2013-5079.pdf

Kortenkamp, A., Martin, O., Faust, M., et al. (2011). *State of the art assessment of endocrine disruptors, final report*. European Commission, Directorate-General for the Environment. http://ec.europa.eu/environment/chemicals/endocrine/pdf/sota_edc_final_report.pdf

- "During the last two decades evidence of increasing trends of many endocrine-related disorders in humans has strengthened…There is good evidence that wildlife populations have been affected, with sometimes widespread effects."

- "Especially where chemicals do not stay for long periods in tissues after exposures have occurred, it is impossible to detect associations when exposure measurements cannot cover periods of heightened sensitivity."

- "For a wide range of endocrine disrupting effects, agreed and validated test methods do not exist. In many cases, even scientific research models that could be developed into tests are missing."

- "Urgently needed are further methods for the identification of endocrine disrupters. Concerted efforts should be undertaken to identify the full spectrum of endocrine disrupters present in the environment and in human tissues."

- This report contains extensive information about the ecotoxicological impact of EDs on fish, birds, arthropods (eg honey bees), invertebrates and sediment dwelling organisms in the European community.

Koziara, J., Lockman, P., Allen, D. and Mumper, R. (2003). In situ blood-brain barrier transport of nanoparticles. *Pharmaceutical Research*. 20. pg. 1772-8. http://www.ncbi.nlm.nih.gov/pubmed/14661921

Kramer, I., Minor, J., Moreno-Bautista, G., et al. (2014). Efficient spray-coated colloidal quantum dot solar cells. *Advanced Materials*. http://onlinelibrary.wiley.com/doi/10.1002/adma.201403281/abstract

Krause, E., Witchels, A., Gimenez, L., et al. (2012). Small changes in pH have direct effects on marine bacterial community composition: A microcosm approach. *PLoS One*. 7(10).

http://www.plosone.org/article/fetchObject.action?uri=info%3Adoi%2F10.1371%2Fjournal.pone.0047035&representation=PDF

Kremen, C., Williams, N., Aizen, M. et al. (2007). Pollination and other ecosystem services produced by mobile organisms: A conceptual framework for the effects of land-use change. *Ecology Letters*. 10(4). http://onlinelibrary.wiley.com/doi/10.1111/j.1461-0248.2007.01018.x/pdf

Kreyling, W., Semmler-Behnke, M. and Moller, W. (2006). Health implications of nanoparticles. *Journal of Nanoparticle Research*. 8. pg. 543-62. http://cms.springerprofessional.de/journals/JOU=11051/VOL=2006.8/ISU=5/ART=9068/BodyRef/PDF/11051_2005_Article_9068.pdf

- "Nanoparticles are so called 'intended' particles intentionally produced for specific use in science, technology, medicine, industries and many day-by-day applications."

Krimsky, Sheldon. (2000). *Hormonal chaos: The scientific and social origins of the environmental endocrine hypothesis*. The Johns Hopkins University Press, Baltimore, MD.

Kuivila, K., Hladik, M., Ingersoll, C., et al. (2012). Occurrence and potential sources of pyrethroid insecticides in stream sediments from seven U.S. metropolitan areas. *Environmental Science and Technology*. 46. pg. 4297-303. http://ca.water.usgs.gov/pubs/KuivilaEtAl2012.pdf

- "Dallas/Fort Worth had the highest pyrethroid detection frequency (89%), the greatest number of pyrethroids (4), and some of the highest concentrations."

- "The variation in pyrethroid concentrations among metropolitan areas suggests regional differences in pyrethroid use and transport processes…pyrethroids commonly occur in urban stream sediments and may be contributing to sediment toxicity."

Kumar, C. (2006). *Nanomaterials: Toxicity, health and environmental issues*. Wiley-VCH.

Kumar, Arun and Xagoraraki, Irene. (2010). Human health risk assessment of pharmaceuticals in water: An uncertainty analysis for meprobamate, carbamazepine, and phenytoin. *Regulatory Toxicology and Pharmacology*. 57. pg. 146-56. http://web.iitd.ac.in/~arunku/files/CEL899_Y12/PharmaRiskAssessment_Kumar.pdf

- "Further research efforts are required to standardize use of acceptable daily intake values to reduce large variability in estimation of hazard quotients."

Kumar, N., Palmer, G., Shah, V. and Walker, V. (2014). The effect of silver nanoparticles on seasonal change in arctic tundra bacterial and fungal assemblages. *PLoS ONE*. 9(6). http://www.plosone.org/article/fetchObject.action?uri=info:doi/10.1371/journal.pone.0099953&representation=PDF

Kusy, R. and Whitley, J. (2005). Degradation of plastic polyoxymethylene brackets and the subsequent release of toxic formaldehyde. *American Journal of Orthodontic Dentofacial Orthopedics*. 127(4). pg. 420-7. http://www.ncbi.nlm.nih.gov/pubmed/15821686

Kwan, C., Takada, H., Mizukawa, K., et al. (2013). PBDEs in leachates from municipal solid waste dumping sites in tropical Asian countries: phase distribution and debromination. *Environmental Science and Pollution Research*. 1-17. http://link.springer.com/article/10.1007/s11356-012-1365-3

- "The unprecedented economic and population growths of some Asian countries over the last decade have led to significant increases in the amount of waste containing PBDEs in that region."

- "A total of 46 PBDE congeners were measured…leachate samples collected, from 2002 to 2010, from ten MSWDS distributed among the eight countries of Lao PDR, Cambodia, Vietnam, India, Indonesia, Thailand, the Philippines, and Malaysia."

- "Municipal solid waste dumping sites (MSWDS) of tropical Asian countries are potential sources of environmental PBDEs, which may be transported to the aquatic environment via dissolution with dissolved organic matter."

La Guardia, M. J., Hale, R. C. and Harvey, E. (2006). Detailed polybrominated diphenyl ether (PBDE) congener composition of the widely used penta-, octa-, and deca-PBDE technical flame-retardant mixtures. *Environ Sci Technol*. 40. pg. 6247-54.

Lang, I., Galloway, T., Scarlett, A., et al. (2008). Association of urinary bisphenol A concentration with medical disorders and laboratory abnormalities in adults. *JAMA*. 300(11). pg. 1303-10. http://www.oehha.org/prop65/public_meetings/pdf/BPAcomment11h.pdf

Larson, D., Zipfel, W., Williams, R., et al. (2003). Water-soluble quantum dots for multiphoton fluorescence imaging in vivo. *Science*. 300. pg. 1434-6. http://www.ncbi.nlm.nih.gov/pubmed/12775841

Latini, G., De Felice, C., Verrotti, A., (2004). Plasticizers, infant nutrition and reproductive health. *Reproductive Toxicology*. 19 (1). 27-33. http://www.ncbi.nlm.nih.gov/pubmed/15336709

- " Di-(2-ethylhexyl)-phthalate (DEHP) is the most widely used plasticizer in PVC formulations."

- Infants are "the most sensitive population, as they may be exposed to several different sources (breast milk, infant formula, baby food, indoor air, and by dermal and oral exposure via indoor dust containing DEHP)."

- "Here, we report a review on dietary phthalate exposure in babies."

Lattin, G., Moore, C., Zellers, Z., et al. (2004). A comparison of neustonic plastic and zooplankton at different depths near the southern California Shore. *Marine Pollution Bulletin*. 49(4). pg. 291-4. http://www.ncbi.nlm.nih.gov/pubmed/15341821

Laursen, D., Danner, B., Stewardson, N. and Thompson, R. (2009). International pellet watch: Global monitoring of persistent organic pollutants (POPs) in coastal waters. 1. Initial phase data on PCBs, DDTs, and HCHs. *Marine Pollution Bulletin*. 58(10). pg. 1437-46. http://www.ncbi.nlm.nih.gov/pubmed/19635625

Law, R. J., Alaee, M., Allchin, C. R., et al. (2003). Levels and trends of polybrominated diphenylethers and other brominated flame retardants in wildlife. *Environ Int*. 29. pg. 757-70.

Law, R. J., Allchin, C., de Boer, J., et al. (2006a). Levels and trends of brominated flame retardants in the European environment. *Chemosphere*. 64. pg. 187-208. http://www.ncbi.nlm.nih.gov/pubmed/16434081

Law, R. J., Bersuder, P., Allchin, C. R. and Barry J. (2006b). Levels of the flame retardants hexabromocyclododecane and tetrabromobisphenol A in the blubber of harbour porpoises (Phocoena phocoena) stranded or bycaught in the UK, with evidence for an increase in HBCD concentrations in recent years. *Environ Sci Technol*. 40. pg. 2177-83.

Ladhar, C., Geffroy, B., Cambier, S., et al. (2013). Impact of dietary cadmium sulphide nanoparticles on Danio rerio zebrafish at very low contamination pressure. *Nanotoxicology*.

- "The food used in the present study mimics the dispersion of NPs in aquatic environments and its transfer along the food web at environmentally realistic concentrations."

- "Different effects, such as gene expression level modifications, mutations and mitochondrial impairment, occurred for both sizes of CdSNPs."

- "NPs ingested through drinking water or food will possibly cross the human intestinal barrier."

LaKind, J., Berlin, C. and Naiman, D. (2001). Infant exposure to chemicals in breast milk in the United States: What we need to learn from a breast milk monitoring program. *Environmental Health Perspectives*. 109. pg. 75-88. http://www.ncbi.nlm.nih.gov/pmc/articles/PMC1242055/

- "A better understanding of an infant's level of exposure to environmental chemicals is essential, particularly in the United States where information is sparse."

- "Two parameters needed to conduct realistic exposure assessments for breast-fed infants: a) levels of chemicals in human milk in the United States (and trends for dioxins/furans); and b) elimination kinetics (depuration) of chemicals from the mother during breastfeeding."

- "Although the data indicate a decrease in breast milk dioxin toxic equivalents over time for several countries, the results for the United States are ambiguous."

- "Previous studies should be extended by testing for an increased number of environmental chemicals in breast milk. In addition to the chemicals discussed in this paper, analytes should include certain heavy metals as well as other chemicals with significant lipid solubility and long biological half-life."

- The bibliography in this review contains a comprehensive list of monitoring programs through 1999.

Lam, C., James, J., McCluskey, R., Arepalli, S. and Hunter, R. (2006). A review of carbon nanotube toxicity ad assessment of potential occupational and environmental health risks. *Critical Reviews in Toxicology*. 36(3). pg. 189-217. http://informahealthcare.com/doi/abs/10.1080/10408440600570233

- "Nanotechnology has emerged at the forefront of science research and technology development. Carbon nanotubes (CNTs) are major building blocks of this new technology. They possess unique electrical, mechanical, and thermal properties, with potential wide applications in the electronics, computer, aerospace, and other industries."

- "CNTs exist in two forms, single-wall (SWCNTs) and multi-wall (MWCNTs). They are manufactured predominantly by electrical arc discharge, laser ablation and chemical vapor deposition processes; these processes involve thermally stripping carbon atoms off from carbon-bearing compounds. SWCNT formation requires catalytic metals."

- "There has been a great concern that if CNTs, which are very light, enter the working environment as suspended particulate matter (PM) of respirable sizes, they could pose an occupational inhalation exposure hazard."

- "CNTs were capable of producing inflammation, epithelioid granulomas (microscopic nodules), fibrosis, and biochemical/toxicological changes in the lungs."

- "They are expected to produce fibers 100 times stronger than steel at only 1/6th the weight—almost certainly the strongest fibers that will ever be made out of anything."

- "Besides inducing histopathological changes, including granuloma formation and fibrosis, both SWCNTs and MWCNTs were found to release toxic cytokines and biomarkers of inflammation, oxidative stress, and cytotoxicity. As discussed above, Shvedova's group further found that SWCNTs produced functional respiratory deficiency, decreased bacterial clearance, and biochemical toxicity in the heart in treated mice."

Lambert, S., Sinclair, C., Bradley, E. and Boxall, A. (2012). Effects of environmental conditions on latex degradation in aquatic systems. *Science of the Total Environment.* 447. pg. 225-34. http://www.resodema.org/publications/publication25.pdf

- "Degradation rate was dependent on light and material thickness. Photooxidation is shown to be the primary degradation pathway. Degradation products include the formation of nanosized particles."

- "The disappearance of the bulk material corresponded to an increase in nanoparticles and dissolved organic material in the test media."

- "The major finding of this study was to identify and quantify the formation of particles in the nanosize range as well as dissolved organic compounds."

Lambert, Scott. (2013). *Environmental risk of polymers and their degradation products.* Thesis submitted to the University of York Environment Department. http://etheses.whiterose.ac.uk/4194/1/EnvironPoly.pdf

- "Analysis of the degradation solutions demonstrated that when the latex polymer degraded, there was an increase in the formation of microscopic latex."

- "Atmosphere is a receiving environmental compartment for polymer degradates though the identification of a range of volatile substances produced during the degradation process."

- "Environments receiving polymer debris are potentially exposed to a mixture of compounds that include the parent polymer, fragmented particles, leached additives, and subsequent degradation products."

Lambert, S., Sinclair, C. and Boxall, A. (2014). Occurrence, degradation, and effect of polymer-based materials in the environment. *Review of Environmental Contamination and Toxicology*. 227. pg. 1-53. http://www.ncbi.nlm.nih.gov/pubmed/24158578

- "Polymer-based materials (PBMs) are exposed to a variety of mechanical and chemical weathering processes. This causes a change to the PBM structure and facilitates the disintegration of the PBM into increasingly smaller fragments."

- "PBMs and their associated degradation products may compromise the viability of organisms at all trophic levels…Non-selective and filter-feeding consumers could be susceptible to ingesting both bulk PBMs and fragmented particles, leading to the potential passage up the food chain to secondary and tertiary consumers."

- Extensive bibliography.

Law, K., Moret-Ferguson, S., Maximenko, N., et al. (2010). Plastic accumulation in the North Atlantic subtropical gyre. *Science*. 329. pg. 1185-88. http://www.grid.unep.ch/FP2011/step1/pdf/015_Law_2010.pdf

- "It is likely that plastic pieces ultimately become small enough to pass through the 335-um mesh net used in this study."

- "The fate of plastic particles that become dense enough to sink below the sea surface is unknown."

Law, K., Moret-Ferguson, S., Goodwin, D., et al. (2014). Distribution of surface plastic debris in the Eastern Pacific Ocean from an 11 year data set. *Environmental Science and Technology*. 48. pg. 4732-8. https://darchive.mblwhoilibrary.org/bitstream/handle/1912/6778/es4053076.pdf?sequence=1&isAllowed=y

- "We present an extensive survey of floating plastic debris in the eastern North and South Pacific Oceans from more than 2500 plankton net tows conducted between 2001 and 2012."

- "Utilizing all available plankton net data collected in the eastern Pacific Ocean (17.4°S to 61.0°N; 85.0 to 180.0°W) since 1999, we estimated a minimum of 21 290 t of floating microplastic."

- So where have all the plastic gone?

Lawton, Sir John. (2008). *Novel materials in the environment: The case of nanotechnology*. Royal Commission on Environmental Pollution. https://www.gov.uk/government/uploads/system/uploads/attachment_data/file/228871/7468.pdf

- Earlier reports include those on subject matters for nuclear pollution, lead, oil pollution in the sea to diesel emissions, waste incinteration, fisheries and chemicals in products.

- "The aggregation behavior of nanomaterials is especially important. Aggregate size, morphology and kinetics alter with nanomaterials type and other environmental factors."

- "Examples of potentially harmful nanomaterials include nanosilver, carbon nanotubes and Buckminsterfullerenes (C60). However, we are very conscious of the extent to which knowledge about the potential health and environmental impacts of nanomaterials lags significantly behind the pace of innovation and these areas of concern could change as new scientific information arises. This is an area of considerable uncertainty."

- "We are also concerned that more sophisticated third and fourth generation nanoproducts may represent a further step change in functionalities and properties...during research for this study, we found a worrying incidence of very myopic views of benefits because a full life cycle assessment of the material had not been considered by its proponents."

Lazar, B. and Gracau, R. (2011). Ingestion of marine debris by loggerhead sea turtles, *Caretta caretta*, in the Adriatic Sea. *Marine Pollution Bulletin*. 62. pg. 43-7. http://www.seaturtle.org/PDF/LazarB_2011_MarPollutBull.pdf

- "Environmental compartments studied comprise the atmosphere, sediments and soils, sewage sludges, and a variety of biological samples and food chains."

- " Findings include that the input of BDEs (especially BDE209) to the Baltic Sea by atmospheric deposition now exceeds that of PCBs by a factor of almost 40 times."

- "The major source is from diffuse leaching from products into wastewater streams from users, households and industries generally."

- "BDEs are found widely distributed in fish, including those from high mountain lakes in Europe, as a consequence of long-range atmospheric transport and deposition."

Lechner, A., Keckeis, H., Lumesberger-Loisl, F., et al. (2014). The Danube so colourful: A potpourri of plastic litter outnumbers fish larvae in Europe's second largest river. *Environmental Pollution*. 188. pg. 177-81. http://www.ncbi.nlm.nih.gov/pubmed/24602762

- "Industrial raw material (pellets, flakes and spherules) accounted for substantial parts (79.4%) of the plastic debris."

- "For several reasons, our values must be regarded as an underestimation of the total plastic load into the Black Sea."

- "1) The amount of filtered microplastics is negatively correlated with the mesh size. Norén (2007) found the abundance of small plastic fibres in a 80 mm net to be five orders of magnitude higher than in a 450 mm net. Therefore we suppose microscopic fragments."

- No mention of the presence of nanoplastics (< 1 um).

Lee, C., Kim, J., Lee, W., et al. (2008). Bactericidal effect of zero-valent iron nanoparticles on Escherichia coli. *Environmental Science & Technology*. 42(13). pg. 4927-33. http://www.ncbi.nlm.nih.gov/pmc/articles/PMC2536719/

- "Nano-Fe^0 showed a strong bactericidal activity comparable to that of silver nanoparticles."

- "Because the toxicity of engineered nanoparticles is one of the key uncertainties associated with widespread application of nanotechnology (33–35), oxidation of the nano-Fe^0 surface may be considered as a means of decreasing the toxicity of the nanoparticles."

Lee, D., Lee, I., Song, K., et al. (2006). A strong dose-response relation between serum concentrations of persistent organic pollutants and diabetes. *Diabetes Care*. 29. pg. 1638-44. http://medicinaycomplejidad.org/pdf/probsal/pops7[1].pdf

- "Diabetes prevalence was strongly positively associated with lipid-adjusted serum concentrations of all six POPs."
- "The association was consistent in stratified analyses and stronger in younger participants, Mexican Americans, and obese individuals."

- This study was sponsored by the Republic of Korea utilizing the National Health and Examination Survey 1999-2002 of 2016 US adults.

Lee, Jongmyoung, Hong, Sunwook, Song, Y., et al. (2013). Relationships among the abundances of plastic debris in different size classes on beaches in South Korea. *Marine Pollution Bulletin.* http://www.sciencedirect.com/science/article/pii/S0025326X13004657

Lee, Kathy E., Langer, Susan K., Barber, Larry B. et al. (2011). *Endocrine active chemicals, pharmaceuticals, and other chemicals of concern in surface water, wastewater-treatment plant effluent, and bed sediment, and biological characteristics in selected streams, Minnesota—Design, Methods, and Data, 2009.* US Geological Survey, Reston, VA. http://pubs.usgs.gov/ds/575/pdf/ds575.pdf

- A detailed summary of pharmaceuticals etc. in waste water effluents in Minnesota.

Lee, O., Takesono, A., Tada, M., et al. (2012). Biosensor zebrafish provide new insights into potential health effects of environmental estrogens. *Environmental Health Perspectives.* 120(7). http://ehp.niehs.nih.gov/wp-content/uploads/120/7/ehp.1104433.pdf

- "Environmental estrogens [estrogenic endocrine-disrupting chemicals (EDCs)] alter hormone signaling in the body that can induce reproductive abnormalities in both humans and wildlife."

Leibold, M.A. (1989). Resource edibility and the effects of predators and productivity on the outcome of trophic interactions. *American Nature.* 134. pg. 922-49. http://www.jstor.org/discover/10.2307/2462017?sid=21105483560761&uid=3739712&uid=2&uid=4&uid=3739256

Leslie, H., van der Meulen, M., Kleissen, F. and Vethaak, A. (2011). *Microplastic litter in the Dutch marine environment.* Dutch Ministry of Infrastructure and Environment, Deltares, Amsterdam, Netherlands. http://www.noordzeeloket.nl/images/Microplastic%20Litter%20in%20the%20Dutch%20Marine%20Environment_851.pdf

- "There is currently a worldwide shortage of dedicated studies on the biological and ecological effects of microplastics."

- "Microplastics represents a new, major, complex global environmental problem that could have great adverse effects on the environment and on humans."

Lever, J., van Nes, E., Scheffer, M. and Bascompte, J. (2014). The sudden collapse of pollinator communities. *Ecology Letters.* 17(3). pg. 350-9.

http://digital.csic.es/bitstream/10261/91808/1/REVISION2_Lever_etal_Collapse_Pollin
ators.pdf

- "Widespread declines in wild and domesticated pollinator populations raise concerns about the future of biodiversity and agricultural productivity."

- "Commonly used insecticides strongly increase pollinator mortality. Habitat destruction, parasites, and disease are also seen as 24 important drivers of pollinator decline."

- Environmental chemicals as a trigger for pollinator declines are referenced but not discussed in detail.

Lewinski, N., Colvin, V. and Drezek, R. (2008). Cytoxocity of nanoparticles. *Small*. 4. pg. 26-49. http://isites.harvard.edu/fs/docs/icb.topic725334.files/LECT7c%2003-24-2010.pdf

- "Exposure to nanoparticles for medical purposes involves intentional contact or administration; therefore, understanding the properties of nanoparticles and their effect on the body is crucial before clinical use can occur."

- "This review presents a summary of the in vitro cytotoxicity data currently available on three classes of nanoparticles."

- "While the number of nanoparticle types and applications continues to increase, studies to characterize their effects after exposure and to address their potential toxicity are few in comparison."

- This article contains an extensive bibliography on nanoparticle physiochemical properties.

Ley, Brian. (1999). *Width of a human hair*. The Physics Factbook. http://hypertextbook.com/facts/1999/BrianLey.shtml

Leys, Sally and Eerkes-Medrano, Dafne. (2006). Feeding in a calcareous sponge: Particle uptake by pseudophobia. *Biological Bulletin*. 211. pg. 157-71. http://www.biolbull.org/content/211/2/157.long

- "Sponges are considered to be filter feeders like their nearest protistan relatives, the choanoflagellates. Specialized 'sieve' cells (choanocytes) have an apical collar of tightly spaced, rodlike microvilli that surround a long flagellum. The beat of the flagellum is believed to draw water through this collar, but how particles caught on the collar are brought to the cell surface is unknown."

- "Of all particles, only 0.1-μm latex microspheres adhered to the collar microvilli in large numbers."

- "Solutions of beads in three sizes (1.0, 0.5, and 1.0 μm) were diluted to 1 x 10^9 in seawater collected from 20-m depth and mixed with 10% bovine serum albumin to prevent clumping."

- "Each chamber is carpeted by a single layer of cuboidal cells called choanocytes. About 10,000 choanocytes 3.5 μm in diameter were estimated to line an average-sized chamber (450 μm long, 100 μm in diameter)."

- "Choanocytes of all sponges fixed 5–10 min after being fed were already filled with phagosomes containing either beads or bacteria, suggesting that uptake occurs within only a few minutes of particles entering the animal."

- "Only 0.1 μm latex beads were found in large numbers on collar microvilli, but these beads adhered even more often to the cell surface."

Li, J., Carlson, B. and Lacis, A. (2015). How well do satellite AOD observations represent the spatial and temporal variability of PM2.5 concentration for the United States? *Atmospheric Environment*. 102. pg. 260-73. http://www.sciencedirect.com/science/article/pii/S1352231014009583

Li, K., Subba Rao, V., Harrison, W., et al. (1983). Autotrophic picoplankton in the tropical ocean. *Science*. 219. pg. 292-5. http://www.cmep.ca/jcullen/publications/1983/Li_et_al_Science.pdf

Li, K. and Fu, S. (2013). Polybrominated diphenyl ethers (PBDEs) in house dust in Beijing, China. *Bulletin of Environmental Contaminant Toxicology*. 91. pg. 382-5. http://www.ncbi.nlm.nih.gov/pubmed/23995797

- "Total PBDEs concentrations ranged from 140 to 1,300 ng g^{-1}. The dominant PBDEs congener identified was BDE 209, which made up more than 70% of all PBDEs congeners."

- "Concentrations of PBDEs in Chinese house dust were lower than in other countries. The most polluted areas were electronics shops and households."

Li, N., Sioutas, C., Cho, A., et al (2003). Ultrafine particulate pollutants induce oxidative stress and mitochondrial damage. *Environmental Health Perspectives*. 111(4). pg. 455-60. http://www.ncbi.nlm.nih.gov/pmc/articles/PMC1241427/

- "UFPs were most potent toward inducing cellular heme oxygenase-1 (HO-1) expression and depleting intracellular glutathione."

- "UFPs also had the highest ROS activity in the DTT assay. Because the small size of UFPs allows better tissue penetration…[they] localize in mitochondria, where they induce major structural damage."

Li, N., Xia, T. and Nel, A. (2008). The role of oxidative stress in ambient particulate matter induced lung diseases and its implications in the toxicity of engineered nanoparticles. *Free Radical Biology & Medicine*. 44(9). pg. 1689-99. http://www.ncbi.nlm.nih.gov/pubmed/18313407

Li, X., Gao, Y., Wang, Y. and Pan, Y. (2014). Emerging persistent organic pollutants in Chinese Bohai Sea and its coastal regions. *The Scientific World Journal*. http://www.hindawi.com/journals/tswj/2014/608231/

- "Emerging persistent organic pollutants (POPs) have widely aroused public concern in recent years. Polybrominated diphenyl ethers (PBDEs) and perfluorooctane sulfonyl fluoride/perfluorooctane sulfonic acid (POSF/PFOS) had been newly listed in Stockholm Convention in 2009, and short chain chlorinated paraffins (SCCPs) and hexabromocyclododecanes (HBCDs) were listed as candidate POPs."

- "The pollution levels of some kinds of emerging organic pollutants (PFOS, HBCDs) in organisms were comparable to some legacy POPs."

- "The emerging pollutants also exhibited ubiquitous existence in both abiotic and biotic media... the average values of PBDEs, PFOS, and PFOA were higher in organism than those in sediments."

Li, Y. (2001). Toxaphene in the United States. *Journal of Geophysical Research*. 106(D16). pg. 17919-27. http://onlinelibrary.wiley.com/doi/10.1029/2000JD900824/pdf

- "The intensive use of toxaphene on croplands was concentrated in the southeastern part of the United States...the state of Alabama was the largest user of toxaphene, reaching as much as 87 kt, followed by Mississippi at 60 kt."

- "Toxaphene... is a mixture of polychlorinated hydrocarbon compounds. Vetter [1993] noted that 177-670 compounds have been identified by chromatography, and predicted that the theoretical maximum number of compounds was over 32,000."

- "Because toxaphene has a high octanol/water partition coefficient, it is lipid-soluble and can be stored in fatty tissues including human milk. Toxaphene is one of the most heavily used organochlorine pesticides globally, with an estimated total usage of 1.33 million tons."

Li, Z., Cai, W. and Chen, X. (2007). Semiconductor quantum dots for in vivo imaging. *Journal of Nanoscience and Nanotechnology*. 7, pg. 2567-81. http://www.ncbi.nlm.nih.gov/pubmed/17685272

Li, Z. (2011). *Relative influences of uncertainty in physical-chemical property data and variability in climate parameters in determining the fate of PCBs.* UMEA University, Sweden. http://www.diva-portal.org/smash/get/diva2:535806/FULLTEXT01.pdf

Lin, M., Lee, C., Lin, Y. and Yang, K. (2011). Potentially toxic trace elements accumulating in marine sediment and bivalves in the outfall area of a desalination plant. *Desalination and Water Treatment.* 25. pg. 106-12. http://www-o.ntust.edu.tw/~khyang/word/Paper/Lin%20et%20al.%20%282011%29,WTD.pdf

Lind, L. and Lind, P. M. (2012). Can persistent organic pollutants and plastic-associated chemicals cause cardiovascular disease?. *J Intern Med.* 271(6). pg. 537-53.

Lianquan, G., Changxian, M., Shuai, W., et al. (2005). Molecular simulation of hydrogen adsorption density in single-walled carbon nanotubes and multilayer adsorption mechanism. *Journal of Material Science and Technology.* 21(1). http://www.jmst.org/fileup/PDF/2004019.pdf

- "Hydrogen is a renewable and unpolluted source of energy and has many advantages. It will bring great structural changes for the sources of energy in the future."

- "Carbon nanotubes were reported to be very promising materials for achieving the requirement of the Department of Energy (DOE) and International Energy Agency (IEA) hydrogen plan."

Limburg, Karin and Waldman, John. (2009). Dramatic declines in North Atlantic diadromous fishes. *Bioscience.* 59(11). pg. 955-65. http://www.esf.edu/efb/limburg/Fisheries/Debate/Limburg_%26_Waldman_BioScience_Dec_09.pdf

Linak, W. P., Perry, E., Williams, R., et al. (1989). Used agricultural plastics as fuel: Chemical and biological characterization of products of incomplete combustion from the simulated field burning of agricultural plastic. *Journal of Air Pollution Control Association.* 39. pg. 836-46. https://www.alfredstate.edu/files/downloads/academics/plastopub3.pdf

Lintelmann, J., Katayama, A., Kurihara, N., et al. (2003). Endocrine disruptors in the environment – (IUPAC Technical Report). *Pure and Applied Chemistry.* 75(5). pg. 631-81. http://www.iupac.org/publications/pac/pdf/2003/7505/7505x0631.html

- "The systems involved are very complex, and the number of possible compounds and organisms affected reach proportions that only large-scale and long time investigations can determine if indeed a serious ecological problem is present."

- "What is also needed is ecological 'common sense', e.g., if the fish in rivers used for drinking water display sex reversal, action should be take immediately."

- "Although the environmental exposure is well documented, there is a scarcity of literature on point source contamination."

- Contains an excellent summary of mechanisms of endocrine disruptors on hormone systems and an extensive bibliography.

Lippiatt, S., Opfer, S. and Arthur, C. (2013). *NOAA technical memorandum NOS-OR&R-46: Recommendations for monitoring debris trends in the marine environment.* National Oceanic and Atmospheric Administration. http://marinedebris.noaa.gov/sites/default/files/Lippiatt%20et%20al%202013.pdf

Lithner, D., Damberg, J., Dave, G. and Larsson, Å. (2009). Leachates from plastic consumer products: Screening for toxicity with Daphnia magna. *Chemosphere.* 74(9). pg. 1195-200. http://www.ncbi.nlm.nih.gov/pubmed/19108869

Lithner, D. (2011a). *Environmental and health hazards of chemicals in plastic polymers and products.* University of Goethenburg, Department of Plant and Environmental Sciences. https://gupea.ub.gu.se/handle/2077/24978

- See the appendices for a copy of the hazard ranked plastic polymers table.

Lithner, D., Halling, M. and Dave, G. (2011b). Toxicity of electronic waste leachates to Daphnia magna: Screening and toxicity identification evaluation of different products, components and materials. *Environmental Contamination Toxicology.* 62(4). pg. 723-4. http://www.ncbi.nlm.nih.gov/pubmed/22193862

- "Electronic waste has become one of the fastest growing waste problems in the world. It contains both toxic metals and toxic organics. The aim of this study was to (1) investigate to what extent toxicants can leach from different electronic products, components, and materials into water and (2) identify which group of toxicants (metals or hydrophobic organics) that is causing toxicity."

- "Toxicity identification evaluations (with C18 and CM resins filtrations and ethylenediaminetetraacetic acid addition) indicated that metals caused the toxicity in the majority of the most toxic leachates. Overall, this study has shown that electronic waste can leach toxic compounds also during short-term leaching with pure water."

Lithner, D., Larsson, Å. and Dave, G. (2011c). Environmental and health hazard ranking and assessment of plastic polymers based on chemical composition. *Science of*

the Total Environment. 409(18). pg. 3309-24.
http://www.sciencedirect.com/science/article/pii/S0048969711004268

- "The knowledge of human and environmental hazards and risks from chemicals associated with the diversity of plastic products is very limited."

- "In this study the environmental and health hazards of chemicals used in 55 thermoplastic and thermosetting polymers were identified and compiled."

- "The polymers were ranked based on monomer hazard classifications...polymers that ranked as most hazardous are made of monomers classified as mutagenic and/or carcinogenic (category 1A or 1B). These belong to the polymer families of polyurethanes, polyacrylonitriles, polyvinyl chloride, epoxy resins, and styrenic copolymers. All have a large global annual production (1-37 million tonnes)."

- "A considerable number of polymers (31 out of 55) are made of monomers that belong to the two worst of the ranking model's five hazard levels, i.e. levels IV-V. The polymers that are made of level IV monomers and have a large global annual production (1-5 million tonnes) are phenol formaldehyde resins, unsaturated polyesters, polycarbonate, polymethyl methacrylate, and urea-formaldehyde resins."

Lithner, D., Nordensvan, I. and Dave, G. (2011d). Comparative acute toxicity of leachates from plastic products made of polypropylene, polyethylene, PVC, acrylonitrile-butadiene-styrene, and epoxy to Daphnia magna. *Environmental Science and Pollution Research.* 19(5). pg. 1763-72.
http://www.ncbi.nlm.nih.gov/pubmed/22183785

- "Purpose: The large global production of plastics and their presence everywhere in the society and the environment create a need for assessing chemical hazards and risks associated with plastic products. The aims of this study were to determine and compare the toxicity of leachates from plastic products made of five plastics types and to identify the class of compounds that is causing the toxicity...Conclusions: Toxic chemicals leached even during the short-term leaching in water, mainly from plasticized PVC and epoxy products."

- "All leachates from plasticized PVC (5/5) and epoxy (5/5) products were toxic...None of the leachates from polypropylene (5/5), ABS (5/5), and rigid PVC (1/1) products showed toxicity... one of the five tested HDPE leachates was toxic...Evaluations indicated that mainly hydrophobic organics were causing the toxicity."

Lithner, D., Nordensvan, I. and Dave, G. (2012). Comparative acute toxicity of leachates from plastic products made of polypropylene, polyethylene, PVC, acrylonitrile-butadiene-styrene, and epoxy to Daphnia magna. *Environmental Science Pollution Research International*. 19(5). pg. 1763-72. http://www.ncbi.nlm.nih.gov/pubmed/22183785

Liu, L. and Xu, W. (2012). Engineering antibody fragment with the quantum dot in cancer cell imaging diagnosis. *Engineering*. 5. pg. 126-8. http://www.scirp.org/journal/PaperInformation.aspx?PaperID=38233

- "The nanoparticles, such as cadmium selenide quantum dots, only 100-100,000 atoms in size, can emit intrinsic fluorescence often many times stronger than that observed for other classes of fluorophores."

- The "nanocarrier-conjugated antibody fragment has potential to become a new therapeutic tool for cancer diagnosis and treatment."

Liu, Runping and Fan, Fenglei. (2014). *Mass concentration variations characteristics of PM10 and PM2.5 in Guangzhou, China*. Third International Workshop on Earth Observation and Remote Sensing Applications. http://ieeexplore.ieee.org/xpl/login.jsp?tp=&arnumber=6927860&url=http%3A%2F%2Fieeexplore.ieee.org%2Fiel7%2F6917166%2F6927831%2F06927860.pdf%3Farnumber%3D6927860

- "$PM_{2.5}$ is the main pollution type and greater than $PM_{2.5-10}$ in Guangzhou."

Liu, X., Tang, K., Harper, S., et al. (2013). Predictive modeling of nanomaterial exposure effects in biological systems. *International Journal of Nanomedicine*. 8(1). pg. 31-43. http://www.ncbi.nlm.nih.gov/pubmed/24098077

- "Rapid growth of nanobiotechnology will obviously result in increased exposure of humans and the environment to nanomaterials. Hence, there is a need to systematically investigate the potential biological and environmental impacts of newly emerging nanomaterials."

- "We found several important attributes that contribute to the 24 hours post-fertilization (hpf) mortality, such as dosage concentration, shell composition, and surface charge. These findings concur with previous studies on nanomaterial toxicity."

Ljubimova, J., Gangalum P., Portilla-Aria, J., et al. (2012). Molecular changes in rat brain due to air nano pollution. *NSTI Nanotech*. 3. pg. 261-4. http://www.researchgate.net/publication/232084778_Molecular_Changes_in_Rat_Brain_Due_to_Air_Nano_Pollution

- "Air pollution is a global problem that has been correlated with cardiovascular and respiratory-related morbidity [1]. Individuals living in urban areas may be exposed to nanoscopic pollution from automobile exhaust and various other industrial byproducts. However, the detrimental effects of air pollution may extend beyond the respiratory and cardiovascular systems. Studies in both animals and humans have shown that high levels of air pollution are associated with chronic neuroinflammation, as well as signs of neurodegeneration."

Llobet, J. M., Falco, G., Bocio, A. and Domingo, J. L. (2007). Human exposure to polychlorinated naphthalenes through the consumption of edible marine species. *Chemosphere*. 66. pg. 1107-13. https://www.ncbi.nlm.nih.gov/pubmed/16890979

Lobelle, D. and Cunliffe, M. (2011). Early microbial biofilm formation on marine plastic debris. *Marine Pollution Bulletin*. 62(1). pg. 197-200. http://www.ncbi.nlm.nih.gov/pubmed/21093883

- "Microbial biofilms developed rapidly on the plastic and coincided with significant changes in the physicochemical properties of the plastic. Submerged plastic became less hydrophobic and more neutrally buoyant during the experiment. Bacteria readily colonized the plastic but there was no indication that plastic-degrading microorganisms were present. This study contributes to improved understanding of the fate of plastic debris in the marine environment."

Lockman, P., Koziara, J., Mumper, R. and Allen, D. (2004). Nanoparticle surface charges alter blood-brain barrier integrity and permeability. *Journal of Drug Targetting*. 12. pg. 635-41. http://www.ncbi.nlm.nih.gov/pubmed/15621689

- "The blood-brain barrier (BBB) presents both a physical and electrostatic barrier to limit brain permeation."

- "This work evaluated: (1) effect of neutral, anionic and cationic charged NPs on BBB integrity and (2) NP brain permeability."

- "Cationic NPs have an immediate toxic effect at the BBB and NP surface charges must be considered for toxicity and brain distribution profiles."

Lohmann, R., E. Jurado, M. E. Q. Pilson and J. Dachs. (2006). Oceanic deep water formation as a sink of persistent organic pollutants. *Geophysical Research Letters*. 33(L12607). http://onlinelibrary.wiley.com/doi/10.1029/2006GL025953/full

Lohmann, R., MacFarlane, J., and Gschwend, P. (2005). Importance of black carbon to sorption of PAHs, PCBs and PCDDs in Boston and New York harbor sediments. *Environmental Science and Technology*. 39. pg. 141-8. http://web.uri.edu/lohmannlab/files/Lohmann-et-al-tumbling-EST05.pdf

- "The solid-water distribution ratios (Kd values) of 'native' PAHs, PCBs, and PCDDs in Boston and New York Harbor sediments were determined using small passive polyethylene samplers incubated for extended times in sedimentwater suspensions."

- "The fate and effects of hydrophobic organic compounds (HOCs) in the environment are largely determined by their sorption to solid phases."

- "When a piece of PE is placed in a sediment-water slurry, the HOCs accumulate within the PE until phase equilibrium occurs."

- "The PEs were removed, extracted, and analyzed for PAHs, PCBs, and PCDDs after three consecutive 14-day sedimentwater exposures."

Lombardo, F., Maggini, M., Gruden, G. and Bruno, G. (2013). Temporal trend in hospitalizations for acute diabetic complications: A nationwide study, Italy, 2001-2010. *PLoS ONE*. 8(5).
http://www.plosone.org/article/fetchObject.action?uri=info:doi/10.1371/journal.pone.0063675&representation=PDF

Lombard, M., Bryce, J., Mao, H. and Talbot, R. (2011). Mercury deposition in southern New Hampshire, 2006-2009. *University of New Hampshire Scholar's Repository*.
http://scholars.unh.edu/cgi/viewcontent.cgi?article=1001&context=earthsci_facpub

- "The lowest Hg wet deposition was measured in the winter with an average total seasonal deposition of 1.56 µg m^{-2} compared to the summer average of 4.71 µg m^{-2}."

- "The seasonal ratios of Hg wet deposition to RGM dry deposition vary by up to a factor of 80."

Long, T., Saleh, N., Tilton, R., et al. (2006). Titanium dioxide (P25) produces reactive oxygen species in immortalized brain microglia (BV2): Implications for nanoparticle neurotoxicity. *Environmental Science and Technology*. 40(14). pg. 4346-52.
http://www.ncbi.nlm.nih.gov/pubmed/16903269

- "Nanosize titanium dioxide (TiO2) is used in air and water remediation and in numerous products designed for direct human use and consumption. Its effectiveness in deactivating pollutants and killing microorganisms relates to photoactivation and the resulting free radical activity. This property, coupled with its multiple potential exposure routes, indicates that nanosize TiO2 could pose a risk to biological targets that are sensitive to oxidative stress damage (e.g., brain)."

Longnecker, M. P. and Michalek, J. E. (2000). Serum dioxin level in relation to diabetes mellitus among Air Force veterans with background levels of exposure. *Epidemiology*. 11. pg. 44-8.

Lopez-Cervantes, J. and Paseiro-Losada, P. (2003). Determination of bisphenol A in, and its migration from, PVC stretch film used for food packaging. *Food Additives and Contamination*. 20. pg. 596-606. http://www.ncbi.nlm.nih.gov/pubmed/12881134

Lorber, M. (2008). Exposure of Americans to polybrominated diphenyl ethers. *Journal of Expsure Science and Environmental Epidemiology*. 18(1). pg. 2-19. http://www.nature.com/jes/journal/v18/n1/pdf/7500572a.pdf

- "The predicted adult body burden of total PBDEs was 33.8 ng/kg lipid weight (lwt), compared to representative measurements in blood and milk at 64.0 and 93.7 ng/g lwt, respectively."

- "Food intake estimate of about 1.3 ng/kg/day (of the 7.7 ng/kg/day total) cannot explain current US body burdens; exposures to PBDEs in house dust accounted for 82% of the overall estimated intakes."

Lorber, M., Patterson, D., Huwe, J. and Kahn, H. (2009). Evaluation of background exposures of Americans to dioxin-like compounds in the 1990s and the 2000s. *Chemosphere*. 77. pg. 640-51. http://cfpub.epa.gov/si/si_public_record_Report.cfm?dirEntryId=213606&CFID=1830 88884&CFTOKEN=58705237&jsessionid=86308cb9faaba47e19c34f1d3e1a3c513f2f

Lorz, P., Towae, F., Enke, W., et. al. (2007). *Ullman's encyclopedia of industrial industry: Phthalic acid and derivatives*. Wiley-VCH Verlag GmbH & Co., Weinheim, Germany.

Lovern, S.B. and Klaper, R. (2006). *Daphnia magna* mortality when exposed to titanium dioxide and fullerene (C_{60}) nanoparticles. *Environmental Toxicological Chemistry*. 25(4). pg. 1132-7. http://www.ncbi.nlm.nih.gov/pubmed/16629153

- "Because compounds in this miniature size range have chemical properties that differ from those of their larger counterparts, nanoparticles deserve special attention."

- "Exposure to filtered C60 and filtered TiO2 caused an increase in mortality with an increase in concentration, whereas fullerenes show higher levels of toxicity at lower concentrations."

Lovrić, J., Cho, S., Winnik, F., and Maysinger, D. (2005). Unmodified cadmium telluride quantum dots induce oxygen species formation leading to multiple organelle damage and cell death. *Chemistry and Biology*. 12. pg. 1227-34. http://ac.els-cdn.com/S107455210500298X/1-s2.0-S107455210500298X-main.pdf?_tid=a54aac7a-cb32-11e4-8153-00000aab0f6c&acdnat=1426438091_9d62840a03b8ac11eac4e37038f1d7d3

Lovrić, J., Bazzi, H., Cuie, Y., et. al. (2006). Differences in subcellular distribution and toxicity of green and red emitting CdTe quantum dots. *Journal of Molecular Medicine*. 83. pg. 377-85. http://www.ncbi.nlm.nih.gov/pubmed/15688234

- "QD-induced cell death was characterized by chromatin condensation and membrane blebbing and was more pronounced with small (2r=2.2+/-0.1 nm), green emitting positively charged QDs than large (2r=5.2+/-0.1 nm), equally charged red emitting QDs."

Lowell Center for Sustainable Production. (2011). *Phthalates and their alternatives: Health and environmental concerns*. University of Massachusetts, Lowell, USA. http://www.sustainableproduction.org/downloads/PhthalateAlternatives-January2011.pdf

- "Phthalates are a class of synthetic chemicals that are widely used in a variety of consumer products including medical devices, food wrap, building materials, packaging, automotive parts, children's toys, and childcare articles made of polyvinyl chloride (PVC)."

- "The annual global production of phthalates is estimated to be 11 billion pounds."

- "The addition of phthalates to PVC makes this brittle plastic more flexible and durable. PVC products may contain up to 50 percent by weight of plasticizers, most commonly phthalates."

- "Phthalates are not chemically bound to the PVC polymer. Thus, over time they leach out of products and diffuse into the air, water, food, house dust, soil, living organisms, and other media, particularly under conditions involving heat."

- "Although six phthalates are now restricted from children's products in the US and European Union (EU), they are unregulated and continue to be used in toy making in many other parts of the world, such as China and India."

- Extensive information about phthalates and their alternatives.

Loyo-Rosales, J., Rosales-Rivera, G., Lynch, A., et al. (2004). Migration of nonylphenol from plastic containers to water and a milk surrogate. *Journal of Agriculture Food Chemistry*. 52(7). pg. 2016-20. http://www.ncbi.nlm.nih.gov/pubmed/15053545

Lu, C. and Su, F. (2007). Adsorption of natural organic matter by carbon nanotubes. *Separation and Purification Technology*. 58. pg. 113-21.

- "The presence of natural organic matter (NOM) in the water is a major concern… [It] not only forms carcinogenic compounds such as trihalomethanes (THMs) during the chlorination of drinking water, but also enhances bacterial regrowth and biofilm formation in the drinking water distribution system."

- "Carbon nanotubes…have been proven to possess good potential as superior adsorbents for removing many kinds of organic and inorganic pollutants such as dioxin and volatile organic compounds from air… or lead, cadmium, fluoride, 1,2-dichlorobenzene, THMs, zinc and nickel from aqueous solutions."

- "Further research works on testing the toxicity of CNTs and CNT-related nanomaterials are needed in order to promote safe and optimized applications of CNTs in water treatment."

Lubick, N. (2006). Still life with nanoparticles. *Environmental Science and Technology*. 40. pg. 4328.

Luef, B., Frischkorn, K., Wrighton, K. et. al. (2015). Diverse uncultivated ultra-small bacterial cells in groundwater. *Nature Communications*. 6. http://www.nature.com/ncomms/2015/150227/ncomms7372/full/ncomms7372.html

Lundbye, A., Berntssen, M., Lie, A., et. al. (2004). Dietary uptake of dioxins (PCDD/PCDFs) and dioxin-like PCBs in Atlantic salmon (*Salmo salar*). *Aquaculture Nutrition*. 10(3). pg. 199-207. http://onlinelibrary.wiley.com/doi/10.1111/j.1365-2095.2004.00299.x/abstract?deniedAccessCustomisedMessage=&userIsAuthenticated=false

- "Even with the most contaminated feed (4.9 pg WHO-TEQ g^{-1} dw) the dioxin concentrations in salmon did not exceed the maximum level set by the European Commission [4 pg WHO-TEQ g^{-1}."

Lundén, A. and Norén, K. (1998). Polychlorinated naphthalenes and other organochlorine contaminants in Swedish human milk, 1972-1992. *Archives of Environmental Contamination and Toxicology*. 34. pg. 414-23.

Lunder, S. and Jacob, A. (2008). *Fire retardants in toddlers and their mothers*. Environmental Working Group. http://www.ewg.org/research/fire-retardants-toddlers-and-their-mothers

- EWG "found that toddlers and preschoolers typically had 3 times as much of these hormone-disrupting chemicals in their blood as their mothers."

- "In total 11 different flame retardants were found in these children."

Lundgren, K., Tysklind, M., Ishaq, R., et al. (2002). Polychlorinated naphthalene levels, distribution and biomagnification in a benthic food chain in the Baltic Sea. *Environmental Science and Technology*. 36. pg. 5005-13.

Luo, Yuzhou, Jorgenson, Brant C., Thuyet, Dang Quoc et al. (2014). Insecticide washoff from concrete surfaces: Characterization and prediction. *Environmental Science & Technology*. 48(1). pg. 234-43.

Lusher, A.L., McHugh, M. and Thompson, R.C. (2012). Occurrence of microplastics in the gastrointestinal tract of pelagic and demersal fish from the English Channel. *Marine Pollution Bulletin*. 67(1-2). pg. 94-9. http://dx.doi.org/10.1016/j.marpolbul.2012.11.028

Luther, W. and Zweck, Z., eds. (2013). *Safety aspects of engineered nanomaterials*. Pan Stanford Publishing Pte. Ltd. http://www.panstanford.com/books/9789814364850.html

- "This book deals with the question regarding the current status of the safety aspects of engineered nanomaterials."

- This text has 12 research reports exploring the application, safety, and health hazards of nanotechnology.

- This text contains extensive, if not comprehensive, bibliographies of the safety issues pertaining to engineered nanomaterials.

- See Schirmer (2013).

Lynch, I., Dawson, K. and Linse, S. (2006). Detecting cryptic epitopes created by nanoparticles. *Science Signaling*. 327. pg. 14. http://www.ncbi.nlm.nih.gov/pubmed/16552091

- "Because of their small size, nanoparticles are easily taken up into cells (by receptor-mediated endocytosis), whereupon they have essentially free access to all cellular compartments."

- "Protein adsorption may result in the exposure at the surface of amino acid residues that are normally buried in the core of the native protein, which are recognized by the cells as 'cryptic epitopes.' These cryptic epitopes may trigger inappropriate cellular signaling events (as opposed to being rejected by the cells as foreign bodies)."

Lyon, F. (2008). *IARC Volume 97: Monographs on the evaluation of carcinogenic risks to humans*. International Agency on Research. http://monographs.iarc.fr/ENG/Monographs/vol97/mono97.pdf

Ma, W., Yun, S., Bell, E., et al. (2013). Temporal trends of polybrominated biphenyl ethers (PBDEs) in the blood of newborns from New York State during 1997 through 2011: Analysis of dried blood spots from the newborn screening program. *Environmental Science and Technology*. 47(14). pg. 8015-21. http://www.ncbi.nlm.nih.gov/pmc/articles/PMC3725776/

- "On a global basis, North American populations are exposed to the highest doses of PBDEs."

- "Some PBDE formulations were phased out from production in the early 2000s. The effectiveness of the phaseout of commercial Penta-BDE and Octa-BDE mixtures in 2004 in the U.S. on human exposure levels is not known."

- "In this study, seven PBDE congeners were determined by gas chromatography-high resolution mass spectrometry (GC-HRMS)...from newborns in New York State (NYS) from 1997 to 2011."

- "The mean concentrations determined during 1997 through 2011 in the whole blood of newborns were 0.128, 0.040, and 0.012 ng/mL for BDEs −47, −99, and −100, respectively."

- "PBDE concentrations were similar during 1997 through 2002 and, thereafter, decreased significantly, which was similar to the trends observed for perfluorinated compounds (PFCs) in DBS samples. Occurrence of PBDEs in the whole blood of newborns confirms that these compounds do cross the placental barrier."

- "Ingestion of indoor dust was reported as a predominant pathway of PBDE exposure for the U.S. general population.16 Human biomonitoring studies have shown that PBDE exposure in the U.S. is 10- to 100-fold higher than those in Europe and Asia."

Mace, G. M., Collar, Nigel J., Gaston, Kevin J., et al. (2008). Quantification of extinction risk: IUCN's system for classifying threatened species. *Conserv. Biol.* 22, pg. 1424-42. http://onlinelibrary.wiley.com/doi/10.1111/j.1523-1739.2008.01044.x/abstract

Madou, M. (2002). *Fundamentals of Microfabrication: The science of miniaturization, 2nd edition*. CRC, Boca Raton, FL.

Magrez, A., Kasa, S., Salicio, V., et al. (2006). Cellular toxicity of carbon-based nanomaterials. *Nanotechnology Letters*. 6(6). pg. 1121-5. http://pubs.acs.org/doi/abs/10.1021/nl060162e

- "These materials are toxic while the hazardous effect is size-dependent, cytotoxicity is enhanced when the surface of the particles is functionalized after an acid treatment."

Maguire, James. (1999). Review of the persistence of nonylphenol and nonylphenol ethoxylates in aquatic environments. *Water Quality Research Journal of Canada*. 34(1). pg. 37-78. http://www.geol.lsu.edu/blanford/NATORBF/12%20Organic%20Chemicals/Maguire_Water%20Research%20Canada_1999.pdf

- "Nonylphenol ethoxylates, are widely used nonionic surfactants that are discharged in high quantities to sewage treatment plants and directly to the environment in areas where there is no sewage or industrial waste treatment."

- "Degradation products [nonylphenol ethoxylates], including nonylphenol, are more persistent than the parent surfactants and they are found in receiving waters of sewage treatment plants. Nonylphenol in particular is found at high concentrations in some sewage sludges that may be spread on agricultural lands."

Main, K.M. (2006). Human breast milk contamination with phthalates and alterations of endogenous reproductive hormones in infants three months of age. *Environmental Health Perspectives*. 114. pg. 270-6. http://www.ncbi.nlm.nih.gov/pmc/articles/PMC1367843/pdf/ehp0114-000270.pdf

- "Phthalates adversely affect the male reproductive system in animals. We investigated whether phthalate monoester contamination of human breast milk had any influence on the postnatal surge of reproductive hormones in newborn boys as a sign of testicular dysgenesis."

- "Our data on reproductive hormone profiles and phthalate exposures in newborn boys are in accordance with rodent data and suggest that human Leydig cell development and function may also be vulnerable to perinatal exposure to some phthalates. Our findings are also in line with other recent human data showing incomplete virilization in infant boys exposed to phthalates prenatally."

- "Sources of phthalate exposures in women can be inhalation, contamination via building materials and furniture, use of consumer products including cosmetics and food items. Thus, exposure to some phthalates such as DEHP and DiNP is likely to be constant rather than episodic, whereas others such as DMP and DEP, through their presence in cosmetics, are more influenced by personal habits."

Main, K., Kiviranta, H., Virtanen, H., et al. (2007). Flame retardants in placenta and breast milk and cryptorchidism in newborn boys. *Environmental Health Perspectives*.

115. pg. 1519-26.
http://www.ncbi.nlm.nih.gov/pmc/articles/PMC2022640/pdf/ehp0115-001519.pdf

- "Because the prevalence of cryptorchidism appears to be increasing, we investigated whether exposure to PBDEs was associated with testicular maldescent."

- "Polybrominated diphenyl ethers (PBDEs) are widely used as flame retardants, and the general population is exposed through products such as upholstery, building materials, insulation, electronic equipment, and contaminated food."

- "Levels in breast milk, but not in placenta, showed an association with congenital cryptorchidism. Other environmental factors may contribute to cryptorchidism."

Malone, T. C. and Ducklow, H. W. (1990). Microbial biomass in the coastal plume of Chesapeake Bay: Phytoplankton-bacterioplankton relationships. *Limnology and Oceanography*. 35(2).
http://onlinelibrary.wiley.com/doi/10.4319/lo.1990.35.2.0296/pdf

Mark, Frank and Vehlow, J. (1999). *Co-Combustion of end of life plastics in municipal solid waste combustors*. Association of Plastic Manufacturers in Europe.
http://www.plasticseurope.fr/Documents/Document/20100312155603-Co-Combustion_of_End_of_Life_Plastics_in_MSW_Combustors.pdf

Markic, Ana and Nicol, Simon. (2014). In a nutshell: Microplastics and fisheries. *SPC Fisheries Newsletter*. 144. pg. 27-9.
http://www.spc.int/DigitalLibrary/Doc/FAME/InfoBull/FishNews/144/FishNews144_27_Markic.pdf

Martina, C., Weiss, B. and Swan, S. (2012). Lifestyle behaviors associated with exposures to endocrine disruptors. *Neurotoxicology*. 33(6). pg. 1427-33.
http://www.ncbi.nlm.nih.gov/pmc/articles/PMC3641683/pdf/nihms-465645.pdf

- "Our data suggest three [lifestyle of an Old Order Mennonite community] practices…may contribute to these lower levels: (1) consuming mostly homegrown produce (ingestion), (2) no cosmetics and limited use of personal care products, and (3) transportation primarily by sources other than automobiles."

Martinović-Weigelt, D., Mehinto, A. C., Ankley, G .T., et al. (2014). Transcriptomic effects-based monitoring for endocrine active chemicals--Assessing relative contribution of treated wastewater to downstream pollution. *Environmental Science and Technology*. 48(4). pg. 2385-94. http://toxics.usgs.gov/highlights/neuroactive.html

Maso, M., Garces, E., Pages, F. and Camp, J. (2003). Drifting plastic debris as a potential vector for dispersing harmful algal bloom species. *Scientia Marina*. 67(1). pg. 107-10. http://www.cmima.csic.es/files/webcmima/docs/biblio-pdf/doc_2696.pdf

Mason, R., Reinfelder, J. and Morel, F. (1996). Uptake, toxicity, and trophic transfer of mercury in a coastal diatom. *Environmental Science and Technology*. 30. pg. 1835-45. http://www.princeton.edu/morel/publications/pdfs/masonEST1996.pdf

Masoner, Jason R., Kolpin, Dana W., Furlong, Edward T., et al. (2014). Contaminants of emerging concern in fresh leachate from landfills in the conterminous United States *Environ. Sci.: Processes Impacts*.16. pg. 2335-54. http://pubs.rsc.org/en/Content/ArticleLanding/2014/EM/C4EM00124A#!divAbstract

- "Industrial and household chemicals were measured in the greatest concentrations, composing more than 82% of the total measured CEC concentrations."

Massart, F., Meucci, V., Saggese, G. and Soldani, G. (2008). High growth rate of girls with precocious puberty exposed to estrogenic mycotoxins. *Journal of Pediatrics*. 152(5). pg. 690-5. http://www.jpeds.com/article/S0022-3476(07)00983-3/pdf

- "Mycoestrogenic zearalenone is suspected to be a triggering factor for CPP (central precocious puberty) development in girls. Because of its chemical resemblance to some anabolic agents used in animal breeding, ZEA may also represent a growth promoter in exposed patients."

Mato, Y., Isobe, T., Takada, H., et al. (2001). Plastic resin pellets as a transport medium for toxic chemicals in the marine environment. *Environmental Science and Technology*. 35(2). pg. 318-24. http://pubs.acs.org/doi/abs/10.1021/es0010498

- "PCBs, DDE, and nonylphenols (NP) were detected in polypropylene (PP) resin pellets collected from four Japanese coasts. Concentrations of PCBs (4−117 ng/g), DDE (0.16−3.1 ng/g), and NP (0.13−16 µg/g) varied among the sampling sites. These concentrations were comparable to those for suspended particles and bottom sediments collected from the same area as the pellets."

- "The source of PCBs and DDE is ambient seawater and that adsorption to pellet surfaces is the mechanism of enrichment. The major source of NP in the marine PP resin pellets was thought to be plastic additives and/or their degradation products."

- "Comparison of PCBs and DDE concentrations in marine PP resin pellets with those in seawater suggests their high degree of accumulation (apparent adsorption coefficient: 105−106). The high accumulation potential suggests that plastic resin pellets serve as both a transport medium and a potential source of toxic chemicals in the marine environment."

Matsumoto, T., Nakajima, S., Tojo, Y., et. al. (2007). Characterization of shredder residues derived from end-of-life vehicles and home electrical appliances. *Journal of the Japan Society of Waste Management Experts*. 18(2). pg. 126-36. http://www.researchgate.net/publication/250304286_Characterization_of_Shredder_Residues_Derived_from_End-of-Life_Vehicles_and_Home_Electrical_Appliances

Mayer, P. and Reichenberg, F. (2006). Can highly hydrophobic organic substances cause aquatic baseline toxicity and can they contribute to mixture toxicity. *Environmental Toxicological Chemistry*. 25(10). pg. 2639-44. http://www.ncbi.nlm.nih.gov/pubmed/17022404

- "Such substances are still expected to contribute to baseline toxicity when part of a complex mixture."

Maynard, Andrew. (2006a). Nanotechnology: Assessing the risks. *Nanotoday*. 1(2). pg. 22-33. http://www.web.pdx.edu/~pmoeck/phy381/nano%20risks.pdf.

- "There is a lack of overall strategy in the current research portfolio, and that without a clear strategic framework, greater resources, and increased partnerships and collaboration, critical questions are unlikely to be answered in a timely manner."

Maynard, Andrew. (2006b). *Nanotechnology: A research strategy for addressing risk*. Woodrow Wilson International Center for Scholars Project on Emerging Nanotechnologies. http://www.nanotechproject.org/file_download/files/PEN3_Risk.pdf

- "The 2004 *Royal Society and Royal Academy of Engineering* report anticipates that the quantities of engineered nanomaterials in use will increase rapidly over the next few years,with an estimated production rate of 58,000 metric tones per year between 2011-2020. It is sobering to think that, if (as we suspect) the number or surface of particles making up these materials determines the hazard they represent, the impact of these materials *might* be the equivalent of between 5 million and 50 billion metric tones of conventional material."

- "Many of the early applications of nanotechnology are by-design intended to achieve high exposure."

Maynard, Andrew D., Aitken, Robert J., Butz, Tilman, et al. (2006c). Safe handling of nanotechnology. *Nature*. 444. pg. 267-9. http://www.nature.com/nature/journal/v444/n7117/full/444267a.html

- "Size, surface area, surface chemistry, solubility and possibly shape all play a role in determining the potential for engineered nanomaterials to cause harm."

- "Effluent from nanomanufacturing processes, use of nanoparticle-containing substances such as sunscreens, and disposal of nanomaterial-containing products, will inevitably lead to increasing quantities of engineered nanomaterials in water systems."

Maynard, A. and Aitken, R. (2007). Assessing exposure to airborne nanomaterials: Current abilities and future requirements. *Nanotoxicology*. 1. pg. 26-41. http://informahealthcare.com/doi/abs/10.1080/17435390701314720

Maynard, A., Warheit, D. and Philbert, M. (2011). The new toxicology of sophisticated materials: Nanotoxicology and beyond. *Toxicological Sciences*. 120(S1). pg. S109-29. http://toxsci.oxfordjournals.org/content/120/suppl_1/S109.full.pdf+html

- "With the rapid rise of the field of nanotechnology and the design and production of increasingly complex nanoscale materials, it has become ever more important to understand how the physical form and chemical composition of these materials interact synergistically to determine toxicology."

McAllister, E., Dhurandhar, N., Keith, S., et. al. (2009). Ten putative contributors to the obesity epidemic. *Critical Reviews in Food Science and Nutrition*. 49. pg. 868-913. http://www.researchgate.net/publication/40447388_Ten_putative_contributors_to_the_obesity_epidemic

- "The obesity epidemic is a global issue and shows no signs of abating, while the cause of this epidemic remains unclear."

- "Marketing practices of energy-dense foods and institutionally-driven declines in physical activity are the alleged perpetrators for the epidemic."

- "Evidence for microorganisms, epigenetics, increasing maternal age, greater fecundity among people with higher adiposity, assortative mating, sleep debt, endocrine disruptors, pharmaceutical iatrogenesis, reduction in variability of ambient temperatures, and intrauterine and intergenerational effects as contributing factors to the obesity epidemic are reviewed."

- "While the evidence is strong for some contributors such as pharmaceutical-induced weight gain, it is still emerging for other reviewed factors."

- "Obesity has not only increased in the United States but also seems to have increased in virtually every country where detailed data are available. The reasons for this increase are incompletely understood."

- "Our understanding of environmental influences on epigenetic processes remains rudimentary."

- "Environmental factors during development can induce permanent alterations in epigenetic gene regulation, and epigenetic dysregulation can contribute to obesity. It is therefore plausible (if not likely) that environmental influences on epigenetic gene regulation contribute to the secular increase in obesity."

- "Endocrine disrupting chemicals (EDCs), such as the flame retardant polybrominated diphenyl ether (PBDE), and the plasticizer bisphenol A (BPA), are very stable in the environment and many have been steadily increasing in levels in humans."

- This article has a very extensive bibliography.

McCauley, D., Pinsky, M., Palumbi, S., et al. (2015). Marine defaunation: Animal loss in the global ocean. *Science*. 347(6219). pg. 247-55. https://oceansciencenow.files.wordpress.com/2014/04/science-2015-mccauley.pdf

- "Current ocean trends, coupled with terrestrial defaunation lessons, suggest that marine defaunation rates will rapidly intensify as human use of the oceans industrializes."

- "Overall, habitat degradation is likely to intensify as a major driver of marine wildlife loss."

- There is no further explanation of the meaning of the "era of global chemical warfare on marine eco-systems."

- There are no references pertaining to the impact of chemical fallout or microplastics.

- This is a frightening article with respect to what is off limits for discussion pertaining to marine wildlife.

McCormick, A., Hoellein, T., Mason, S., Schluep, J. and Kelly, J. (2014). Microplastic is an abundant and distinct microbial habitat in an urban river. *Environmental Science and Technology*. 48(20). pg. 11863-71. http://pubs.acs.org/doi/abs/10.1021/es503610r

- "We demonstrated that wastewater treatment plant effluent was a point source of microplastic."

McDermid, K. J. and McMullen, T. L. (2004). Quantitative analysis of small-plastic debris on beaches in the Hawaiian archipelago. *Marine Pollution Bulletin*. 48. pg. 790-4. http://www.ncbi.nlm.nih.gov/pubmed/15041436

- "Greatest quantity was found at three of the most remote beaches on Midway Atoll and Moloka'i. Of the debris analyzed, 72% by weight was plastic."

McGuire, V. L. (2014). *Water-level changes and change in water in storage in the high plains aquifer, predevelopment to 2013 and 2011–13*. U.S. Geological Survey Scientific Investigations Report 2014–5218. USGS, Lincoln, NB. http://pubs.usgs.gov/sir/2014/5218/

Meeker, J., Calafat, A. and Hauser, R. (2007). Di(2-ethylhexyl) phthalate metabolites may alter thyroid hormone levels in men. *Environmental Health Perspectives*. 115. pg. 1029-34.

Meeker, J., Sathyanarayana, S. and Swan, S. (2009). Pthalates and other additives in plastic: Human exposure and associated health outcomes. *Philosophical Transcripts of the Royal Society*. 364(1526). pg. 2115-26.

- "Concern exists over whether additives in plastics to which most people are exposed, such as phthalates, bisphenol A or polybrominated diphenyl ethers, may cause harm to human health by altering endocrine function or through other biological mechanisms."

- "Certain chemicals, used in plastics to provide beneficial physical qualities, may also act as endocrine-disrupting compounds (EDCs) that could lead to adverse reproductive and developmental effects."

- "It is well known that humans are exposed to all these compounds simultaneously, and to many other chemicals."

- "There are limited data on the interactions between chemicals within a class or across classes. Chemicals may interact additively, multiplicatively or antagonistically in what is commonly referred to as the 'cocktail effect'."

- "Finally, human research is needed on potential latent and transgenerational effects (e.g. epigenetic modifications) of exposure to plastic additives and other EDCs."

Meier, A., Tsaloglou, N., Mowlem, M., et al. (2013). Hyperbaric biofilms on engineering surfaces formed in the deep sea. *Biofouling*. 29(9). pg. 1029-42. http://www.ncbi.nlm.nih.gov/pubmed/23964799

Meijer, S., Ockenden, W., Sweetman, A., et al. (2003). Global distribution and budget of PCBs and HCB in background surface soils: Implications for sources and environmental processes. *Environment Science and Technology*. 23. pg. 667-72. http://pubs.acs.org/doi/abs/10.1021/es0258091

Melnick, D., Pearl, M. and Warfield, J. (January 20, 2015). Make forests pay. *The New York Times.*

la Merrill, M. and Birnbaum, L. S. (2011). Childhood obesity and environmental chemicals. *Mt. Sinai Journal of Medicine.* 78(1). http://www.ncbi.nlm.nih.gov/pmc/articles/PMC3076189/pdf/nihms-253603.pdf

- The article contains detailed exposure tables, listings, and bibliographies.

Meyer-Reil, Lutz-Arend. (1994). Microbial life in sedimentary biofilms: The challenge to microbial ecologists. *Marine Ecology Progress Series.* 112. pg. 303-11. http://www.int-res.com/articles/meps/112/m112p303.pdf

Meyers, R. and Worm, B. (2003). Rapid worldwide depletion of predatory fish communities. *Nature.* 423. pg. 280-3. http://faculty.wwu.edu/~shulld/ESCI%20432/Myers&Worm2003.pdf

Michalet, X., Pinaud, F., Bentolila, L., et al. (2005) Quantum dots for live cells, in vivo imaging, and diagnostics. *Science.* 396. pg. 538-44. http://www.ncbi.nlm.nih.gov/pmc/articles/PMC1201471/pdf/nihms3260.xml.fixed.pdf

- "The new generations of qdots have far-reaching potential for the study of intracellular processes at the single-molecule level, high resolution cellular imaging, long-term in vivo observation of cell trafficking, tumor targeting, and diagnostics."

Michaels, David. (2008). *Doubt is their product:How industry's assault on science threatens your health.* Oxford University Press, New York, NY.

Michel, C., Herzog, S., de Capitani, C., Burkhardt-Holm, P. and Pietsch, C. (2014). Natural mineral particles are cytotoxic to rainbow trout gill epithelial cells *In Vitro*. *PLoS ONE.* 9(7). http://www.plosone.org/article/fetchObject.action?uri=info:doi/10.1371/journal.pone.0100856&representation=PDF

Mierzykowski S. E. (2012). *Contaminants in Atlantic sturgeon and shortnose sturgeon recovered from the Penobscot and Kennebec Rivers, Maine.* USFWS. Spec. Proj. Rep. FY09-MEFO-3-EC. Maine Field Office, Orono, ME. http://www.fws.gov/northeast/mainecontaminants/pdf/MaineSturgeonContaminantsReport2012.pdf

- "Total PCB in sturgeon fillets ranged from below the detection limit (< 5.00 parts-per-billion, ppb) to 1,900.00 ppb wet weight. Five shortnose sturgeon had PCB fillet concentrations that would exceed suggested criteria for protecting fish-eating wildlife (120 ppb) and aquatic life (400 ppb)."

- "Total PBDE in five shortnose sturgeon fillets ranged from 4.4 ppb to 39.1 ppb. Congener BDE #47 was the greatest contributor to Total PBDE."

Miller, E., Vanarsdale, A., Keeler, G., et al. (2004). Estimation and mapping of wet and dry mercury deposition aross Northeastern North America. *Ecotoxicology*. 14. pg. 53-70. http://citeseerx.ist.psu.edu/viewdoc/download?doi=10.1.1.457.3359&rep=rep1&type=pdf

Miller, G., Senjen, R., Cameron, P., et al. (2008). *Out of the laboratory and on to our plates: Nanotechnology in food and agriculture.* Friends of the Earth. http://static.sdu.dk/mediafiles//Files/Om_SDU/Fakulteterne/Teknik/NANO/nano_food_report.pdf

- "Nanotechnology has been provisionally defined as relating to materials, systems and processes which exist or operate at a scale of 100 nanometres (nm) or less. It involves the manipulation of materials and the creation of structures and systems at the scale of atoms and molecules, the nanoscale. The properties and effects of nanoscale particles and materials differ significantly from larger particles of the same chemical composition."

- "Nanoparticles of silver, titanium dioxide, zinc and zinc oxide, materials now used in nutritional supplements, food packaging and food contact materials, have been found to be highly toxic to cells in test tube studies."

- "Nanoparticles have a very large surface area which typically results in greater chemical reactivity, biological activity and catalytic behavior."

- "Nanotechnology is introducing a new array of potentially more toxic pesticides, plant growth regulators and chemical fertilizers."

- "Nanoparticles are also more adhesive than larger particles to surfaces within our bodies."

- "Chemical release nanopackaging enables food packaging to interact with the food it contains. The exchange can proceed in both directions. Packaging can release nanoscale antimicrobials, antioxidants, flavours, fragrances or nutraceuticals into the food or beverage to extend its shelf life or to improve its taste or smell."

Miller, R., Bennett, S., Keller, A., Pease, S. and Lenihan, H. (2012). TiO_2 nanoparticles are phototoxic to marine phytoplankton. *PLoS ONE*. 7(1). http://www.plosone.org/article/fetchObject.action?uri=info:doi/10.1371/journal.pone.0030321&representation=PDF

- "Relatively low levels of ultraviolet light, consistent with those found in nature, can induce toxicity of TiO_2 nanoparticles to marine phytoplankton, the most important primary producers on Earth."

Minchin, Rodney and Martin, Darren. (2010). Minireview: Nanoparticles for molecular imaging—an overview. *Endocrinology*. 151. pg. 474-81. http://press.endocrine.org/doi/pdf/10.1210/en.2009-1012

- "Nanoprobes can be used to image specific cells and tissues within a whole organism."

- "Perhaps the least understood characteristic of nanoparticles designed for molecular imaging is their potential to induce toxicity after administration. Toxicity can result from the nanoparticles themselves or the individual components of the nanoparticles that can be released during degradation *in vivo*."

- "As more complex and innovative nanoparticle probes are designed, their adverse effects, both as intact particles and as separate chemical entities, will need to be evaluated."

- "Progress toward the clinical adaptation of this technology may be slow, particularly if the adverse effects of the nanoparticles are not investigated."

Mizukawa, K., Takada, H., Takeuchi, I., et al. (2009). Bioconcentration and biomagnification of polybrominated diphenyl ethers (PBDEs) through lower-trophic-level coastal marine food web. *Marine Pollution Bulletin*. 58. pg. 1217-24. http://www.ncbi.nlm.nih.gov/pubmed/19376538

Moghadam, B., Hou, W., Corredor, C., et al. (2012). Role of nanoparticle surface functionality in the disruption of model cell membranes. *Langmuir*. 26(47). pg. 16318-26. http://pubs.acs.org/doi/abs/10.1021/la302654s

Mohan, Durga, Thiyagarajan, Devasena and Murthy, Prakhya Balakrishna. (2013). Toxicity of exhaust nanoparticles. *African Journal of Pharmacy and Pharmacology*. 7(7). pg. 318-31. http://www.academicjournals.org/article/article1380800058_Mohan%20et%20al.pdf

- "The main four sources of air pollution are emissions from vehicles, thermal power plants, industries and refineries."
- "In 2007, the Blacksmith Institute listed the top ten polluted areas in the world as Azerbaijan, China, India, Peru, Russia, Ukraine, and Zambia"
- "Exhaust from vehicles [in India] has increased eight-folds over a period of twenty years; industrial pollution has risen four times over the same period."

Mokdad, A., Ford, E., Bowman, B., et. al. (2003). Prevalence of obesity, diabetes, and obesity-related health risk factors. *JAMA*. 289. pg. 76-9. http://jama.jamanetwork.com/article.aspx?articleid=195663

- "Increases in obesity and diabetes among US adults continue in both sexes, all ages, all races, all educational levels, and all smoking levels."

- "The prevalence of obesity and diabetes among US adults increased substantially from 1990 to 2000."

Moore, C., Moore, S., Leecaster, M. and Weisberg, S. (2001). A comparison of plastic and plankton in the North Pacific Central Gyre. *Marine Pollution Bulletin*. 42(12). pg. 1297-1300. http://5gyres.org/media/Moore_2001_plastic_in_North_Pacific_Gyre.pdf

- "Neuston samples were collected at 11 random sites, using a manta trawl lined with 333um mesh. The abundance and mass of neustonic plastic was the largest recorded anywhere in the Pacific Ocean at 334,271 pieces km^2 and 5,114 g/km^2 respectively. Plankton abundance was approximately five times higher than that of plastic, but the mass of plastic was approximately six times that of plankton."

Moore, C., Moore, S., Leecaster, M. and Weisberg, S. (2002a). A comparison of neustonic plastic and zooplankton abundance in Southern California's coastal waters. *Marine Pollution Bulletin*. 44. pg. 1035-38. http://ftp.sccwrp.org/pub/download/DOCUMENTS/AnnualReports/2001_02AnnualReport/06_ar02-shelly.pdf

Moore, M.N. (2002b). Biocomplexity: The post-genome challenge in ecotoxicology. *Aquatic Toxicology*. 59(1-2). pg. 1-15. http://www.ncbi.nlm.nih.gov/pubmed/12088630

Moore, M. N. (2006). Do nanoparticles present ecotoxicological risks for the health of the aquatic environment? *Environment International*. 32. pg. 967-76. http://www.sciencedirect.com/science/article/pii/S0160412006000857

- "Possible nanoparticle association with naturally occurring colloids and particles is considered together with how this could affect their bioavailability and uptake into cells and organisms."

- "Uptake by endocytotic routes are identified as probable major mechanisms of entry into cells; potentially leading to various types of toxic cell injury."

Moore, Charles James. (2008). Synthetic polymers in the marine environment: A rapidly increasing, long-term threat. *Environmental Research*. 108(2). pg. 131-9. http://www.sciencedirect.com/science/article/pii/S001393510800159X

von Moos, Nadia. (2012). *Cellular effects of plastic particles on the blue mussel Mytilus edulis*. Out to Sea: The Plastic Garbage Project. http://www.plasticgarbageproject.org/en/plastic-garbage/problems/microplastics/nadia-von-moos-the-effects-of-plastic-on-mussels/

- "The results obtained showed that the irregularly shaped polyethylene particles were ingested by the blue mussel during filter feeding and that the particles triggered an immune reaction (significant increase in granulocytoma formation) and negatively affected mussel health status (significant destabilization of lysosomes)."

von Moos, N., Burkhardt-Holm, P., and Kahler, A. (2013). Uptake and effects of microplastics on cells and tissue of the blue mussel *Mytilus edulis L.* after an experimental exposure. *Environmental science & Technology.* 46. pg. 11327-35. http://www.plasticgarbageproject.org/fileadmin/plasticgarbage/Themen/Downloads/Texte/EN/von_Moos_Effects_on_Mussels_MfGZ_EN.pdf

Mora, C., Tittensor, D. P., Adl, S., et al. (2011). How many species are there on Earth and in the ocean? *PLOS Biol.* 9.

Mora, C., Wei, C-L., Rollo, A., Amaro, T., Baco, A. R., et al. (2013). Biotic and human vulnerability to projected changes in ocean biogeochemistry over the 21st century. *PLOS Biology.* 11(10). http://journals.plos.org/plosbiology/article?id=10.1371/journal.pbio.1001682

Morales, R., Pulido-Villena, E. and Reche, I. (2013). Chemical signature of Saharan dust on dry and wet atmospheric deposition in the south-western Mediterranean region. *Tellus.* 65. http://www.tellusb.net/index.php/tellusb/article/view/18720

Morgan, A. and Wilkie, C. (2007). *Flame retardant polymer nanocomposites*. John Wiley & Sons. Hoboken, NJ.

Morgan, Marisa. (2014). *Exposure to endocrine disrupting compounds and reproductive toxicity in women*. FIU Electronic Theses and Dissertations. http://digitalcommons.fiu.edu/etd/1586

- "In this cross-sectional study of women 20-85 years of age, we separately evaluated 6 individual PCB cogeners, the sum of dioxin-like PCBs, the sum of non-dioxin-like PCBs, eight phthalate metabolites, the sum of DEHP, the sum of total phthalates, and BPA in association with reproductive cancers (breast, cervical, ovarian, and uterine) in women…PCBs showed the most significant associations with all reproductive cancers."

- "Separate analyses showed geometric mean levels of individual PCB congeners to be significantly higher among women with breast cancer, ovarian cancer, and uterine cancer when compared to the rest of the study population…monoethyl phthalate (MEP) was found to be significantly higher among women with uterine cancer compared to women never diagnosed with cancer…we found PCB 138 to be significantly associated with breast cancer, cervical cancer, and uterine cancer, and PCB 74 and 118 to be significantly associated with ovarian cancer…in sum, the major findings from the research in this dissertation indicate that exposure to PCBs may increase the risk of endometriosis, uterine leiomyomas, and cancers of the breast, cervix, ovaries, and uterus. Our findings are consistent with the proven contributions of unopposed estrogens to the risk for reproductive toxicity."

Morland, K., Landrigan, P., Sjodin, A., et. al. (2005). Body burdens of polybrominated diphenyl ethers among urban anglers. *Environmental Health Perspectives*. 113(12). pg. 1689-92. http://www.ncbi.nlm.nih.gov/pmc/articles/PMC1314906/pdf/ehp0113-001689.pdf

Morris, S., Allchin, C. R., Zegers, B. N., et al. (2004). Distributon and fate of HBCD and TBBPA brominated flame retardants in North Sea estuaries and aquatic food webs. *Environ Sci Technol*. 38. pg. 5497-5504.

Morrissey, S. (2006). Evaluating the Safety of Nanotechnology. *Chemical & Engineering News*. 84(41).

Möller, A., Sturm, R., Xie, Z., et al. (2012). Organophosphorus flame retardants and plasticizers in airborne particles over the Northern Pacific and Indian Ocean toward the Polar regions: Evidence for global occurrence. *Environmental Science & Technology*. 46(6). pg. 3127-34. http://pubs.acs.org/doi/abs/10.1021/es204272v?journalCode=esthag

- "TCEP and TCPP were the predominating compounds, both over the Asian seas as well as in the polar regions, with concentrations from 19 to 2000 pg m^3 and 22 to 620 pg m^3, respectively."

Montie, E., Reddy, C., Gebbink, W., et. al. (2009). Organohalogen contaminants and metabolites in cerebrospinal fluid and cerebellum gray matter in short-beaked common dolphins and Atlantic white-sided dolphins from the western North Atlantic. *Environmental pollution*. 157(8-9). pg. 2345-58. http://www.ncbi.nlm.nih.gov/pubmed/19375836

Montie, E., Letcher, R., Reddy, C., et. al. (2010). Brominated flame retardants and organochlorine contaminants in winter flounder, harp and hooded seals, and North

Atlantic right whales from the Northwest Atlantic Ocean. *Marine Pollution Bulletin*. 60(8). pg. 1160-9. http://www.ncbi.nlm.nih.gov/pubmed/20434733

Mrakovcic, M., Absenger, M., Riedl, R., et al. (2013). Assessment of long-term effects of nanoparticles in a microcarrier cell culture system. *PLoS ONE*. 8(2). http://www.plosone.org/article/fetchObject.action?uri=info:doi/10.1371/journal.pone.0056791&representation=PDF

- "Properties of nanomaterials (NMs) change as their size approaches the nanoscale [3]. Because of quantum size and large surface area, NMs have unique properties compared with their larger counterparts. Even when made of inert elements (e.g. gold), NMs become highly (re)active or even catalytic at nanometer dimensions [4], mostly because of their high surface to volume ratio."

Muir, D. and Howard, P. (2006). Are there other persistent organic pollutants? A challenge for environmental chemists. *Environmental Science and Technology*. 40. pg. 7157-66. http://www.researchgate.net/publication/6619243_Are_there_other_persistent_organic_pollutants_A_challenge_for_environmental_chemists

- "The past 5 years have seen some major successes in terms of global measurement and regulation of persistent, bioaccumulative, and toxic (PB&T) chemicals and persistent organic pollutants (POPs). The Stockholm Convention, a global agreement on POPs, came into force in 2004."

Munke, J. (2009). Exposure to endocrine disrupting compounds via the food chain: Is packaging a relevant source? *Science of the Total Environment*. 407(16). pg. 4549-59. http://www.ncbi.nlm.nih.gov/pubmed/19482336

- "This article reviews…50 known or potential EDCs used in food contact materials and examined data of EDCs leaching from packaging into food…especially…bisphenol A."

- "The widespread legal use of EDCs in food packaging requires dedicated assessment."

Murphy, C. J. (2002). Materials science: Nanocubes and nanoboxes. *Science*. 298(5601). pg. 2139-41. https://www.sciencemag.org/content/298/5601/2139?related-urls=yes&legid=sci;298/5601/2139

Murray, F. and Cowie, P. (2011) Plastic contamination in the decapod crustacean *Nephrops norvegicus*. *Marine Pollution Bulletin*. 62(6). pg. 1207-17. http://www.wildlifeextra.com/resources/doc/misc/scampi.pdf

Myers, R. A. and Worm, B. (2003). Rapid worldwide depletion of predatory fish communities. *Nature*. 423. pg. 280-3. http://faculty.wwu.edu/~shulld/ESCI%20432/Myers&Worm2003.pdf

Naidenko, O., Nneka, L., Sharp, R. and Houlihan, J. (2008). *Bottled water quality investigation: 10 major brands, 38 pollutants*. Environmental Working Group. http://courses.washington.edu/h2owaste/assignments/ExtraBottled%20Water.pdf

Nalbone, J. (2014). *Unseen threat: How microbeads harm New York waters, wildlife, health, and environment*. Environmental Protection Bureau of the New York State Attorney General's Office. http://ag.ny.gov/pdfs/Microbeads_Report_5_14_14.pdf

Nakada, N., Nyunoya, H., Nakamura, M., et al. (2004). Identification of estrogenic compounds in wastewater effluent. *Environmental Toxicological Chemistry*. 23. pg. 2807-15. http://www.ncbi.nlm.nih.gov/pubmed/15648753

- "[Estrone] and [17beta-estradiol] were the dominant environmental estrogens in the STP effluent, but a significant contribution to estrogenic activities stems from unidentified components in the effluents."

- "Sewage treatment plant (STP) secondary effluent discharged to the Tamagawa River in Tokyo, Japan."

Nakashima, E., Isobe, A., Kako, S., et al. (2012). Toxic metals derived from plastic litter on a beach. In: *Interdisciplinary Studies on Environmental Chemistry—Environmental Pollution and Toxicology*. Kawaguchi, M., Misaki, H., Sato, T., et al. eds. pg. 321-8. http://www.terrapub.co.jp/onlineproceedings/ec/06/pdf/PR639.pdf

Nakata, H., Kawazoe, M., Arizono, K., et al. (2002). Organochlorine pesticides and polychlorinated biphenyl residues in foodstuffs and human tissues from China: Status of contamination, historical trend, and human dietary exposure. *Environmental Contamination and Toxicology*. 43(4). pg. 473-80. http://link.springer.com/article/10.1007/s00244-002-1254-8

- "Among the organochlorines analyzed, DDT and its metabolites were prominent compounds in most of the foodstuffs. In particular, mussels contained noticeable residues of DDTs (34,000 ng/g lipid weight), which are one to three orders greater than those reported levels in bivalves from other Asian countries."

- "Very high concentrations of DDTs and HCHs were detected in human tissues from Shanghai, with the maximum values as high as 19,000 ng/g lipid weight (mean: 7,600 ng/g) and 17,000 ng/g (mean: 7,400 ng/g), respectively."

Nanoaction. (2008). *Principles for the oversight of nanotechnologies and nanomaterials.* International Center for Technology Assessment. http://www.icta.org/files/2012/04/080112_ICTA_rev1.pdf

National Institute of Environment Health Sciences. (2010) *Endocrine disruptors.* NIEHS. http://www.niehs.nih.gov/health/materials/endocrine_disruptors_508.pdf

National Institute for Occupational Safety and Health. (2009). *Approaches to safe nanotechnology: Managing the health and safety concerns associated with engineered nanomaterials.* NIOSH (DHHS) Publication 2009-125. http://www.cdc.gov/niosh/docs/2009-125/

- "Nanoparticles can enter the blood stream, and translocate to other organs."

- "Research is needed to determine the key physical and chemical characteristics of nanoparticles that determine their hazard potential."

- "In the case of nanomaterials, the uncertainties are great because the characteristics of nanoparticles may be different from those of larger particles with the same chemical composition."

- "Inhalation is the most common route of exposure to airborne particles in the workplace."

- "Experimental studies in rats have shown that at equivalent mass doses, insoluble ultrafine particles of similar composition…[cause] pulmonary inflammation, tissue damage, and lung tumors."

National Institute for Occupational Safety and Health. (2010). *Current intelligence bulletin 65: Occupational exposure to carbon nanotubes and nanofibers.* Centers for Disease Control. http://www.cdc.gov/niosh/docs/2013-145/pdfs/2013-145.pdf

National Institute for Occupational Safety and Health. (2012). *World Trade Center chemicals of potential concern and selected other chemical agents.* Centers for Disease Control. http://www.cdc.gov/niosh/docs/2012-115/pdfs/2012-115.pdf

- "A total of 287 chemicals or chemical groups were identified from the [Contaminants of Potential Concern (COPC) Committee] report [2003]."

- "Because of the potential exposure of rescue and recovery workers, as well as workers and residents in area buildings, to soot, biomass fuel, and diesel particulates, these are included with the COPC agents identified by EPA."

National Institute for Occupational Safety and Health. (2014). *NIOSH list of antineoplastic and other hazardous drugs in healthcare settings, 2014.* Publication no.

2014-138. Centers for Disease Control and Prevention.
http://www.cdc.gov/niosh/docs/2014-138/pdfs/2014-138_v3.pdf

- "The National Institute for Occupational Safety and Health (NIOSH) Alert: Preventing Occupational Exposures to Antineoplastic and Other Hazardous Drugs in Health Care Settings was published in September 2004, http://www.cdc.gov/niosh/docs/2004-165/. In Appendix A of the Alert, NIOSH identified a sample list of major hazardous drugs."

- "The current update (2014) adds 27 drugs and includes a review of the 2004 list and the consequent removal of 12 drugs that did not meet the NIOSH criteria for hazardous drugs."

- Small quantities of all of the hundreds of drugs listed in this publication eventually become nanocomponents of riverine washout to marine ecosystems via the vector of sewage treatment and landfill washout events.

National Institute for Occupational Safety and Health. (2015). *NIOSH numbered publications*. Centers for Disease Control and Prevention.
http://www.cdc.gov/niosh/pubs/all_date_desc_nopubnumbers.html

- "A complete list of all publications issued by NIOSH."

- Out of over 400 publications issued from April 9, 2009, other than "Skin notation profiles" (26 chemicals, 26 reports), NIOSH issued one report on "Methylene chloride hazards for bathtub refinishers," a flavoring-related lung disease report (May 2012), two hazardous chemicals reports, and one pocket guide to hazardous chemicals.

- Most of the hundreds of reports related to memory, noise, stress, falls, motor vehicles, and pathogens. Eight reports were on nanotechnology safety diseases.

- None of the reports referenced POPs, HDCs, PBTs, or any other form of chemical fallout.

- No mention is made of dioxin or toxic emissions from landfills or municipal waste incinerators.

National Institute for Public Health and the Environment. (2014). *Size and amount of microplastics in toothpastes*. National Institute for Public Health and the Environment, New Zealand Ministry of Health, Welfare and Sport.
http://www.beatthemicrobead.org/images/pdf/andere_pdfs/007514_Microplastics_poster_V1-2.pdf

National Oceanic and Atmospheric Administration. (2012a). *Gulf of Mexico 'dead zone' predictions feature uncertainty*. NOAA. http://www.noaanews.noaa.gov/stories2012/20120621_deadzone.html

National Oceanic and Atmospheric Administration. (2012b). *Proceedings of the Second Workshop on Microplastic Marine Debris: November 5-6, 2010*. Technical Memorandum NOS-OR&R-39. NOAA Marine Debris Program, Silver Spring, MD. http://marinedebris.noaa.gov/sites/default/files/microplastics_workshop_2012.pdf

- This report contains no new information on, or research results, about the fate of microplastics in the marine environment.

- "The risk of microplastic particles to marine systems is still unclear." This quote summarizes the evasion by NOAA of any attempt to track the evolution of microplastics into toxin sorbing plastic nanoparticles (PNP) and their movement through the marine food chain.

National Oceanic and Atmospheric Administration. (2014a). *NOAA's marine debris blog*. NOAA. https://marinedebrisblog.wordpress.com/

- "Arctic sea ice from remote locations contains concentrations of microplastics at least two orders of magnitude greater than those that have been previously reported in highly contaminated surface waters, such as those of the Pacific Gyre."

- "One main point of agreement across the board was the evolving role of marine plastics in the transfer of persistent organic pollutants to organisms."

National Oceanic and Atmospheric Administration. (2014b). *Historical overview (1800s-present): How has the red snapper fishery changed over time*? NOAA Southeast Regional Office. http://sero.nmfs.noaa.gov/sustainable_fisheries/gulf_fisheries/red_snapper/overview/

National Oceanic and Atmospheric Administration Marine Debris Program. (2014c). *2014 report on the occurrence and health effects of anthropogenic debris ingested by marine organisms*. NOAA. Silver Spring, MD. http://marinedebris.noaa.gov/sites/default/files/mdp_ingestion.pdf

- "Physical and chemical processes can degrade some types of plastic in as little as a few weeks, while other pieces last for decades."

- "It has been estimated in the North Sea that, eventually, 15% of plastic debris washes ashore, 15% floats at the surface, and 70% will sink to the sea floor over an extended amount of time."

- "Future work should identify the location and concentration of debris at depths where organisms are actively feeding in order to assess the potential impacts on those animals actively feeding in these areas."

- "Debris that has already made its way into the ocean will continue to impact the organisms therein for decades to come."

National Research Council. (2008). *Tackling marine debris in the 21st century*. The National Academies Press, Washington, DC. http://docs.lib.noaa.gov/noaa_documents/NOAA_related_docs/marine_debris_2008.pdf

National Resources Defense Council. (2005a). *New Orleans environmental quality test results*. NRDC. Available online at: http://www.nrdc.org/health/effects/katrinadata/contents.asp.

National Resources Defense Council. (2005b). *Sampling results: Bywater/Marigny including agriculture street landfill*. NRDC. Available online at: http://www.nrdc.org/health/effects/katrinadata/bywater.asp.

National Resources Defense Council. (2011). *Your computer's lifetime journey*. NRDC. http://www.nrdc.org/living/stuff/your-computers-lifetime-journey.asp

National Toxicology Program. (2014). *Report on carcinogens, 13th edition*. US Department of Health and Human Services. http://ntp.niehs.nih.gov/pubhealth/roc/roc13/index.html

Needham, L. L., Barr, D. B., Caudill, S. P., et al. (2005). Concentrations of environmental chemicals associated with neurodevelopmental effects in U.S. population. *Neurotoxicology*. 26. pg. 531-45.

Nel, A., Xia, T., Madler, L. and Li, N. (2006). Toxic potential of materials at the nanolevel. *Science*. 311(5761). pg. 622-7. http://www.ncbi.nlm.nih.gov/pubmed/16456071

Nelson, J., Carney, L., Wille-Irmiter, J., et al. (2011). *Life-cycle impact of toner and ink for CU-boulder*. Sustainable Solutions Consulting, Spring. http://www.colorado.edu/envs/sites/default/files/attached-files/Printer%20Project%20Chapter%203%20%20Life%20Cycle%20Impact%20of%20Toner%20and%20Ink.pdf

Nevison, Cynthia D. (2014). A comparison of temporal trends in United States autism prevalence to trends in suspected environmental factors. *Environmental Health*. 13(73). http://www.ncbi.nlm.nih.gov/pmc/articles/PMC4177682/

- "Children's exposure to most of the top ten toxic compounds has remained flat or decreased over this same time frame."

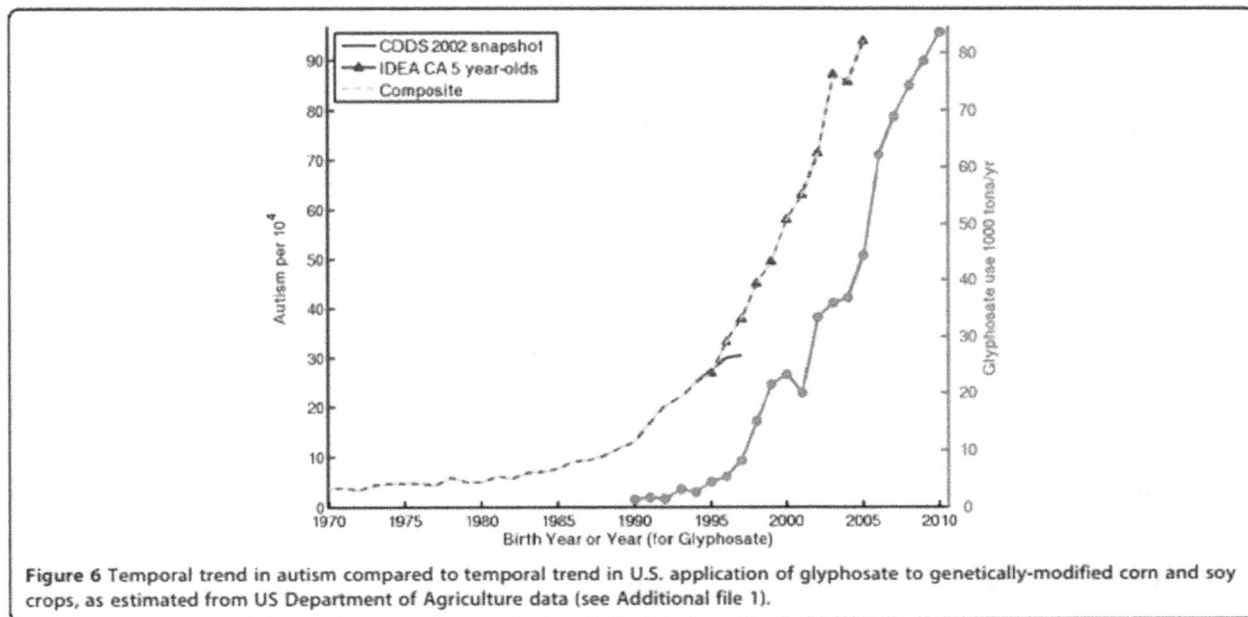

Figure 6 Temporal trend in autism compared to temporal trend in U.S. application of glyphosate to genetically-modified corn and soy crops, as estimated from US Department of Agriculture data (see Additional file 1).

Newbold, R. R., Padilla-Banks, E., Snyder, R. J., et al. (2007a). Perinatal exposure to environmental estrogens and the development of obesity. *Mol Nutr Food Res.* 51(7). pg. 912-7.

Newbold, R. R., Padilla-Banks, E., Snyder, R. J., et al. (2007b). Developmental exposure to endocrine disruptors and the obesity epidemic. *Reprod. Toxicol.* 23(3). pg. 290–6. http://www.ncbi.nlm.nih.gov/pmc/articles/PMC1931509

Newbold, R. R., Padilla-Banks, E., Jefferson, W. and Heindall, J. (2008). Effects of endocrine disruptors on obesity. *International Journal of Andrology.* 31(2). pg. 201-8. http://onlinelibrary.wiley.com/doi/10.1111/j.1365-2605.2007.00858.x/pdf

- "Environmental chemicals with hormone-like activity can disrupt the programming of endocrine signalling pathways that are established during perinatal life and result in adverse consequences that may not be apparent until much later in life."

- "In this review, we summarize the literature reporting an association of EDCs and the development of obesity…and provide evidence that support the scientific term 'the developmental origins of adult disease.'"

- "The data summarized in this review support the idea that brief exposure early in life to environmental endocrine disrupting chemicals, especially those with oestrogenic activity like DES, increases body weight with age."

- "Public health risks can no longer be based on the assumption that overweight and obesity are just personal choices involving the quantity and kind of foods we eat combined with inactivity, but rather that complex events including exposure to environmental chemicals during development may be contributing to the obesity epidemic."

Newbold, Retha. (2010). Impact of environmental endocrine disrupting chemicals on the development of obesity. *Hormones*. 9(3). pg. 206-17. http://www.hormones.gr/pdf/HORMONES%202010%20206-217.pdf

- "Recent reports link exposure to environmental endocrine disrupting chemicals during development with adverse health consequences, including obesity and diabetes."

- "These particular diseases are quickly becoming significant public health problems and are fast reaching epidemic proportions worldwide."

- "Over the last 2 to 3 decades, the prevalence of obesity has risen dramatically in wealthy industrialized countries."

- "In the United States, the Center for Disease Control (CDC) reported in 2008 that obesity has reached epidemic proportions with more than 60% of U.S. adults being either obese or overweight."

- Also data "show a link between exposure to environmental chemicals (such as estrogenic chemicals, BPA, PCBs, DDE, and persistent organic pollutants and heavy metals) and the development of obesity."

Newton, I. (1979). *Population ecology of raptors*. T. & A.D. Poyser, Berhamsted, UK.

Nfon, E., Cousins, I. T. and Broman, D. (2008). Biomagnification of organic pollutants in benthic and pelagic marine food chains from the Baltic Sea. *Science of the Total Environment*. 397. pg. 190-204. http://www.sciencedirect.com/science/article/pii/S0048969708001976

Ng, K. and Obbard, J. (2006). Prevalence of microplastics in Singapore's coastal marine environment. *Marine Pollution Bulletin*. 52(7). pg. 761-7. http://www.ncbi.nlm.nih.gov/pubmed/16388828

Ng, C., Scheringer, M., Fenner, K., and Hungerbuhler, K. (2011). A framework for evaluating the contribution of transformation products to chemical persistence in the

environment. *Environmental Science and Technology*. 45. pg. 111-17.
http://pubs.acs.org/doi/pdf/10.1021/es1010237

- "Evidence indicates that transformation products can be as or even more persistent than their parent compounds and thus must be included in chemical assessment."

- "Identification and evaluation of all possibly relevant TPs is a formidable problem that cannot be solved routinely for thousands of chemicals."

NHANES. (2006). *1999–2000 public data release file documentation [article online], 2005*. http://www.cdc.gov/nchs/about/major/nhanes/nhanes99 – 00.htm

NHANES. (2006). *2000–2001 public data release file documentation [article online], 2005*. http://www.cdc.gov/nchs/about/major/nhanes/nhanes01– 02.htm

Nilsson, R. (2000). Endocrine modulators in the food chain and environment. *Toxicological Pathology*. 28(3). pg. 420-31.
http://tpx.sagepub.com/content/28/3/420.long

- "In view of the extremely low potency of most xenobiotic endocrine disruptors, in combination with very low exposure levels with respect to the general population, a role for such chemicals as a cause of various adverse endocrine effects in humans is biologically implausible."

- A dissenting opinion.

Norén, F. (2008). *Small plastic particles in coastal Swedish waters*. N-Research report, commissioned by KIMO Sweden.
http://www.kimointernational.org/WebData/Files/Small%20plastic%20particles%20in%20Swedish%20West%20Coast%20Waters.pdf

- "The amount of plastic particles, concentrated with a 80μm mesh net, was in the range of $150 - 2400$ per m^3."

- "A very high concentration, 102,000 per m^3 of plastic particles (diam. ~0.5 - 2mm) was found locally in the harbour outside a polyethene production plant."

Norén, K. and Meironyté, D. (2000). Certain organochlorine and organobromine contaminants in Swedish human milk in perspective of past 20–30 years. *Chemosphere*. 40. pg. 1111-23. http://www.ncbi.nlm.nih.gov/pubmed/10739053

- "The present study summarises the investigations of polychlorinated biphenyls (PCBs), naphthalenes (PCNs), dibenzo-p-dioxins (PCDDs), dibenzofurans (PCDFs), polybrominated diphenyl ethers (PBDEs) and pesticides (DDT, DDE, hexachlorobenzene, dieldrin) as well as methylsulfonyl metabolites of PCBs and DDE in human milk."

- "A decrease to the half of the original concentration was attained in the range of 4-17 yr periods. On the contrary to the organochlorine compounds, the concentrations of PBDEs have increased during the period 1972-1997."

North American Bird Conservation Initiative. (2013). *The state of the birds 2013 report on private lands, United States of America.* State of the Birds. http://www.stateofthebirds.org/2013/2013%20State%20of%20the%20Birds_low-res.pdf

Nowack, B. and Bucheli, T. (2007). Occurrence, behavior and effects of nanoparticles in the environment. *Environmental Pollution.* 150. pg. 5-22. http://www.pnl.gov/nano/pdf/science_direct.pdf

- "The increasing use of engineered nanoparticles in industrial and household applications will very likely lead to the release of such materials into the environment. Assessing the risks of these nanoparticles in the environment requires an understanding of their mobility, reactivity, ecotoxicity and persistency."

- "Today, nanoscale materials find use in a variety of different areas such as electronic, biomedical, pharmaceutical, osmotic, energy, environmental, catalytic and material applications."

- "Nanotechnology is defined as the understanding and control of matter at dimensions of roughly 1e100 nm, where unique physical properties make novel applications possible (EPA, 2007). NP are therefore considered substances that are less than 100 nm in size in more than one dimension. They can be spherical, tubular, or irregularly shaped and can exist in fused, aggregated or agglomerated forms."

- "A special class of unintentionally produced NP is composed of platinum and rhodium containing particles produced from automotive catalytic converters."

- "Of the large family of fullerenes, the buckminsterfullerene C60 is by far the most widely investigated. Fullerenes are mainly proposed to be used in fullerene polymer combinations, as thin films, in electro-optical devices and in biological applications."

- "Carbon nanotubes (CNT) are considered as the hottest topic in physics. Depending on the synthesis method, the technique used for the separation from the amorphous by-products, subsequent cleaning steps, and finally different functionalizations, a variety of different CNT are obtained that have very different properties. Especially biological and medical applications explore the potential of modifying the properties of CNT."

- "These NP are taken up by a wide variety of cells and are studied for their ability to cross the blood brain barrier."

- "Engineered inorganic NP cover a broad range of substances including elemental metals, metal oxides and metal salts. Elemental silver is used in many products as bactericide, whereas elemental gold is explored for many possible applications and its catalytic activity. The use of nanoscale zero-valent iron (nZVI) for groundwater remediation ranks as the most widely investigated environmental nanotechnological application. Metallic iron is very effective in degrading a wide variety of common contaminants such as chlorinated methanes, brominated methanes, trihalomethanes, chlorinated ethanes, chlorinated benzenes, other polychlorinated hydrocarbons, pesticides and dyes."

- "Nano-sized particles are taken up by a wide variety of mammalian cell types, are able to cross the cell membrane and become internalized. The uptake on NP is size-dependent. Aggregation and size-dependent sedimentation onto the cells or diffusion towards the cell were the main parameters determining uptake. The uptake occurs via endocytosis or by phagocytosis in specialized cells. One hypothesis is that the coating of the NP by protein in the growth medium results in conformational changes of the protein structure, which triggers the uptake into the cell by specialized structures, limiting uptake to NP below about 120 nm. Within the cells NP are stored in certain locations (e.g. inside vesicles, mitochondria) and are able to exert a toxic response. The small particle size, a large surface area and the ability to generate reactive oxygen species play a major role in toxicity of NP. Inflammation and fibrosis are effects observed on an organism level, whereas oxidative stress, antioxidant activity and cytotoxicity are observed effects on a cellular level. Several respiratory and cardiovascular diseases in humans are caused by BC. Ultra-fine soot globules migrate deep into the lungs and carry very toxic, often carcinogenic compounds such as polycyclic aromatic hydrocarbons (PAH)."

- "Ecotoxicological studies show that NP are also toxic to aquatic organisms, both unicellular (e.g. bacteria or protozoa) and animals (e.g. Daphnia or fish). CNT induced a dose-dependent growth inhibition in a protozoan."

- "Release of NP may come from point sources such as production facilities, landfills or wastewater treatment plants or from nonpoint sources such as wear from materials containing NP. Accidental release during production or transport is also possible."

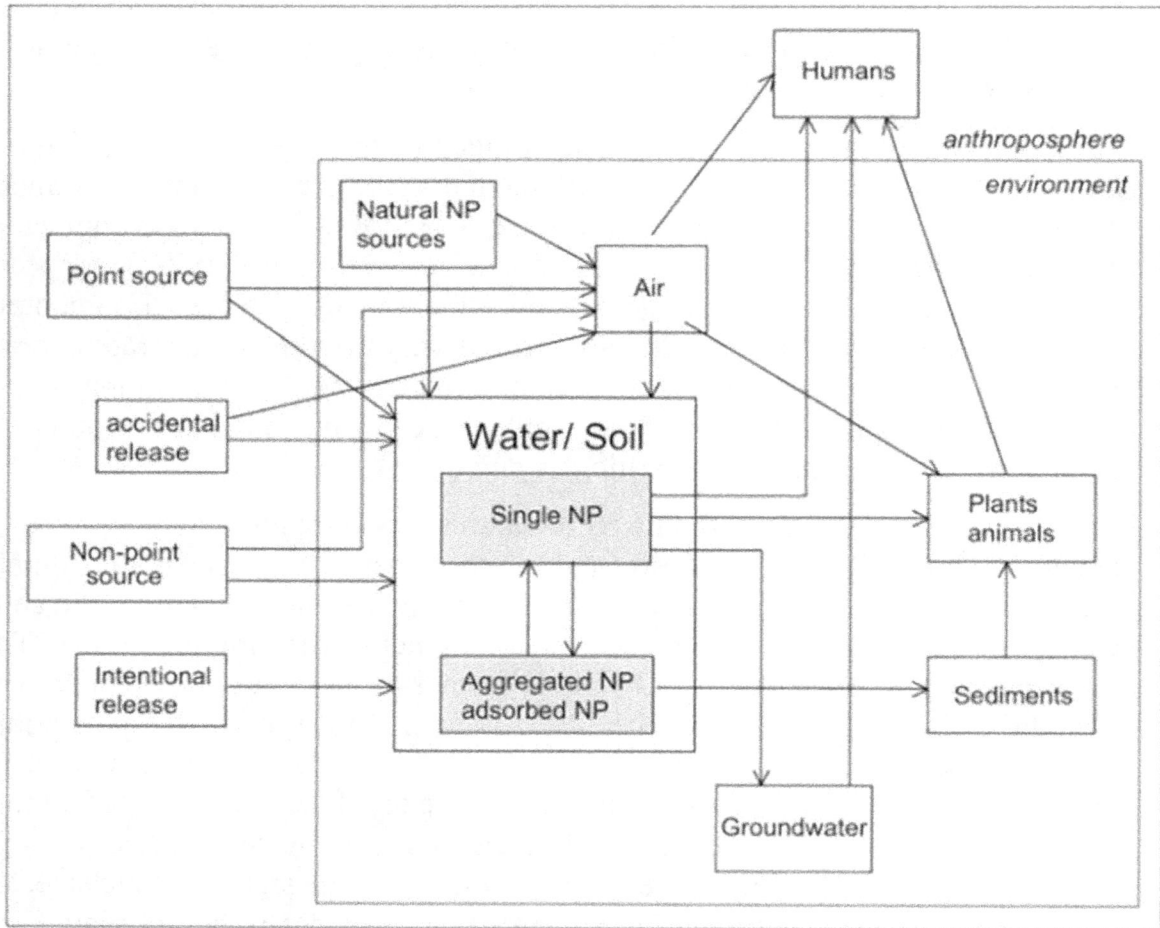

Fig. 2. Nanoparticle pathways from the anthroposphere into the environment, reactions in the environment and exposure of humans.

Nyman, M., Koistenen, J., Fant, M., et. al. (2002). Current levels of DDT, PCB and trace elements in the Baltic ringed seals (*Phoca hispida baltica*) and grey seals (*Halichoerus grypus*). *Environmental Pollution*. 119. pg. 399-412. http://www.ncbi.nlm.nih.gov/pubmed/12166673

Oberbeckmann, S., Loeder, M., Gerdts, G. and Osborn, M. (2014). Spatial and seasonal variation in diversity and structure of microbial biofilms on marine plastics in Northern European waters. *FEMS Microbiology Ecology*. 90. pg. 478-92. http://femsec.oxfordjournals.org/content/femsec/90/2/478.full.pdf

Oberdörster, Eva. (2004). Manufactured nanomaterials (fullerenes, C_{60}) induce oxidative stress in the brain of juvenile largemouth bass. *Environmental Health Perspectives*. 112. pg. 1058-62. http://infohouse.p2ric.org/ref/52/51402.pdf

- "Fullerenes are lipophilic and localize into lipid-rich regions such as cell membranes in vitro, and they are redox active."

- "Fullerenes (C_{60}) can form aqueous suspended colloids (nC_{60})."

- "Further research needs to be done to evaluate the potential toxicity of manufactured nanomaterials, especially with respect to translocation into the brain."

- "Engineered nanomaterials are useful because of their large surface area:mass ratio, which makes them important as catalysts in chemical reactions, and they have desirable properties as drug delivery devices, as imaging agents in medicine, and in consumer products such as sunscreens and cosmetics."

- "Quantum dots with cadmium selenium cores were initially rendered nontoxic with coatings, but if the quantum dots were either exposed to air or ultraviolet radiation for as little as 30 min, they became extremely cytotoxic."

Oberdörster, G. Sharp, Z., Atudorei, V., et al. (2004). Translocation of inhaled ultrafine particles to the brain. *Inhalation Toxicology*. 15. pg. 437-45. http://www.ncbi.nlm.nih.gov/pubmed/15204759

- "Ultrafine particles (UFP, particles <100 nm) are ubiquitous in ambient urban and indoor air from multiple sources and may contribute to adverse respiratory and cardiovascular effects of particulate matter (PM). Depending on their particle size, inhaled UFP are efficiently deposited in nasal, tracheobronchial, and alveolar regions due to diffusion."

Oberdörster, G., Maynard, A., Donaldson, K., et al. (2005a). Principles for characterising the potential human health effects from exposure to nanomaterials: Elements of a screening strategy. *Particle Fibre Toxicology*. 2(8). http://www.particleandfibretoxicology.com/content/pdf/1743-8977-2-8.pdf

Oberdörster, G., Oberdörster, E. and Oberdörster, J. (2005b). Nanotoxicology: An emerging discipline evolving from studies of ultrafine particles. *Environmental Health Perspectives*. 113(7). http://www.ncbi.nlm.nih.gov/pmc/articles/PMC1257642/pdf/ehp0113-000823.pdf

- "Although humans have been exposed to airborne nanosized particles (NSPs; < 100 nm) throughout their evolutionary stages, such exposure has increased dramatically over the last century due to anthropogenic sources. The rapidly developing field of nanotechnology is likely to become yet another source through inhalation, ingestion, skin uptake, and injection of engineered nanomaterials."

- "When inhaled, specific sizes of NSPs are efficiently deposited by diffusional mechanisms in all regions of the respiratory tract."

- "Access to the central nervous system and ganglia via translocation along axons and dendrites of neurons has also been observed."

- "Endocytosis and biokinetics are largely dependent on NSP surface chemistry (coating) and in vivo surface modifications. The greater surface area per mass compared with larger-sized particles of the same chemistry renders NSPs more active biologically."

Table 2. Particle number and particle surface area per 10 μg/m^3 airborne particles.

Particle diameter (μm)	Particle no. (cm^{-3})	Particle surface area (μm^2/cm^3)
5	153,000,000	12,000
20	2,400,000	3,016
250	1,200	240
5,000	0.15	12

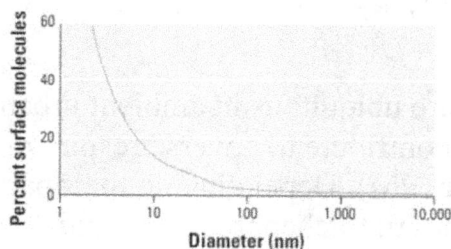

Figure 2. Surface molecules as a function of particle size. Surface molecules increase exponentially when particle size decreases < 100 nm, reflecting the importance of surface area for increased chemical and biologic activity of NSPs. The increased biologic activity can be positive and desirable (e.g., antioxidant activity, carrier capacity for therapeutics, penetration of cellular barriers), negative and undesirable (e.g., toxicity, induction of oxidative stress or of cellular dysfunction), or a mix of both. Figure courtesy of H. Fissan (personal communication).

Oberdörster, G., Stone, V. and Donaldson, K. (2007). Toxicology of nanoparticles: A historical perspective. *Nanotoxicology*. 1(1). pg. 2-25.
http://www.researchgate.net/publication/216213018_Toxicology_of_nanoparticles_A_historical_perspective

- "The propensity of NP to cross cell barriers, enter cells and interact with subcellular structures is well established, as is the induction of oxidative stress as a major mechanism of nanoparticle effects."

- A comprehensive review of the history of Nanotoxicology, defined as "dealing with effects and potential risks of particulate structures <100 nm in size."

Oberg, K., Warman, K. and Oberg, T. (2002). Distribution and levels of brominated flame retardants in sewage sludge. *Chemosphere*. 48. pg. 805-9. http://www.ncbi.nlm.nih.gov/pubmed/12222774

Öberg T. and Iqbal M. (2012). The chemical and environmental property space of REACH chemicals. *Chemosphere*. 87(8). pg. 975-81. http://www.researchgate.net/publication/221884857_The_chemical_and_environmental_property_space_of_REACH_chemicals

- "Long environmental half-lives were linked to halogenated substances in general and chlorinated and brominated aromatics and perfluorinated aliphatics in particular."

O'Brine, T. and Thompson, R. (2010). Degradation of plastic carrier bags in the marine environment. *Marine Pollution Bullein*. 60(12). pg. 2279-83. http://www.ncbi.nlm.nih.gov/pubmed/20961585

- "Our data indicate that compostable plastics may degrade relatively quickly compared to oxo-biodegradable and conventional plastics. While degradable polymers offer waste management solutions, there are limitations to their effectiveness in reducing hazards associated with plastic debris."

- Is this degradation chemical, or structural, as in metaplastic > mesoplastics > microplastics > nanoplastics > picoplastics?

- Where have all the plastics gone, since much of it is no longer visible?

Occupational Safety & Health Administration and National Institute for Occupational Safety and Health. (2015). *OSHA NIOSH hazard alert: Worker exposure to silica during hydraulic fracturing*. OSHA and NIOSH. https://www.osha.gov/dts/hazardalerts/hydraulic_frac_hazard_alert.html

- "NIOSH's recent field studies show that workers may be exposed to dust with high levels of respirable crystalline silica (called 'silica' in this Hazard Alert) during hydraulic fracturing."

- "*Respirable crystalline silica* is the portion of crystalline silica that is small enough to enter the gas-exchange regions of the lungs if inhaled; this includes particles with aerodynamic diameters less than approximately 10 micrometers (μm)."

OECD. (2014). *Obesity update: June 2014*. Organization for Economic Cooperation and Development. http://www.medicosypacientes.com/gestor/plantillas/articulos/archivos/imagenes/www.oecd.org_els_health-systems_Obesity-Update-2014.pdf

- "The majority of the population, and one in five children, are overweight or obese in the OECD area."

- "Overweight and obese people are a majority today in the OECD area. The obesity epidemic continues to spread, and no OECD country has seen a reversal of trends since the epidemic began. Until 1980, fewer than one in ten people were obese in OECD countries. In the following decades, rates doubled or tripled, and are continuing to grow."

Oehlmann, J., Schulte-Oehlmann, U., Kloas, W., et al. (2009). A critical analysis of the biological impacts of plasticizers on wildlife. *Philosophical Transcripts of the Royal Society*. 364(1526). pg. 2047-62. http://rstb.royalsocietypublishing.org/content/364/1526/2047

- "This review provides a critical analysis of the biological effects of the most widely used plasticizers, including dibutyl phthalate, diethylhexyl phthalate, dimethyl phthalate, butyl benzyl phthalate and bisphenol A (BPA), on wildlife, with a focus on annelids (both aquatic and terrestrial), molluscs, crustaceans, insects, fish and amphibians."

- "The most striking gaps in our current knowledge on the impacts of plasticizers on wildlife are the lack of data for long-term exposures to environmentally relevant concentrations and their ecotoxicity when part of complex mixtures."

- "In the 1950s, BPA was rediscovered when a Bayer chemist, Hermann Schnell, reacted BPA with phosgene to produce polycarbonate plastic, and its use in plastics has subsequently become its primary commercial application."

Officer, R., Thompson, R. and O'Connor, I. (2013). *Microplastic distribution and ecological interactions across latitudinal gradients*. MARES Doctoral Programme on Marine Ecosystem Health and Conservation. http://www.mares-eu.org/index.asp?p=1861&a=1853&mod=phd&id=172

Ogden, C., Flegal, K., Carroll, M. and Johnson, C. (2002). Prevalence and trends in overweight among US children and adolescents, 1999-2000. *JAMA*. 288. pg. 1728-32. http://jama.jamanetwork.com/article.aspx?articleid=195387

- "The prevalence of overweight was 15.5% among 12- through 19-year-olds, 15.3% among 6- through 11-year-olds, and 10.4% among 2- through 5-year-olds, compared with 10.5%, 11.3%, and 7.2%, respectively, in 1988-1994 (NHANES III). The prevalence of overweight among non-Hispanic black and Mexican-American adolescents increased more than 10 percentage points between 1988-1994 and 1999-2000."

Olson, Erik. (2005). *Statement of Erik D. Olson, Senior Attorney, Natural Resources Defense Council (on Hurricane Katrina)*. NRDC. https://www.nrdc.org/legislation/katrina/leg_05100601A.pdf

Ogata, Y., Takada, H., Mizukawa, K., et al. (2009). International pellet watch: Global monitoring of persistent organic pollutants (POPs) in coastal waters. 1. Initial phase data on PCBs, DDTs, and HCHs. *Marine Pollution Bulletin*. 58(10). pg. 1437-46. http://www.ncbi.nlm.nih.gov/pubmed/19635625

Orthodox Union. (2004). *OU fact sheet on NYC water: August 13, 2004*. Kosher Orthodox Union. https://oukosher.org/blog/consumer-kosher/ou-fact-sheet-on-nyc-water-august-13-2004/

OSPAR. (2010). *Quality status report 2010*. OSPAR. http://qsr2010.ospar.org/en/media/chapter_pdf/QSR_Ch01_EN.pdf

OSPAR. (2012). *CEMP 2011 Assessment Report*. OSPAR. http://www.ospar.org/documents/dbase/publications/p00563/p00563_cemp_2011_assessment_report.pdf

- In general, OSPARs (The Convention for the Protection of the Marine Environment for the North East Atlantic, 1992) coordinated environmental monitoring program (CEMP) documents only a miniscule component of the thousands of POPs (persistent organic pollutants) and PBTs (persistent bioaccumulative and toxic) chemicals now known to be present in the biotic environment.

- OSPAR has been issuing annual monitoring reports since 2005. Due to restrictions on the use of some POPs, contamination trends for some biologically significant contaminants have been declining. Almost all input reduction occurred before 2000.

- "The ban on the use of *gamma*-HCH in Europe is still resulting in declining concentrations in the environment, but concentration factors to fish are higher than to shellfish, on average a factor of 7 higher in fish liver compared to shellfish. One major exception is the Portuguese Praia da Barra, where concentrations are around 25 times higher than ordinary mussel concentrations (13 compared to 0.5 µg/kg)."

- "PCBs are however among the most prevalent pollutants in the Arctic and are widely distributed by long-range atmospheric transport… Concentrations are decreasing at a high proportion of the fish/shellfish stations, particularly along the continental coast of the North Sea, the west of the UK, and Ireland. A small number of stations showed increasing trends."

- "Polycyclic aromatic hydrocarbons (PAHs) are natural components of coal and oil and are also formed during the combustion of fossil fuels and organic material. They are one of the most widespread organic pollutants in the marine environment of the OSPAR area, entering the sea from offshore activities, operational and accidental oil spills from shipping, river discharges and the air. Long-range atmospheric transport is an important pathway for PAHs within and to the OSPAR area and is of regional and global concern."

- "PAHs are toxic and, since they are hydrophobic, bioconcentrate particularly in fatty tissues. They can adversely affect reproduction, and may affect immune systems so as to make disease epidemics worse."

- "Trends in PAH concentrations in fish and shellfish are predominantly downward, especially in Region III, but concentrations are still at levels which pose a risk of pollution effects in many estuaries and urbanised and industrialised locations."

Ostiguy, C., Soucy, B., Lapointe, G., et al. (2008). *Health effects of nanoparticles: Second edition*. Insitut de recherché Robert-Sauve en Sante et en Securite du Travail. http://www.irsst.qc.ca/en/-irsst-publication-health-effects-of-nanoparticles-second-edition-r-589.html

- "Insoluble or low-solubility nanoparticles in biological fluid are the greatest cause for concern. Because of their tiny size, several studies have shown behaviour unique to NP. Some of them can pass through our various defence mechanisms and be transported through the body in insoluble form."

- "These properties, extensively studied in pharmacology, could allow NP to be used as vectors to carry drugs to targeted body sites, including the brain. The corollary is that undesirable NP could be distributed through the bodies of exposed workers and has deleterious effects."

- "In the case of NP, it has been clearly shown that the measured effects are not linked to the mass of the product…at equal mass, NP are more toxic than products of the same chemical composition but of greater size."

- "Although major trends may emerge and show numerous toxic effects related to certain NP, it can be seen that each product, and even each synthesized NP batch, can have its own toxicity. Any process or surface modification can have an impact on the toxicity of the resulting product."

- "Chemical reactivity increases rapidly as particle size diminishes…Surface energy also rises by a factor of one million as size decreases from millimetres to nanometers."

- "On the nano-scale, factors such as specific surfaces, surface modifications, number of particles, surface properties…concentration, dimensions, structure are all factors that must be considered in toxicity assessment."

Oterhals, A. and Nygard, E. (2008). Reduction of persistent organic pollutants in fishmeal: A feasibility study. *Journal of Agricultural and Food Chemistry*. 56(6). pg. 2012-20. http://www.ncbi.nlm.nih.gov/pubmed/18284205

- "Dioxins and DL-PCBs are fat-soluble compounds…Soybean oil extraction of the press cake [fish meal] reduced the dioxin and DL-PCB content by 97%."

- "A new integrated fishmeal and fish oil production and decontamination process line is proposed."

Oterhals, A. (2011). *Decontamination of persistent organic pollutants in fishmeal and fish oil*. Dissertation for the degree philosophiae doctor (PhD) at the University of Bergen. https://bora.uib.no/bitstream/handle/1956/5129/Dr.thesis_Aage_Oterhals.pdf?sequence=1

- "POPs comprise pesticides, industrial chemicals and unwanted by-products…are fat soluble chemical substances that persist in the environment and bioaccumulate in the food chain."

- "Fish caught in some of the North-European fishing areas contain high dioxin and PCB levels resulting in fishmeal and oil with WHO-PCDD/F-PCB-TEQ levels above the maximum permitted. To meet the new industrial and socialeconomic challenges there is a need for development of cost-effective decontamination technologies."

Owen, R. and Depledge, M. (2005). Nanotechnology and the environment: Risks and rewards. *Marine Pollution Bulletin*. 50. pg. 609-612.
http://www.sciencedirect.com/science/article/pii/S0025326X05002043

Owen, R. and Handy, M. (2007). Formulating the problems for environmental risk assessment of nanomaterials. *Environmental Science and Technology*. 42. pg. 5582-8.
http://pubs.acs.org/doi/pdf/10.1021/es072598h

- "When nanomaterials and the environment are generally considered, in most cases the potential connectivity is currently unclear, and the supporting evidence is indirect or simply absent."

Pacheco, S., Tapia, J., Medina, M. and Rodriguez, R. (2006). Cadmium ions adsorption in simulated wastewater using structured alumina-silica nanoparticles. *Journal of NonCrystalline Solids*. pg. 5475-81.
http://www.sciencedirect.com/science/article/pii/S0022309306011495

Padmanabhan, V., Siefert, K., Ransom, S., et. al. (2008). Maternal bisphenol-A levels at delivery: A looming problem. *Journal of Perinatology*. 28(4). pg. 258-63.
http://www.ncbi.nlm.nih.gov/pmc/articles/PMC4033524/pdf/nihms559013.pdf

- "Maternal levels of unconjugated BPA ranged between 0.5 to 22.3 ng/mL in Southeastern Michigan mothers…this is the first study to document measurable levels of BPA in maternal blood of a U.S. population."

- "Every year, over 6 billion pounds of BPA are used in the manufacture of epoxy resins and polycarbonate plastics used in a wide variety of domestic products…[including] plastics…in dental fillings, plastic food and commercially available water containers, baby bottles, and food wrap, as well as in the lining of beverage and food cans."

- "BPA is particularly potent during fetal and neonatal development because the liver has limited capacity to deactivate BPA in fetuses and newborns, especially in humans."

Paget, Vincent, Dekali, Samir, Kortulewski, Thierry, et al. (2015). Specific uptake and genotoxicity induced by polystyrene Nanobeads with distinct surface chemistry on human lung epithelial cells and macrophages. *PLoS ONE*. 10(4). pg. 1-20.
http://journals.plos.org/plosone/article?id=10.1371/journal.pone.0123297

- "Nanoparticle surface chemistry is known to play a crucial role in interactions with cells and their related cytotoxic effects."

- "Our results strongly support the primordial role of nanoparticles surface chemistry on cellular uptake and related biological effects."

Painter, M., Buerkley, M., Julius, M., et al. (2009). Antidepressants at environmentally relevant concentrations affect predator avoidance behavior of larval fathead minnows (*Pimephales promelas*). *Environmental Toxicology and Chemistry*. 28(12). pg. 2677-84. http://onlinelibrary.wiley.com/doi/10.1897/08-556.1/pdf

Pakrashi, S., Dalai, S., Humayun, A., et al. (2013). *Ceriodaphnia dubia* as a potential bio-indicator for assessing acute aluminum oxide nanoparticle toxicity in fresh water environment. *PLoS ONE*. 8(9). http://www.plosone.org/article/fetchObject.action?uri=info:doi/10.1371/journal.pone.0074003&representation=PDF

Pan, J., Buffet, P., Poirier, L., et al. (2012). Size dependent bioaccumulation and ecotoxicity of gold nanoparticles in an endobenthic invertebrate: The Tellinid clam *Scrobicularia plana*. *Environmental Pollution*. 168. pg. 37-43. http://www.ffcsa-club.org/userfiles/File/Publication%20PAN%20Jinfen%20et%20al%202012.pdf

Pan, Z., Lee, W., Slutsky, L., et al. Adverse effects of titanium dioxide nanoparticles on human dermal fibroblasts and how to protect cells. *Small*. 5(4). pg. 511-20. http://www.ncbi.nlm.nih.gov/pubmed/19197964

- "The exposure to nanoparticles decreases cell area, cell proliferation, mobility, and ability to contract collagen. Individual particles are shown to penetrate easily through the cell membrane in the absence of endocytosis."

Panseri, S., Biondi, P. A., Vigo, D., et al. (2013). Occurrence of organochlorine pesticides residues in animal feed and fatty bovine tissue. In: Muzzalupo, Innocenzo, ed. *Food industry*.InTech. pg. 261-83. http://www.intechopen.com/books/food-industry/occurrence-of-organochlorine-pesticides-residues-in-animal-feed-and-fatty-bovine-tissue

- "Low volatility and high stability, together with lipophilic behaviour, are responsible critical factor for [organochlorine pesticides residues] persistence in the environment (air, water and soil) and subsequent concentration in fatty tissues through the food chain."

- Organochlorine pesticides (OCP) residues "… are present in fatty foods, both foods of animal origin such as meat, eggs and milk, and of plant origin such as vegetable oil, nuts, oat and olives."

Pardue, J. H., Moe, W. M. McInis, D. et al. 2005. Chemical and microbiological parameters in New Orleans floodwater following hurricane Katrina. *Environmental Science and Technology*. 39(22). pg. 8591-9.

Park, E., Yi, J., Chung, K.-H., et al. (2008). Oxidative stress and apoptosis induced by titanium dioxide nanoparticles in cultured BEAS-2B cells. *Toxicology Letters*. 180(3). pg. 222-9. http://www.ncbi.nlm.nih.gov/pubmed/18662754

- "Cytotoxicities of titanium dioxide nanoparticles of different concentrations (5, 10, 20 and 40 microg/ml) were evaluated…Exposure of the cultured cells to nanoparticles led to cell death, reactive oxygen species (ROS) increase, reduced glutathione (GSH) decrease, and the induction of oxidative stress-related genes."

- "Uptake of the nanoparticles into the cultured cells was observed and titanium dioxide nanoparticles seemed to penetrate into the cytoplasm and locate in the peri-region of the nucleus as aggregated particles."

Parner, E., Schendel, D. and Thorsen, P. (2008). Autism prevalence trends over time in Denmark. *Arcives of Pediatric and Adolescent Medicine*. 162(12). pg. 1150-6. http://archpedi.jamanetwork.com/article.aspx?articleid=380557

Pascall, M., Zabik, M., Zabik, M. and Hernandez, R. (2005). Uptake of polychlorinated biphenyls (PCBs) from an aqueous medium by polyethylene, polyvinyl chloride, and polystyrene films. *Journal of Agricultural Food Chemistry*. 53. pg. 164-9. http://pubs.acs.org/doi/abs/10.1021/jf048978t

Paul, A., Jones, K. and Sweetman, A. (2009). A first global production, emission, and environmental inventory for perfluorooctane sulfonate. *Environmental Science and Technology*. 43(2). pg. 385-92. http://pubs.acs.org/doi/abs/10.1021/es802216n

- "Direct emissions from POSF-derived products are the major source to the environment…primarily through losses from stain repellent treated carpets, waterproof apparel, and aqueous fire fighting foams…environmental monitoring from the 1970s onward shows strong upward trends in biota."

Pauly, D., Christensen, V., Dalsgaard, J., et al. (1998). Fishing down marine food webs. *Science*. 279(5352). pg. 860-3. http://umanitoba.ca/institutes/natural_resources/pdf/pauly_fishing_down_marine_food_webs.pdf

- "The mean trophic level of the species groups report in Food and Agricultural Organization global fisheries statistics declined from 1950 to 1994. This reflects a gradual transition in landings from long-lived, high trophic level, piscivorous bottom fish toward short-lived, low trophic level invertebrates and planktivorous pelagic fish."

- "Fishing down food webs (that is, at lower trophic levels) leads at first to increasing catches, then to a phase transition associated with stagnating or declining catches. These results indicate that present exploitation patterns are unsustainable."

- "Continuation of present trends will lead to widespread fisheries collapses."

Pauly, Daniel and Palomares, Maria-Lourdes. (2005). Fishing down marine food web: It is more pervasive than we thought. *Bulletin of Marine Science*. 76(2). pg. 197-211. http://www.seaaroundus.org/researcher/dpauly/PDF/2005/JournalArticles/FishingDown MarineFoodWebItisFarMorePervasivethanWeThought.pdf

Pelletier, C., Imbeault, P. and Tremblay, A. (2003). Energy balance and pollution by organochlorines and polychlorinated biphenyls. *Obesity Review*. 4(1). pg. 17-24. http://www.ncbi.nlm.nih.gov/pubmed/12608524

Pelley, J. and Saner, M. (2009). *International approaches to the regulatory governance of nanotechnology*. Regulation Paper, Regulatory Governance Initiative, Carleton University. http://www.nanowerk.com/nanotechnology/reports/reportpdf/report127.pdf

- "Our survey focuses on five key jurisdictions: the United States (US), the United Kingdom (UK), the European Union (EU), Australia, and Canada."

- "For each jurisdiction, we provide descriptions of the policy, regulatory and stewardship approaches undertaken to date in response to the emergence of nanotechnology onto the marketplace."

- "In the case of nanotechnology, there is currently only limited knowledge available regarding the potential health, safety, and environmental impacts of this technology."

- "In an article published in Wired magazine in April 2000, Sun Microsystems co-founder and prominent American scientist Bill Joy argued that technological advances in the fields of genetic engineering, robotics, and nanotechnology posed 4 significant risks that threatened the very existence of the human species."

- "Some dimensions of nanotechnology may pose potential risks to human health, worker safety, and the environment. However, at this time we cannot yet fully appreciate the precise nature, magnitude or frequency of such risks."

- No mention is made of the necessity of life cycle assessments of nanoparticles, especially their end of life behavior and fate after nano product disposal.

Pelley, J., Daar, Abdallah and Saner, M. (2009). State of academic knowledge on toxicity and biological fate of quantum dots. *Toxicological Sciences*. 112(2). pg. 276-96. http://toxsci.oxfordjournals.org/content/112/2/276.full.pdf+html

- "Quantum dots (QDs), an important class of emerging nanomaterial, are widely anticipated to find application in many consumer and clinical products in the near future... parameters that determine exposure (e.g., dosage, transformation, transportation, and persistence) are just as important as inherent toxicity…we summarize the state of academic knowledge on QDs pertaining not only to toxicity, but also their physicochemical properties, and their biological and environmental fate."

- "The properties of QDs make them potentially useful in a wide variety of settings, including electronics, computing, and various biomedical and clinical imaging applications."

- "The luminescent properties of QDs are being explored for use in next generation versions of light-emitting diodes and diode lasers. QDs are also being explored for potential applications in the emerging field of quantum computing."

- "Biomedical applications exploit the fluorescent properties of QDs, and particularly their advantages over traditional organic dyes, for both diagnostic and clinical applications. The *in vitro* biomedical and diagnostic applications of QDs include such techniques as the multicolor fluorescent labeling of cell surface molecules and cellular proteins in microscopy and other applications, detection of pathogens and toxins, DNA and RNA technologies, and fluorescence resonance energy transfer."

- "The results from a number of studies have indicated that the placement of molecules such as proteins onto the surface of QDs can greatly impact their pharmacokinetics and biodistribution."

Perkel, Jeffrey M. (2003). Nanoscience is out of the bottle. *The Scientist*. http://www.the-scientist.com/?articles.view/articleNo/14955/title/Nanoscience-is-Out-of-the-Bottle/

Petersen, E. J. and Nelson, B. C. (2010). Mechanisms and measurements of nanomaterial-induced oxidative damage to DNA. http://www.ncbi.nlm.nih.gov/pubmed/20563891

- "The potential of nanomaterials to directly or indirectly promote the formation of reactive oxygen species is one of the primary steps in their genotoxic repertoire."

Pflieger-Bruss, S., Schuppe, H. and Schill, W. (2004). The male reproductive system and its susceptibility to endocrine disrupting chemicals. *Andrologia*. 36. pg. 337-45. http://www.ncbi.nlm.nih.gov/pubmed/15541049

- "In the past years, there has been increased interest in assessing the relationship between impaired male fertility and environmental factors…recent studies suggest a correlation between pesticide exposure and standard semen parameters."

Pham, C., Ramirez-Llodra, E., Alt, C., et. al. (2014). Marine litter distribution and density in European seas, from the shelves to deep basins. *PLoS ONE*. 9(4). http://www.plosone.org/article/fetchObject.action?uri=info:doi/10.1371/journal.pone.0095839&representation=PDF

- "The highest litter density occurs in submarine canyons, whilst the lowest density can be found on continental shelves and on ocean ridges. Plastic was the most prevalent litter item found on the seafloor."

- There is no mention of microplastics or their evolution into plastic nanoparticles (PNPs).

Phan, D., Jansson, S. and Boily, J. (2014). Link between fly ash properties and polychlorinated organic pollutants formed during simulated municipal solid waste incineration. *Energy and Fuels*. 28(4). pg. 2761-9. http://www.researchgate.net/publication/263947392_Link_between_Fly_Ash_Properties_and_Polychlorinated_Organic_Pollutants_Formed_during_Simulated_Municipal_Solid_Waste_Incineration

Phillips, P. J. and Chalmers, A. T. (2009). Wastewater effluent, combined sewer overflows, and other sources of organic compounds to Lake Champlain. *Journal of the American Water Works Association*. 45(1). pg. 45-57. http://onlinelibrary.wiley.com/doi/10.1111/j.1752-1688.2008.00288.x/abstract;jsessionid=822DB9C3351425AB0CD670A2EFC26237.f02t03

Pignatello, J. and Xing, B. (1996). Mechanisms of slow sorption of organic chemicals to natural particles. *Environmental Science and Technology*. 30. pg. 1-11. http://ag.udel.edu/soilchem/Pignatello95EST.pdf

- "Sorption kinetics are extremely important in modeling the transport of contaminants in the subsurface."

Pimm, S. L., Russell, G. J., Gittleman, J. L. and Brooks, T. M. (1995). The future of biodiversity. *Science*. 269. pg. 347-50. http://www.montana.edu/screel/Webpages/conservation%20biology/pimm%20et%20al%20-%20biodiversity.pdf

Pinsky, M. L., Jensen, O. P., Ricard, D. et al. (2011). Unexpected patterns of fisheries collapse in the world's oceans. *Proc. Natl. Acad. Sci. U.S.A*. 108. pg. 8317-22.

Plastics Europe. (2013). *Plastics: The facts, 2013*. Association of Plastic Manufacturers. http://www.plasticseurope.org/documents/document/20131014095824-final_plastics_the_facts_2013_published_october2013.pdf

- "There are various types of plastics featuring different properties. The international recycling codes (ranging from 1 to 7) which are featured on most plastic products are meant to make (unmixed) separation easy."

Plumer, Brad. (2015). *We dump 8 million tons of plastic into the ocean each year. Where does it all go?* http://www.vox.com/2015/2/12/8028267/plastic-garbage-patch-oceans

Poland, C., Duffin, R., Kinloch, I., et al. (2008). Carbon nanotubes introduced into the abdominal cavity of mice show asbestos-like pathogenicity in a pilot study. *Nature Nanotechnology*. 3. pg. 423-8. http://www.nature.com/nnano/journal/v3/n7/full/nnano.2008.111.html

Polder, A., Savinova, T., Tkachev, A., et al. (2010). Levels and patterns of persistent organic pollutants (POPS) in selected food items from Northwest Russia (1998-2002) and implications for dietary exposure. *Science of the Total Environment*. 408(22). pg. 5352-61. http://www.ncbi.nlm.nih.gov/pubmed/20719362

- "Contamination of animal feed and agricultural practice were assumed the most important causes for the results in the present study."

- "Dairy products, meat products and fish were the main sources for the human exposure to POPs."

Pongratz, I. and Vikstrom, B. (2011). *Hormone-disruptive chemical contaminants in food*. Royal Society of Chemistry, Cambridge, UK.

Porter, A., Gass, M., Muller, K., et al. (2007). Direct imaging of single-walled carbon nanotubes in cells. *Nature Nanotechnology*. 2. pg. 713-7. http://www.nature.com/nnano/journal/v2/n11/full/nnano.2007.347.html

- "The development of single-walled carbon nanotubes for various biomedical applications is an area of great promise. However, the contradictory data on the toxic effects of single-walled carbon nanotubes highlight the need for alternative ways to study their uptake and cytotoxic effects in cells."

- "Single-walled carbon nanotubes have been shown to be acutely toxic in a number of types of cells, but the direct observation of cellular uptake of single-walled carbon nanotubes has not been demonstrated previously due to difficulties in discriminating carbon-based nanotubes from carbon-rich cell structures."

Posner, Richard A. (2006). *Catastrophe: Risk and response*. Oxford University Press, Oxford, UK.

Possatto, F., Barletta, M., Costa, M., et al. (2011). Plastic debris ingestion by marine catfish: An unexpected fisheries impact. *Marine Pollution Bulletin*. 62(5). pg. 1098-102. http://www.ncbi.nlm.nih.gov/pubmed/21354578

Presley, S. M., Rainwater, T. R. Austin, G. P. et al. (2006). Assessment of pathogens and toxicants in New Orleans, LA following Hurricane Katrina. *Environmental Science and Technology*. 40(2). pg. 468-74.

Preton Ltd. (2010). *Environmental issues associated with toner and ink usage*. Preton LTD, Tel Aviv, Israel.
http://www.preton.com/pdf/PretonSaver_envi_whitePaperFinal_1403010.pdf

- "Office printers have a major influence on indoor air quality. Studies show a clear rise in the concentration of ozone, VOCs and ultrafine particles during operation of printers as compared to idle mode. GHG emissions caused by manufacturing a single mono toner cartridge have been calculated to approximately 4.8 Kg of CO_2…each cartridge becomes 3.5 pounds of solid waste sitting in a landfill and can take up to 450 to 1000 years to decompose, as it includes mixed resin, one of the most difficult plastics to recycle."

- "Burning cartridges emits dioxins and polycyclic aromatic hydrocarbons (PAHs), both cancerous pollutants that pollute local rivers and lands, make their way into the food chain and affect all levels of species."

Price, H. J. (1988). Feeding mechanisms in marine and freshwater zooplankton. *Bulletin of Marine Science*. 43(3). pg. 327-43.

Priester, John H., Ge, Yuan, Mielke, Randall E., et al. (2012). Soybean susceptibility to manufactured nanomaterials with evidence for food quality and soil fertility interruption. *PNAS*. 109(37). pg. 14734-5.
http://www.pnas.org/content/109/37/E2451.full.pdf

- "Manufactured nanomaterial (MNM) environmental buildup could profoundly alter soil-based food crop quality and yield."

- "MNMs can enter soil through atmospheric routes, e.g., by nano-CeO_2 In fuel additives being released with diesel fuel combustion exhaust. Another route of entry is from biosolids treated in conventional wastewater treatment plants (WWTPs). Given that half of US biosolids are disposed to land, biosolids with MNMs will enter soils."

- "Juxtaposed against widespread land application of wastewater treatment biosolids to food crops, these findings forewarn of agriculturally associated human and environmental risks from the accelerating use of MNMs."

Purcell, S., Polidoro, B., Harnel, J., et al. (2014). The cost of being valuable: Predictors of extinction risk in marine invertebrates exploited as luxury seafood. *Proceedings of the Royal Society*.
http://www.esf.edu/efb/parry/Invert_Cons_14_Readings/Purcell_etal_2014.pdf

- "77 known species of sea cucumber…are exploited worldwide as luxury seafood for Asian markets. Extinction risk was primarily driven by high market value, compounded by accessibility and familiarity (well known) in the marketplace."

- "Our finding that low-income countries have many threatened species to manage illustrates a dilemma that exacerbates those constraints—threats to biodiversity loss are greatest where capacity is weakest to manage them."

- "Whereas the USA, the European Union and Japan are the main final destinations of most biodiversity-implicated commodities, harvested sea cucumbers are mostly destined for China."

Radjenovic, J., Petrovic, M. and Barcelo, D. (2009). Fate and distribution of pharmaceuticals in wastewater and sewage sludge of the conventional activated sludge (CAS) and advanced membrane bioreactor (MBR) treatment. *Water Research*. 43(3). pg. 831-41. http://www.sciencedirect.com/science/article/pii/S0043135408005642

- "The most ubiquitous contaminants in the sewage water were analgesics and anti-inflammatory drugs ibuprofen (14.6–31.3 µg/L) and acetaminophen (7.1–11.4 µg/L), antibiotic ofloxacin (0.89–31.7 µg/L), lipid regulators gemfibrozil (2.0–5.9 µg/L) and bezafibrate (1.9–29.8 µg/L), β-blocker atenolol (0.84–2.8 µg/L), hypoglycaemic agent glibenclamide (0.12–15.9 µg/L) and a diuretic hydrochlorothiazide (2.3–4.8 µg/L)."

- "None of the residual pharmaceuticals initially detected in the sewage sludge were degraded during the anaerobic digestion. Out of the 26 pharmaceutical residues passing through the WWTP, 20 were ultimately detected in the treated sludge that is further applied on farmland."

Rahman, F., Langford, K., Scrimshaw, M. and Lester, J. (2001). Polybrominated diphenyl ether (PBDE) flame retardants. *Science of the Total Environment*. 275(1). pg. 1-17. http://www.ncbi.nlm.nih.gov/pubmed/11482396

Ramakrishna, S., Fujihara, K., Teo, W.-E., et al. (2006). Electrospun nanofibers: Solving global issues. *Materials Today*. 9(3). pg. 40-50. http://ac.els-cdn.com/S136970210671389X/1-s2.0-S136970210671389X-main.pdf?_tid=5c026100-8c6e-11e4-8d6d-00000aacb35d&acdnat=1419536814_36338799abd3f364e9c17569d6a8b4b5

- "Nanofibers are able to form a highly porous mesh and their large surface-to-volume ratio improves performance for many applications. Electrospinning has the unique ability to produce nanofibers of different materials in various fibrous assemblies. The relatively high production rate and simplicity of the setup makes electrospinning highly attractive to both academia and industry. A variety of nanofibers can be made for applications in energy storage, healthcare, biotechnology, environmental engineering, and defense and security."

Ramirez-Llodra, E., Tyler, P., Baker, M., et al. (2011). Man and the last great wilderness: Human impact on the deep sea. *PLoS ONE*. 6(7). http://www.plosone.org/article/fetchObject.action?uri=info:doi/10.1371/journal.pone.0022588&representation=PDF

Ramu, K., Kajiwara, N., Lam, P., et al. (2006). Temporal variation and biomagnification of organohalogen compounds in finless porpoises (*Neophocaena phocaenoides)* from the South China Sea. *Environmental Pollution*. 144. pg. 516-23. https://swfsc.noaa.gov/uploadedFiles/Divisions/PRD/Publications/Ramuetal.06(91).pdf

- "Levels of organochlorine compounds (OCs) in various environmental matrices show declining trends due to the ban on their use and production."

- "PBDE concentrations in the environment increased substantially since the large scale PBDE production began in the early 1970s."

- "Lack of regulation on the use of PBDEs and the dumping of e-waste from the developed nations along the South China coast for recycling suggest that the situation may continue to deteriorate in this region."

Rani, B., Singh, U., Chuhan, A., et al. (2011). Photochemical smog pollution and its mitigation measures. *Journal of Advanced Scientific Research*. 2(4). pg. 28-33. http://www.sciensage.info/journal/1359642715JASR_2609111.pdf

Raun, A. P. and Preston, R. L. (2002). *History of diethylstilbestrol use in cattle*. American Society of Animal Science, Savoy, IL.

Ray, P., Yu, H. and Fu, P. (2009). Toxicity and environmental risks of nanomaterials: Challenges and future needs. *Journal of Environmental Science and Health*. 27. pg. 1-35. http://www.ncbi.nlm.nih.gov/pmc/articles/PMC2844666/pdf/nihms181469.pdf

- "There are still huge gaps in knowledge about the nature of interaction of nanoparticles with the environmental system…the potential for human exposure to the nanoscale components of commercially available products, as well as future products."

- "Many of the surfactants used to control size and shape of the nanomaterials are toxic."

- "To assess the safety of complex multi-component and multi-functional nanomaterials, scientists will need to develop validated models capable of predicting the release, transport, transformation, accumulation, and uptake of engineered nanomaterials in the environment."

- "Developing structure-activity relationships is needed to predict biological impacts, ecological impacts, and degradation at end-of-life."

- "Advances in information technology and sensor design should lead to the development of smart sensors that detect nanoparticle concentrations and determine their potential toxicity, possibly providing early indications of harm."

- "Have we learned anything new in biological toxicity mechanism because of these toxicological studies? The answer, with several exceptions, is not really."

- Poorly written, disorganized, and redundant, but with a very comprehensive bibliography.

Rayne, S., Ikonomou, M. G., Ross, P. S., et al. (2004). PBDEs, PBBs, and PCNs in three communities of free-ranging killer whales (*Orcinus orca*) from the Northeastern Pacific Ocean. *Environmental Science and Technology*. 38. pg. 4293-9.

Raz, R., Roberts, A., Lyall, K., et al. (2014). Autism spectrum disorder and particulate matter air pollution before, during, and after pregnancy: A nested case–control analysis within the Nurses' Health Study II cohort. *Environmental Health Science*. http://ehp.niehs.nih.gov/wp-content/uploads/advpub/2014/12/ehp.1408133.acco.pdf

Reed, J., Koenig, C., Shepard, A. and Gilmore, R. (2007). Long term monitoring of a deep-water coral reef: Effects of bottom trawling. In: Pollock, N. W. and Godfrey, J. M. eds. *Diving for Science 2007. Proceedings of the American Academy of Underwater Sciences 26th Symposium*. Dauphin Island, AL. http://archive.rubicon-foundation.org/xmlui/bitstream/handle/123456789/7004/AAUS_2007_18.pdf?sequence=1

- "Recent quantitative analyses… reveal drastic loss of live coral cover between 1975 and present."

Reible, D., Haas, C., Pardue, J. and Walsh, W. (2006). Toxic and contaminant concerns generated by Hurricane Katrina. *The Bridge*. https://www.nae.edu/File.aspx?id=7393

- "Several chemical plants, petroleum refining facilities, and contaminated sites, including Superfund sites, were covered by floodwaters. In addition, hundreds of commercial establishments, such as service stations, pest control businesses, and dry cleaners, may have released potentially hazardous chemicals into the floodwaters."

- "Figure 1 shows potential petroleum-related release points, including refineries, oil and gas wells, and service stations near the city. Figure 2 shows the major hazardous-materials storage locations, Superfund sites, and Toxic Release Inventory reporting facilities."

- "Adding to the potential sources of toxics and environmental contaminants are metal-contaminated soils typical of old urban areas and construction lumber preserved with creosote, pentachlorophenol, and arsenic. Compounding these concerns is the presence of hazardous chemicals commonly stored in households and the fuel and motor oil in approximately 400,000 flooded automobiles."

- "In the confusion immediately after the flooding, the amount of contamination was not known."

- "Initial concerns in the city were focused on acute exposures for stranded residents and relief workers. Subsequent efforts have been focused on acute exposures for returning residents and initial assessments of chronic exposures."

- This report contains no media specific data about contamination released during the flooding.

Reijnders, L. (2006). Cleaner nanotechnology and hazard reduction of manufactured nanoparticles. Journal of Cleaner Production. 14. pg. 124-33. http://www.sciencedirect.com/science/article/pii/S0959652605000867

Reisser, J., Shaw, J., Wilcox, C., et al. (2013). Marine plastic pollution in waters around Australia: Characteristics, concentrations, pathways. *PLoS ONE*. 8(11). http://marinedebris.info/sites/default/files/archives/Marine%20Plastic%20Pollution%20in%20Waters%20around%20Australia%20Characteristics,%20Concentrations,%20and%20Pathways.pdf

Reisser, J., Proietti, M., Shaw, J. and Pattiaratchi, C. (2014a). Ingestion of plastics at sea: Does debris size really matter? *Frontiers in Marine Science*. 1(70). http://www.researchgate.net/publication/268516584_Ingestion_of_plastics_at_sea_does _debris_size_really_matter

- "Some of these plastic surface textures are feeding marks produced by invertebrates grazing upon the plastic biofilm."

- "Copepods are an abundant planktivorous group and possess strong feeding apparatuses to feed upon organisms such as diatoms."

- "Copepods could also feed upon biofilm of plastic debris, which is often rich in 'epiplastic' diatoms."

- "Our hypotheses are that (1) plastic biofouling induces plastic ingestion, and (2) plastic pieces must not necessarily be smaller than the organism for a feeding interaction to occur."

Reisser, J., Shaw, J., Hallegraeff, G., et al. (2014b). Millimeter-sized marine plastics: A new pelagic habitat for microorganisms and invertebrates. *PLoS ONE*. 9(6). http://www.plosone.org/article/fetchObject.action?uri=info:doi/10.1371/journal.pone.01 00289&representation=PDF

Remillard, R. B. and Bunce, N. J. (2002). Linking dioxins to diabetes: Epidemiology and biologic plausibility. *Environ Health Perspect*. 110. pg. 853-8.

Renner, R. (2000). Increasing levels of flame retardants found in North American environment. *Environmental Science and Technology*. 34(21). pg. 452A-3. http://www.ncbi.nlm.nih.gov/pubmed/21662258

Ribic, C., Sheavly, S., Rugg, D. and Erdmann, E. (2010). Trends and drivers of marine debris on the Atlantic coast of the United States 1997-2007. *Marine Pollution Bulletin*. 60. pg. 1231-42. http://www.sciencedirect.com/science/article/pii/S0025326X10001116

Ribic, C., Sheavly, S. and Rugg, D. (2011). Trends in marine debris in the U.S. Caribbean and the Gulf of Mexico. *Journal of Integrated Coastal Zone Management*. 11(1). pg. 7-19. http://www.aprh.pt/rgci/pdf/rgci-181_Ribic.pdf

- "Debris loads on beaches in the Gulf of Mexico are likely affected by Gulf circulation patterns, reducing loads in the eastern Gulf and increasing loads in the western Gulf."

Ribic, C., Sheavly, S., Klavitter, J. (2012). Baseline for beached marine debris on Sand Island, Midway Atoll. *Marine Pollution Bulletin*. 64. pg. 1726-9. http://digitalcommons.unl.edu/cgi/viewcontent.cgi?article=1437&context=usfwspubs

Ribic, C., Sheavly, S., Rugg, D. and Erdmann, E. (2012). Trends in marine debris along the U.S. Pacific Coast and Hawaii 1998-2007. *Marine Pollution Bulletin*. 64. pg. 994-1004. http://www.ncbi.nlm.nih.gov/pubmed/22385753

- "Hawai'i had the highest debris loads; the North Pacific Coast region had the lowest debris loads."

- "Debris loads decreased over time for all source categories in all regions except for land-based and general-source loads in the North Pacific Coast region, which were unchanged."

- How much of the debris was plastics? Where have all the plastics gone?

Rice, M. R. and Gold, H. S. (1984). Polypropylene as an adsorbent for trace organics in water. *Analytical Chemistry*. 56. pg. 1436-40. http://www.researchgate.net/publication/244455372_Polypropylene_as_an_adsorbent_for_trace_organics_in_water

Rico-Martinez, R., Snell, T. and Shearer, T. (2013). Synergistic toxicity of Macondo crude oil and dispersant Corexit 9500A® to the Brachionus plicatilis species complex (Rotifera). *Environmental Pollution*. 173. pg. 5-10. http://www.ncbi.nlm.nih.gov/pubmed/23195520

Rietman, E. A. (2001). *Molecular engineering of nanosystems*. Springer, Berlin. Heidelberg, Germany.

Rillig, Matthew. (2012). Microplastic in terrestrial ecosystems and the soil? *Environmental Science and Technology*. 46(12). pg. 6453-4. http://pubs.acs.org/doi/full/10.1021/es302011r

- "These microplastics present a new set of issues, because of two main reasons: (i) they are small enough to be taken up by biota and thus can accumulate in the food chain; and (ii) they can sorb pollutants on their surfaces, thus enriching them on these particles."

- "Secondary microplastics could result from abrasion of plastic debris at soil surfaces (where UV light could render the material brittle) or inside the soil profile. In agricultural fields where plastic mulching is practiced, an abundant source of plastic material would be available; in other cases, incidental plastic debris would be the starting material. Curiously, even washing machines can produce secondary microplastic fibers; via water treatment plants these could end up on agricultural fields."

- "Microplastic could be ingested also by micro- and mesofauna, such as mites, collembola, or enchytraeids, and thus accumulate in the soil detrital food web."

Rios, L., Moore, C. and Jones, P. (2007). Persistent organic pollutants carried by synthetic polymers in the ocean environment. *Marine Pollution Bulletin*. 54(8). pg. 1230-7. http://5gyres.org/media/Persistent_organic_pollutants.pdf

- "Thermoplastic resin pellets are melted and formed into an enormous number of inexpensive consumer goods, many of which are discarded after a relatively short period of use, dropped haphazardly onto watersheds and then make their way to the ocean where some get ingested by marine life."

- "The total concentration of PCBs ranged from 27 to 980 ng/g; DDTs from 22 to 7100 ng/g and PAHs from 39 to 1200 ng/g, and aliphatic hydrocarbons from 1.1 to 8600 lg/g."

- "The results of this study confirm that plastic debris is a trap for POPs…One pound of the most common pellets costs about $1US and contains approximately 25,000 pellets."

Rios, L., Jones, P., Moore, C. and Narayan, U. (2010). Quantification of persistent organic pollutants adsorbed on plastic debris from the Northern Pacific Gyre's "eastern garbage patch." *Journal of Environmental Monitoring*. 12. pg. 2226-36. http://www.ncbi.nlm.nih.gov/pubmed/21042605

Risom, L., Møller, P. and Loft, S. (2005). Oxidative stress-induced DNA damage by particulate air pollution. *Mutation Research*. 592(1-2). pg. 119-37. http://www.ncbi.nlm.nih.gov/pubmed/16085126

Robinson, Brett. (2009). E-waste: An assessment of global production and environmental impacts. *Science of the Total Environment*. 408. pg. 183-91. http://kiwiscience.com/JournalArticles/STOTEN2009.pdf

- "E-waste contains valuable metals (Cu, platinum group) as well as potential environmental contaminants, especially Pb, Sb, Hg, Cd, Ni, polybrominated diphenyl ethers (PBDEs), and polychlorinated biphenyls (PCBs)."

- "Most E-waste is disposed in landfills. Effective reprocessing technology, which recovers the valuable materials with minimal environmental impact, is expensive."

List of common Waste Electrical and Electronic Equipment (WEEE) items, including those normally considered as E-waste.

Item	Wt of Item (kg)	Typical life (year)
WEEE normally considered E-waste		
Computer[a]	25	3
Facsimile machine	3	5
High-fidelity system[b]	10	10
Mobile telephone[b]	0.1	2
Electronic games[b]	3	5
Photocopier	60	8
Radio[b]	2	10
Television[c]	30	5
Video recorder and DVD player[b]	5	5
WEEE not normally considered E-waste		
Air conditioning unit	55	12
Dish washer[b]	50	10
Electric cooker[b]	60	10
Electric heaters[b]	5	20
Food mixer[b]	1	5
Freezer[b]	35	10
Hair dryer[b]	1	10
Iron[b]	1	10
Kettle[b]	1	3
Microwave[b]	15	7
Refrigerator[b]	35	10
Telephone[b]	1	5
Toaster[b]	1	5
Tumble dryer[b]	35	10
Vacuum cleaner[b]	10	10
Washing machine[b]	65	8

Potential environmental contaminants arising from E-waste disposal or recycling.

Contaminant	Relationship with E-waste	Typical E-waste concentration (mg/kg)[a]	Annual global emission in E-waste (tons)[b]
Polybrominated diphenyl ethers (PBDEs) polybrominated biphenyls (PBBs) tetrabromobisphenol-A (TBBPA)	Flame retardants		
Polychlorinated biphenyls (PCB)	Condensers, transformers	14	280
Chlorofluorocarbon (CFC)	Cooling units, insulation foam		
Polycyclic aromatic hydrocarbons (PAHs)	Product of combustion		
Polyhalogenated aromatic hydrocarbons (PHAHs)	Product of low-temperature combustion		
Polychlronated dibenzo-p-dioxins (PCDDs), polychlorinated dibenzofurans (PCDFs)	Product of low-temperature combustion of PVCs and other plastics		
Americium (Am)	Smoke detectors		
Antimony	Flame retardants, plastics (Ernst et al., (2003))	1700	34,000
Arsenic (As)	Doping material for Si		
Barium (Ba)	Getters in cathode ray tubes (CRTs)		
Beryllium (Be)	Silicon-controlled rectifiers		
Cadmium (Cd)	Batteries, toners, plastics	180	3600
Chromium (Cr)	Data tapes and floppy disks	9900	198,000
Copper (Cu)	Wiring	41,000	820,000
Gallium (Ga)	Semiconductors		
Indium (In)	LCD displays		
Lead (Pb)	Solder (Kang and Schoenung, (2005)), CRTs, batteries	2900	58,000
Lithium (Li)	Batteries		
Mercury (Hg)	Fluorescent lamps, batteries, switches	0.68	13.6
Nickel (Ni)	Batteries	10,300	206,000
Selenium (Se)	Rectifiers		
Silver (Ag)	Wiring, switches		
Tin (Sn)	Solder (Kang and Schoenung, (2005)), LCD screens	2400	48,000
Zinc (Zn)		5100	102,000
Rare earth elements	CRT screens		

Rochman, Chelsea. (2013a). Plastics and priority pollutants: A multiple stressor in aquatic habitats. *Environmental Science and Technology*. 47. pg. 2439-40. http://pubs.acs.org/doi/pdf/10.1021/es400748b

Rochman, Chelsea, Browne, M., Halpern, B., et al. (2013b). Policy: Classify plastic waste as hazardous. *Nature*. 494(7436). pg. 169-71. http://www.ncbi.nlm.nih.gov/pubmed/23407523

Rochman, Chelsea, Hoh, E., Hentschel, B. and Kaye, S. (2013c). Long-term field measurement of sorption of organic contaminants to five types of plastic pellets: Implications for plastic marine debris. *Environmental Science and Technology*. 47(3). pg. 1646-54. http://pubs.acs.org/doi/abs/10.1021/es303700s?journalCode=esthag

Rochman, Chelsea, Hoh, E., Kurobe, T. and Teh, S. (2013d). Ingested plastic transfers hazardous chemicals to fish and induces hepatic stress. *Scientific Reports*. 3(3263). http://www.nature.com/srep/2013/131121/srep03263/full/srep03263.html

- "Concerns regarding marine plastic pollution and its affinity for chemical pollutants led us to quantify relationships between different types of mass-produced plastic and organic contaminants in an urban bay."

- "At five locations in San Diego Bay, CA, we measured sorption of polychlorinated biphenyls (PCBs) and polycyclic aromatic hydrocarbons (PAHs) throughout a 12-month period to the five most common types of mass-produced plastic: polyethylene terephthalate (PET), high-density polyethylene (HDPE), polyvinyl chloride (PVC), low-density polyethylene (LDPE), and polypropylene (PP)."

- "During this long-term field experiment, sorption rates and concentrations of PCBs and PAHs varied significantly among plastic types and among locations. Our data suggest that for PAHs and PCBs, PET and PVC reach equilibrium in the marine environment much faster than HDPE, LDPE, and PP. Most importantly, concentrations of PAHs and PCBs sorbed to HDPE, LDPE, and PP were consistently much greater than concentrations sorbed to PET and PVC."

- "These data imply that products made from HDPE, LDPE, and PP pose a greater risk than products made from PET and PVC of concentrating these hazardous chemicals onto fragmented plastic debris ingested by marine animals."

Rochman, Chelsea, Lewison, R., Eriksen, M., et al. (2014a). Polybrominated diphenyl ethers (PBDEs) in fish tissue may be an indicator of plastic contamination in marine habitats. *Science of the Total Environment*. 476-477. pg. 622-33. http://www.ciagent-stormwater.com/documents/watermonitoring/Field%20Studies/Rochman%20et%20al%20%202014%20PBDEs%20S%20Atlantic.pdf

- "Mesopelagic lanternfishes, sampled from each station and analyzed for bisphenol A (BPA), alkylphenols, alkylphenol ethoxylates, polychlorinated biphenyls (PCBs) and polybrominated diphenyl ethers (PBDEs), exhibited variability in contaminant levels, but this variability was not related to plastic debris density for most of the targeted compounds with the exception of PBDEs."

- "This research…is to the best of our knowledge the first study to examine concentrations of chemical burdens in myctophid from the South Atlantic Gyre."

- "PBDEs can transfer from plastic to organisms, including fish, upon ingestion."

Rochman, C., Hentschel, B. and Teh, S. (2014b). Long-term sorption of metals is similar among plastic types: Implications for plastic debris in aquatic environments. *PLoS ONE*. 9(1). http://www.plosone.org/article/info%3Adoi%2F10.1371%2Fjournal.pone.0085433

- "Plastic debris may accumulate greater concentrations of metals the longer it remains at sea…a complex mixture of metals, including those listed as priority pollutants by the US EPA (Cd, Ni, Zn and Pb) can be found on plastic debris."

Rockström, J., Steffen, W., Noone, K., et al. (2009a). A safe operating space for humanity. *Nature*. 461. Pg. 472-5. http://pubs.giss.nasa.gov/docs/2009/2009_Rockstrom_etal_1.pdf

- "Since the Industrial Revolution, a new era has arisen, the Anthropocene, in which human actions have become the main driver of global environmental change."

Rockström, J., Steffen, W., Noone, K. et al. (2009b). Planetary boundaries: Exploring the safe operating space for humanity. *Ecology and Society*. 14(2). pg. 32. http://www.ecologyandsociety.org/vol14/iss2/art32/

Roper, W., Weiss, K. and Wheeler, J. (2006). *Water quality assessment and monitoring in New Orleans following Hurricane Katrina*. U.S. Environmental Protection Agency, Washington, DC. http://www.epa.gov/oem/docs/oil/fss/fss06/roper_3.pdf

- "Following the passage of Hurricane Katrina, New Orleans was left with eighty percent of its land area flooded. In some locations the flood waters were over thirty feet deep. In the heat and stagnation that followed the waters quickly became heavily polluted with petroleum products, industrial chemicals, raw sewage, dead animals etc. In addition, a super fund cleanup site was flooded and contributed to the water pollution problem."

- "Lake Pontchartrain was the primary receiving water body with the Mississippi River as a secondary receiver."

- "Working twenty four hour days throughout September New Orleans was close to being pumped out when Rita swept through on September 24, reflooding almost forty percent of the area again, causing more damage to Southeast Louisiana. But even with the Rita set back, the New Orleans dewater effort was completed on October 11. Dewatering was completed in just forty three days, less than half the time estimated by some."

- "In emergency situations such as this, EPA serves as the lead Agency for water quality including the cleanup of hazardous materials such as oil and gasoline. EPA national and regional Emergency Operations Centers were activated 24 hours a day. The Corps had employees embedded with the EPA/LDEQ team in Baton Rouge and onsite teams locally in New Orleans for rapid and effective communication regarding water quality issues."

- "The large amounts of water resulting from the hurricanes may have actually helped some of the water pollution concerns particularly in Lake Pontchartrain due to dilution. The large volumes of water from outside the city also provided dilution to the flood waters, and the time consuming process of dewatering allowed many solids to settle, likely removing some sorbed constituents from the water column."

- "The floodwaters removed from New Orleans proper consisted of a mixture of rain, Gulf of Mexico and Lake Pontchartrain waters, and other constituents."

- "Chlorinated hydrocarbons are 100 times or more resistant to bacterial breakdown than PAHs in the natural environment. However, the earlier inventory showed that these compounds, too were well below toxic levels in sediments, except for isolated occurrences in the immediate vicinity of New Orleans' canals. In the decades since the early 1970s, PCBs and long-lived pesticides like DDT, dieldrin, and aldrin have been systematically removed from use in the U.S., and none were detected in the initial DEQ samplings."

- "The flooding in New Orleans resulted in the potential for unparalleled exposure to toxics and contaminants. Initial concerns about a 'toxic gumbo,' however, have not been supported by sampling and analyses to date."

- Flood waters "did not contain chemical toxicants at levels that are expected to lead to long-term impacts on the surroundings beyond the impacts expected of a similar volume of stormwater from the city."

- "The most serious continuing issue facing most residents is the presence of high concentrations of mold and airborne mold spores. However, respiratory protection during the removal of all mold-contaminated materials and reconstruction can mitigate the risk."

- This is a total cover-up of pre- and post-Katrina contamination levels for several thousand PBT chemicals.

Rorije, E., Verbruggen, E., Hollander, A., et. al. (2011). *Identifying potential POP and PBT substances: Development of a new persistence/bioaccumulation score*. National Institute for Public Health and the Environment, Bilthoven, Netherlands. http://www.rivm.nl/dsresource?objectid=rivmp:24659&type=org&disposition=inline&ns_nc=1

- "A new methodology has been developed which indicates whether substances will persist in the environment and/or bioaccumulate in biological organisms… [and] which have inherent chemical properties that potentially make them a long-term hazard for the environment."

- "The Persistence/Bioaccumulation score presented in this report can be used as a tool to quickly screen data-poor substances for their potential environmental persistence and bioaccumulation in the food chain."

Rose, M. and Fernandes, A. (2013). *Persistent organic pollutants and toxic metals in foods*. Woodhead Publishing, Oxford, England.

http://www.worldcat.org/title/persistent-organic-pollutants-and-toxic-metals-in-foods/oclc/820110817

- A detailed survey (476 pp, 18 chapters) of POPs and toxic metals in food by researchers from England, Belgium, Italy, Ireland, Germany, Austria and Sweden. Only 1 chapter on mercury was written by an American (E.M. Sunderland, Harvard University School of Public Health).

Ross, P. (2000). Marine mammals as sentinels in ecological risk assessment. *HERA*. 6. pg. 29-46.
http://www.whoi.edu/science/B/people/mhahn/Ross_weight_of_evidence.pdf

- "The number of contaminants to which marine mammals are exposed is staggering."

- "Marine mammals ultimately provide information on the chemicals which present the greatest risk to consumers at the top of the food chain."

Rossi, M. and Blake, A. (2014). *The plastics scorecard: Evaluating the chemical footprint of plastics.* Clean Production Action.
http://www.ksat.com/content/dam/pns/ksat/news/defenders/2014/07/Plastics-Scorecard.pdf

- "Reducing the chemical footprint of plastics is a significant challenge. Starting from their feedstock base of fossil fuels, CoHCs [Chemicals of High Concern] litter the plastics pathway from primary chemicals to intermediates to monomer to final product compounded with additives. Exposure to a wide array of CoHCs during manufacturing, usage, and disposal poses a significant risk to the health of workers, communities, and the global environment."

- "In terms of the chemicals in products, additives are the key driver affecting the Chemical Footprint of Plastic Products. Residing in the product in the greatest concentrations beyond the polymer, additives dictate the concentration of CoHCs in plastic products. Companies are reducing CoHCs in plastic products by eliminating the need for the additive, changing additives, or changing polymers to avoid the need for the additive in the first place."

- "Among the challenges of effectively evaluating the hazards of additives include the absence of relevant publically available data for the various additive chemistries."

- "The plastics economy, from cradle to grave, remains largely based on CoHCs. The Plastics Scorecard v1.0 presents a novel method for evaluating the chemical footprint of plastics, selecting safer alternatives, and measuring progress away from CoHCs."

Plastics and the Chemicals they Consume

Steps in Polymer Manufacturing	Plastic Polymers							
Core Chemical Inputs	ABS	PC	PE	PET	PLA	PP	PS	PVC
Primary Chemical Inputs								
1,3-Butadiene	●							
Benzene	●	●					●	
Chlorine		●						●
Ethylene	●		●				●	●
Glucose					●			
Methanol				●				
Propylene	●	●				●		
Xylenes (p-Xylene)				●				
Intermediate Chemical Inputs								
Acetic acid				●				
Acetone		●						
Ammonia	●							
Cumene		●						
Dimethyl terephthalate / Terephthalic acid				●				
Ethylbenzene	●						●	
Ethylene dichloride								●
Ethylene glycol				●				
Lactic Acid					●			
Phenol		●						
Monomer Inputs								
1,3-Butadiene	●							
Acrylonitrile	●							
bis(2-hydroxyethyl) terephthalate				●				
Bisphenol A (BPA)		●						
Ethylene			●					
Lactide					●			
p-tert-Butylphenol		●						
Propylene						●		
Styrene	●						●	
Vinyl chloride monomer								●

ABS = Acrylonitrile Butadiene Styrene
PC = Polycarbonate
PE = Polyethylene
PET = Polyethylene Terephthalate

PLA = Polylactic Acid
PP = Polypropylene
PS = Polystyrene
PVC = Polyvinyl Chloride

▨ Chemical of High Concern to human health or the environment

● Chemical is an input in the manufacture of the indicated

Plastics and the Chemicals of High Concern they Consume

Chemicals of High Concern (plastics)	Total Global Consumption (million metric tons)	Consumed by Plastics (%)	Consumed by Plastics (million metric tons)
Ethylene dichloride (PVC)[b]	43.45	97%	42.14
para-Xylene (PET)[b]	42.89	88%	37.62
Benzene (PS)[b]	39.67	85%	33.52
Vinyl chloride monomer (PVC)[b]	32.79	97%	31.80
Ethylbenzene (ABS, PS)[b]	27.57	99%	27.29
Styrene (ABS, PS, SAN, SBR)[b]	23.63	91%	21.38
Ethylene glycol (PET, Nylon)[a]	21.00	80%	16.80
Cumene (PC)[b]	12.23	84%	10.27
Butadiene (ABS, SBR)[b]	9.28	94%	8.75
Acrylonitrile (ABS)[a]	5.35	96%	5.16
Phenol (PC)[c]	8.90	55%	4.88
Bisphenol A (PC, epoxy resins)[c]	4.04	96%	3.86
Acetone (PC)[d]	5.67	45%	2.53
Total	**270.79**	**90%**	**243.48**

"Chemicals of High Concern" to human health or the environment = carcinogen, mutagen, reproductive / developmental toxicant; persistent, bioaccumulative, toxicant (PBT); endocrine disruptor; or chemical of equivalent concern.

ABS = Acrylonitrile Butadiene Styrene   PLA = Polylactic Acid   SAN = Styrene Acrylonitrile
PC = Polycarbonate   PP = Polypropylene   SBR = Styrene Butadiene Rubber
PE = Polyethylene   PS = Polystyrene
PET = Polyethylene Terephthalate   PVC = Polyvinyl Chloride

Rotander, A., van Bavel, B., Polder, A., et al. (2012). Polybrominated diphenyl ethers (PBDEs) in marine mammals from Arctic and North Atlantic regions, 1986-2009. *Environment International*. 40. pg. 102-9.

Rotander, A., van Bavel, B., Riget, F., et al. (2012). Polychlorinated naphthalenes (PCNs) in sub-Arctic and Arctic marine mammals, 1986-2009. *Environmental Pollution*. 164. pg. 118-24.
https://events.iwc.int/index.php/workshops/ISPEPR2013/paper/viewFile/64/43/

- "A selection of PCN congeners was analyzed in pooled blubber samples… covering a time period of more than 20 years (1986-2009). A large geographical area of the North Atlantic and Arctic areas was covered."

- "PCN congeners 48, 52, 53, 66 and 69 were found in the blubber samples between 0.03 and 5.9 ng/glw... total PCN content accounted for 0.2% or less of the total non-planar PCB content. No statistically significant trend in contaminant levels could be established."

Rotllant, G., Holgado, A., Sarda, F., et al. (2006). Dioxin compounds in the deep-sea rose shrimp *Aristeus antennatus* throughout the Mediterranean Sea. *Deep Sea Research*. 53. pg. 1895-1906.
http://www.sciencedirect.com/science/article/pii/S0967063706002354

- "Polychlorodibenzo-*p*-dioxins (PCDDs) and polychlorodibenzofurans (PCDFs) are among the more toxic anthropogenic contaminants. They are fat-soluble and accumulate in animal tissues."

- "Specimens of *Aristeus antennatus* were collected from depths of 600–2500 m at different points in the Mediterranean Sea."

- "PCDD/F levels detected in the edible parts (muscle) of the commercial shrimp...PCDD/Fs were found in the shrimp *A. antennatus* throughout the Mediterranean Sea. Total PCDD/Fs burdens were higher in shrimps caught in the western Mediterranean than in those caught at eastern Mediterranean sites. There was a tendency for higher levels of PCDD/F contamination in samples obtained from deeper (2500 m) than from shallower sites (600 m)."

Rousk, J., Ackermann, K., Curling, S. and Jones, D. (2012). Comparative toxicity of nanoparticulate CuO and ZnO to soil bacterial communities. *PLoS ONE*. 7(3).
http://www.plosone.org/article/fetchObject.action?uri=info:doi/10.1371/journal.pone.0034197&representation=PDF

- "The increasing industrial application of metal oxide Engineered Nano-Particles (ENPs) is likely to increase their environmental release to soils."

- "Our findings suggested that the principal mechanism of toxicity was dissolution of metal oxides and sulphates into a metal ion form known to be highly toxic to bacteria."

Roux, S. (2008). *Nanopollutants, their effects on the environment, and potential impacts on the South African water treatment and supply infrastructure*. Research Space Technical Report.
http://researchspace.csir.co.za/dspace/bitstream/10204/3289/1/Roux3_2008.pdf

- "Nanotechnology research and development is rapidly expanding into a full-scale industry that includes building materials, automotive parts, sports equipment and consumer products."

- "Due to economic pressures the availability of nanotechnology-based consumer products are increasing rapidly while the development of regulatory frameworks and environmental impact assessment tools regarding nanotoxicity to the environment are lagging behind."

- "The study of nanotoxicity and its effects on the future sustainability of the environment is a relatively new field in science, unfortunately hampered by a lack of funding."

- "Although the industrial sector has been quick to take advantage of the financial implications of NT, research regarding the environmental impacts of NT has not kept pace. Government sponsored research into the safe and environmentally friendly use and release of nano-products has not been funded adequately and a much bigger effort in this regard from the industrialised nations in Europe and North America must be considered a priority. Future research into the environmental impact of NT could ultimately be the determining factor regarding the successful and sustainable utilisation of NT."

- "The large-scale release of NPs into the environment is considered to be inevitable based on the increasing industrial exploitation of NT. Personal care products (PCP) like sunscreens and cosmetics as well as food products and packaging are examples of the increasing use of nanomaterials associated with future impact on the environment."

- "Fullerenes also have a very low adsorption potential in soil and are therefore more easily absorbed by earthworms which causes concern regarding their environmental fate after release or disposal. The lipophilicity of these molecules allows relatively easy penetration of cell membranes and collection areas in the cell include mitochondria, cytoplasm, lysosomes and nuclei. Organic solvents like tetrahydrofuran may be trapped inside the fullerene structure and has been linked to observed cytotoxicity. "

- "Nobody will argue against the potential of nanotechnology to assist in the mitigation and elimination of many of the industry-related environmental pollution problems currently faced by modern society. Cleaner production processes, more economic manufacturing and fewer waste products are all promises made by the supporters of NT. Direct intervention in current polluted environments has been demonstrated. Future increase in life quality and even life expectancy due to medical NT intervention for a larger fraction of the human population are factors in favour of the development of NT."

Rowe, Amy A. (2011). *Assessment of water quality of runoff from sealed asphalt surfaces*. US EPA. http://nepis.epa.gov/Exe/ZyPDF.cgi?Dockey=P100ECC8.txt

Roy, P. K., Hakkarainen, M., Varma, I. K. and Albertsson, A. (2011). Degradable polyethylene: Fantasy or reality. *Environmental Science and Technology*. pg. 4217-27.

Royal Society and Royal Academy of Engineering. (2004). *Nanoscience and nanotechnologies: Opportunities and uncertainties*. Royal Society and Royal Academy of Engineering. http://www.raeng.org.uk/publications/reports/nanoscience-and-nanotechnologies-opportunities

- "Until more is known about environmental impacts of nanoparticles and nanotubes, we recommend that the release of manufactured nanoparticles and nanotubes into the environment be avoided as far as possible."

- "Our conclusions have been based on incomplete information about the toxicology and epidemiology of nanoparticles and their behaviour in air, water and soil, including their explosion hazard. If nanotechnologies are to expand and nanomaterials become commonplace in the human and natural environment, it is important that research into health, safety and environmental impacts keeps pace with the predicted developments."

Rudel, R., Camann, D., Spengler, J., Korn, L. and Brody, J. (2003). Phthalates, alkylphenols, pesticides, polybrominated diphenyl ethers, and other endocrine disrupting compounds in indoor air and dust. *Environmental Science and Technology*. 37. 4543-53. http://www.rachel.org/files/document/Phthalates_Alkylphenols_Pesticides_Polybromina.pdf

- "We sampled indoor air and dust in 120 homes, analyzing for 89 organic chemicals identified as EDCs. Fifty-two compounds were detected in air and 66 were detected in dust."

- "The number of compounds detected per home ranged from 13 to 28 in air and from 6 to 42 in dust."

- "The banned pesticides heptachlor, chlordane, methoxychlor, and DDT were also frequently detected, suggesting limited indoor degradation."

Rudel, R., Camann, D., Spengler, J., Korn, L. and Brody, J. (2008). PCB-containing wood floor finish is a likely source of elevated PCBs in residents' blood, household air and dust: A case study of exposure. *Environmental Health*. 7. pg. 2. http://www.ncbi.nlm.nih.gov/pmc/articles/PMC2267460

Ruder, A., Hein, M., Nilsen, N., et al. (2006). Mortality among workers exposed to polychlorinated biphenyls (PCBs) in an electrical capacitor manufacturing plant in Indiana: An update. *Environmental Health Perspectives*. 114(1). pg. 18-23. http://www.ncbi.nlm.nih.gov/pmc/articles/PMC1332650/

Rutkowski, Joseph and Levin, Barbara. (1986). Acrylonitrile-Butadiene-Styrene copolymers (ABS): Pyrolysis and combustion products and their toxicity: A review of the literature. *Fire and Materials*. 10. pg. 93-105. http://fire.nist.gov/bfrlpubs/fire86/PDF/f86017.pdf

Ryan, P., Connell, A. and Gardner, B. (1988). Plastic ingestion and PCBs in seabirds: Is there a relationship? *Marine Pollution Bulletin*. 19(4). pg. 174-6. http://www.mindfully.org/Plastic/Ocean/Plastic-Ingestion-PCBs1apr88.htm

Ryan, P., Moore, C., van Franeker, J. and Moloney, C. (2009). Monitoring the abundance of plastic debris in the marine environment. *Philosophical Transcriptions of the Royal Society*. 364(1526). pg. 1999-2012.

- "Monitoring is crucial to assess the efficacy of measures implemented to reduce the abundance of plastic debris, but it is complicated by large spatial and temporal heterogeneity in the amounts of plastic debris and by our limited understanding of the pathways followed by plastic debris and its long-term fate."

Ryman-Rasmussen, J., Riviere, J. and Monteiro-Riviere, N. (2006). Penetration of intact skin by quantum dots with diverse physiochemical properties. *Toxicological Sciences*. 91(1). pg. 159-65. http://toxsci.oxfordjournals.org/content/91/1/159

- "Quantum dots of different sizes, shapes, and surface coatings can penetrate intact skin at an occupationally relevant dose within the span of an average-length work day…skin is surprisingly permeable to nanomaterials with diverse physicochemical properties and may serve as a portal of entry for localized, and possibly systemic, exposure of humans to QD and other engineered nanoscale materials."

Ryu, J., Yoon, Y. and Oh, J. (2011). Occurrence of endocrine disrupting compounds and pharmaceuticals in 11 WWTPs in Seoul, Korea. *Ksce Journal of Civil Engineering*. 15(1). pg. 57-64. http://www.researchgate.net/publication/227198374_Occurrence_of_endocrine_disrupting_compounds_and_pharmaceuticals_in_11_WWTPs_in_Seoul_Korea/links/0c960529dad64efaba000000

- "Results of this study can provide evidence that WWTP effluent is one of the major sources contaminating the Han River."

- "Numerous recent studies have shown that conventional wastewater treatment plants cannot completely remove many Endocrine Disrupting Compounds (EDCs) and pharmaceuticals and personal care products (PPCPs)."

Saber, A., Lamson, J., Jacobsen, N., et. al. (2013). Particle-induced pulmonary acute phase response correlates with neutrophil influx linking inhaled particles and cardiovascular risk. *PLoS ONE*. 8(7). http://www.plosone.org/article/fetchObject.action?uri=info:doi/10.1371/journal.pone.0069020&representation=PDF

Sadiq, M. (2002). Metal contamination in sediments from a desalination plant effluent outfall area. *Science of the Total Environment*. 287(1-2). pg. 37-44. http://www.ncbi.nlm.nih.gov/pubmed/11883759

Safe S. (1998). Hazard and risk assessment of chemical mixtures using the toxic equivalency factor approach. *Environmental Health Perspectives*. 106. pg. 1051-8.

Sahoo, S., Parveen, S. and Panda, J. (2007). The present and future of nanotechnology in human health care. *Nanomedicine: Nanotechnology, Biology and Medicine*. 3. pg. 20-31. http://www.ncbi.nlm.nih.gov/pubmed/17379166

- "Applications of nanotechnology to medicine and physiology imply materials and devices designed to interact with the body at subcellular (i.e., molecular) scales with a high degree of specificity."

Sakai, S., Urano, S. and Takatsuki, R. (2000). Leaching behavior of PCBs and PCDDs/DFs from some waste materials. *Waste Management*. 20. pg. 241-7. http://www.researchgate.net/publication/32137359_Leaching_behavior_of_PCBs_and_PCDDsDFs_from_some_waste_materials

- "Little attention has been paid to the leaching behavior of these chemicals because of their low solubility."

- "Shredder residues from car/electrical goods recycling and fly ash from a municipal solid waste (MSW) incinerator were used in these leaching tests."

- "The results indicate that surfactant-like substances increase the leaching concentration of POPs, and fine particles related closely to the transporting behavior of POPs."

Sajiki, J. and Yonekubo, J. (2003). Leaching of bisphenol A (BPA) to seawater from polycarbonate plastic and its degradation by reactive oxygen species. *Chemosphere*. 51(1). pg. 55-62. http://www.ncbi.nlm.nih.gov/pubmed/12586156

- "BPA was degraded in both control water and seawater in the presence of radical oxygen species, but the degradation rate was lower in seawater than in control water, suggesting that anti-oxidative system exists in seawater."

Sajwan, K. S., Kumar, K. S., Nune, S., et al. G. (2008). Persistent organochlorine pesticides, polychlorinated biphenyls, polybrominated diphenyl ethers in fish from coastal waters off Savannah, GA, USA. *Toxicol Environ Chem*. 90. pg. 81-96.

Salykina, Y., Zherdeva, V., Dezhurov, S., et al. (2011). Biodistribution and clearance of quantum dots in small animals. *SPIE Proceedings*. 7999. http://www.uvp.com/pdf/SPIESaratovManuskript.pdf

- "Many in vitro studies have suggested that nanomaterials induce toxic responses through mechanisms such as particle breakdown and the subsequent release of toxic metals."

- "Nanoparticle surface chemistry, chemical composition, size, shape and aggregation have been shown to influence toxicity of nanomaterials."

- "All of the reports in the literature were unanimous in concluding that QDs show a preference for deposition in organs and tissues and that they do not remain circulating in the bloodstream."

Sanchez, P., Maso, M., Saez, R., et al. (2013). Baseline study of the distribution of marine debris on soft-bottom habitats associated with trawling grounds in the northern Mediterranean. *Scientia Marina*. 77. pg. 247-55. http://digital.csic.es/bitstream/10261/78396/1/SciMar77(2)194Sanchez.pdf

Sancivens, N. and Marco, M. (2008). Multifunctional nanoparticles – properties and prospects for their use in human medicine. *Trends in Biotechnology*. 26(8). pg. 425-33.http://www.ncbi.nlm.nih.gov/pubmed/18514941

Sandberg, A. and Bostrom, N. (2008). *Global catastrophic risks survey*. Technical Report # 2008-1. Future of Humanity Institute, Oxford University, Oxford, UK. http://www.fhi.ox.ac.uk/gcr-report.pdf

Sanders, Robert, Caron, David and Beringer, Ulrike. (1992). Relationships between bacteria and heterotrophic nanoplankton in marine and fresh waters: An inter-ecosystem comparison. *Marine Ecology Progress Series*. 86. pg. 1-14. http://www.int-res.com/articles/meps/86/m086p001.pdf

- "There are broad similarities in microbial food webs across systems. Relative abundances of bacteria and nanoplanktonic protozoa (HNAN, primarily heterotrophic flagellates) are similar in marine and freshwater environments."

- "Densities of bacteria and heterotrophic nanoplankton, therefore, are strongly related to the degree of eutrophication…Data from the literature is compiled to demonstrate a remarkably consistent numerical relationship (ca 1000 bacteria: 1 HNAN) between bacterioplankton and HNAN from the euphotic zones of a variety of marine and freshwater systems."

- "Nanoplanktonic protists are the major consumers in most (if not all) aquatic ecosystems…ciliates and copepods consume nanoplanktonic flagellates in freshwater and marine systems."

Sanderson, H., Johnson, D., Reitsma, T., et. al. (2004). Ranking and prioritization of environmental risks of pharmaceuticals in surface waters. *Regulatory Toxicological Pharmacology*. 39(2). pg. 158-83. http://www.ncbi.nlm.nih.gov/pubmed/15041147

Santos, S., Dinis, A., Rodrigues, D., et al. (2013). Studies on the toxicity of an aqueous suspension of C_{60} nanoparticles using a bacterium (gen. *Bacillus*) and an aquatic plant (*Lemna gibba*) as in vitro model systems. *Aquatic toxicology*. 142-143. pg. 347-54. https://estudogeral.sib.uc.pt/bitstream/10316/25592/1/1-s2.0-S0166445X13002282-main(1).pdf

- "The increasing use of C_{60} nanoparticles and the diversity of their applications in industry and medicine has led to their production in a large scale. C_{60} release into wastewaters and the possible accumulation in the environment has raised concerns about their ecotoxicological impact."

- "C_{60} aqueous dispersions must be viewed as an environmental pollutant, potentially endangering the equilibrium of aquatic ecosystems."

- "C_{60} is released into the environment and activated by sunlight, inducing the production of ROS, it may endanger live organisms due to the reaction of ROS with proteins, nucleic acids and the double bonds of membrane phospholipid hydrocarbon chains, which will lead to downstream detrimental effects, such as protein and DNA adduction, lipid peroxidation, membrane rupture and, eventually, cell death."

- "The paucity of toxicological data on carbon nanoparticles makes it difficult to forecast the risks arising from exposure to those nanomaterials, whose widespread application may lead to an extensive environmental contamination."

- "C_{60} nanoparticles affected the photosynthetic activity of *L. gibba* and inhibited its growth. Considering that *L. gibba* is a primary producer in aquatic food chains, the contamination of natural environment with C60 nanoparticles may have serious implications in the equilibrium of aquatic ecosystems."

Saskatchewan Ministry of Environment. (2012). *EPB 433: Health and environmental effects of burning waste plastics*. Saskatchewan Ministry of Environment. http://www.saskwastereduction.ca/assets/upload/pdf/plastics-pdf/effects-of-buring-plastics.pdf

- "The composition of byproducts of plastic combustion…depends on the combustion temperature and the flame residence time."

- "A study of the combustion of PE (both low (LDPE) and high density (HDPE) polyethylene) at different operating conditions detected more than 230 VOCs and semi-VOCs especially olefins, paraffin, aldehydes and light hydrocarbons. Amongst VOCs, benzene is a known carcinogen and has been observed to be released in significant quantity during plastic combustion. Some of the toxic semi-VOCs including benzo(a)pyrene and 1,3,5 trimethylbenzene have also been observed in significant quantities in the emissions from plastic combustion."

Sathyanarayana, S., Karr, C., Lozano, P., et al. (2008). Baby care products: Possible sources of infant phthalate exposure. *Pediatrics*. 121. pg. e260-8. http://www.nonadjavid.com/research/phthalate.pdf

Sayes, C., Fortner, L., Guo, W., et al. (2004). The differential cytotoxicity of water soluble fullerenes. *Nanoletters*. 4. pg. 1881-7. http://ytao.rice.edu/pdfs/08-nanoletters2004.pdf

- "The cytotoxicity of water-soluble fullerene species is a sensitive function of surface derivatization: In two different human cell lines, the lethal dose of fullerene changed over 7 orders of magnitude with relatively minor alterations in fullerene structure."

Sayes, C., M., Grobin, A., Ausman, K., et al. (2005). Nano-C60 cytotoxicity is due to lipid peroxidation. *Biomaterials*. 26. pg. 7587-95. http://nanonet.rice.edu/publications/2005/Sayes_Nano-C60_cytotoxicity.pdf

- "This water-soluble nano-C60 colloidal suspension disrupts normal cellular function through lipid peroxidation; reactive oxygen species are responsible for the membrane damage."

- "With the addition of an antioxidant, L-ascorbic acid, the oxidative damage and resultant toxicity of nano-C60 was completely prevented."

Scanlon, B., Faunt, C., Lonquevergne, L., et. al. (2012). Groundwater depletion and sustainability of irrigation in the US high plains and Central Valley. *Proceedings of the National Academy of Sciences, USA*. 109. pg. 9320-5. http://www.pnas.org/content/109/24/9320.full.pdf

- "Groundwater depletion in the irrigated High Plains and California Central Valley accounts for ~50% of groundwater depletion in the United States since 1900."

- "Extrapolation of the current depletion rate suggests that 35% of the southern High Plains will be unable to support irrigation within the next 30 years."

Scarascia-Mugnozza, G., Sica, C. and Russo, G. (2011). Plastic materials in European agriculture: Actual use and perspectives. *Journal of Agricultural Engineering*. 3. pg. 15-28. http://www.agroengineering.org/jae/article/viewFile/jae.2011.3.15/26

Schafer, K. and Kegley, S. (2002). Persistent toxic chemicals in the US food supply. *Journal of Epidemiological Health*. 56. pg. 813-7. http://jech.bmj.com/content/56/11/813.full.pdf+html

Schauffler, Marina. (2015). *Plastic microbeads: A small piece of a vast problem.* http://naturalchoices.com/plastic-microbeads-a-small-piece-of-a-vast-problem/

Schecter, A., Pavuk, M., Papke, O., et al. (2003). Polybrominated diphenyl ethers (PBDEs) in U.S. mothers' milk. *Environmental Health Perspectives*. 111. pg. 1723-9. https://www.ocf.berkeley.edu/~cfp/EHP2003%20Scheter.pdf

- "The PBDE levels in breast milk from Texas were similar to levels found in U.S. blood and adipose tissue lipid from California and Indiana and are 10-100 times greater than human tissue levels in Europe."

- "There are particular concerns especially about infant health because the fetus and the developing child are more sensitive than adults to the effects of exogenous chemical compounds, including PBDEs in breast milk or diet."

Schecter, A., Papke, O., Tung, K., et. al. (2004). Polybrominated diphenyl ethers contamination of United States Food. *Environmental Science and Technology*. 38(20). pg. 5306-11. http://pubs.acs.org/doi/abs/10.1021/es0490830

- "Elevated levels of polybrominated diphenyl ethers (PBDEs), a type of brominated flame retardant, were recently detected in U.S. nursing mothers' milk...whose route of intake is almost exclusively through food of animal origin."

- " Levels of PBDEs are highest in fish, then meat, and lowest in dairy products; median levels were 1725 (range 8.5–3078), 283 (range 0.9–679), and 31.5 (0.2–1373), parts per trillion (ppt), or pg/g, wet weight, respectively."

- No mention is made of the fact that house dust is a major vector for PBDE ingestion.

Schecter, A., Päpke, O., Tung, K. C., et al. (2005). Polybrominated diphenyl ether flame retardants in the U.S. population: Current levels, temporal trends, and comparison with dioxins, dibenzofurans, and polychlorinated biphenyls. *Journal of Occupational Environmental Medicine*. 47. pg. 199-211.

Schecter, A., Birnbaum, L., Ryan, J. and Constable, J. (2006a). Dioxins: An overview. *Environmental Research*. 101. pg. 419-28.
http://www.researchgate.net/publication/7327824_Dioxins_an_overview

Schecter, A., Papke, O., Harris, T., et al. (2006b). Polybrominated diphenyl ether (PBDE) levels in an expanded market basket survey of U.S. food and estimated PBDE dietary intake by age and sex. *Environmental Health Perspectives*. 114. pg. 1515-20.
http://www.ncbi.nlm.nih.gov/pmc/articles/PMC1626425/pdf/ehp0114-001515.pdf

- "Fish were highest in PBDEs (mean, 1,120 pg/g; median, 616 pg/g; range, 11.14–3,726 pg/g."

- "We estimated PBDE intake from food to be 307 ng/kg/day for nursing infants and from 2 ng/kg/day at 2–5 years of age for both males and females to 0.9 ng/kg/day in adult females."

- "Dietary exposure alone does not appear to account for the very high body burdens measured. The indoor environment (dust, air) may play an important role in PBDE body burdens in addition to food."

- "The high level of PBDE contamination in the U.S. population and food is cause for concern because these compounds are chemically similar to polychlorinated biphenyls (PCBs)."

- This article contains a detailed market basket survey of PBDEs in food.

Schecter, A., Johnson-Welch, S., Tung, K., et al. (2007). Polybrominated diphenyl ether (PBDE) levels in livers of U.S. human fetuses and newborns. *Journal of Toxicological Environmental Health*. 70. pg. 1-6. http://www.ncbi.nlm.nih.gov/pubmed/17162494

- "U.S. human breast milk and blood levels of PBDEs are presently the highest in the world."

- "All samples were contaminated with PBDEs. Levels varied from 4 to 98 ppb, lipid… These data document the transfer of PBDEs from maternal to fetal tissue."

Schecter, A., Papke, O., Tung, K., et. al. (2008). Brominated flame retardants in US food. *Molecular Nutritional Food Research*. 52. pg. 266-72.

http://www.researchgate.net/profile/Arnold_Schecter/publication/5806824_Brominated _flame_retardants_in_US_food/links/0deec53c41c2bdc2c6000000.pdf

- "All US women's milk samples were contaminated with PBDEs from 6 to 419 ng/g, lipid, orders of magnitude higher than levels reported in European studies, and are the highest reported worldwide."

- "Fish were most highly contaminated (median 616 pg/g), then meat (median190 pg/g) and dairy products (median 32.2 pg/g)."

Scheringer, M., Strempel, S., Hukari, S., et al. (2012). How many persistent organic pollutants should we expect? *Atmospheric Pollution Research*. 3. pg. 383-91. http://www.atmospolres.com/articles/Volume3/issue4/APR-12-045.pdf

- "Under the Stockholm Convention on Persistent Organic Pollutants (POPs), currently 22 chemicals or groups of chemicals are regulated as POPs...another five are currently under review by the POP Review Committee (POPRC) of the Convention."

- "We apply the screening criteria for persistence, bioaccumulation and long-range transport potential of the Stockholm Convention to a set of 93,144 organic chemicals."

- "510 chemicals that exceed all four critieria and can be considered potential POPs. Ninety eight percent of these chemicals are halogenated."

- "Ten substances are high-production volume chemicals and 249 are pre-registered in the EU."

Schirmer, Kristin, Behra, Renata, Sigg, Laura and Suter, Marc J.-F. (2013). Ecotoxicological aspects of nanomaterials in the aquatic environment. In: Luther, W. and Zweck, Z., eds. *Safety aspects of engineered nanomaterials*. Pan Stanford Publishing. http://www.panstanford.com/books/9789814364850.html

- "A proper evaluation of risks of NP [nanoplastics] to the aquatic environment requires synthesis of knowledge covering physical, chemical and biological aspects of NP behavior and NP-biota interaction."

- "A useful framework is to acknowledge the similarity of processes taking place in either the abiotic or biotic environment."

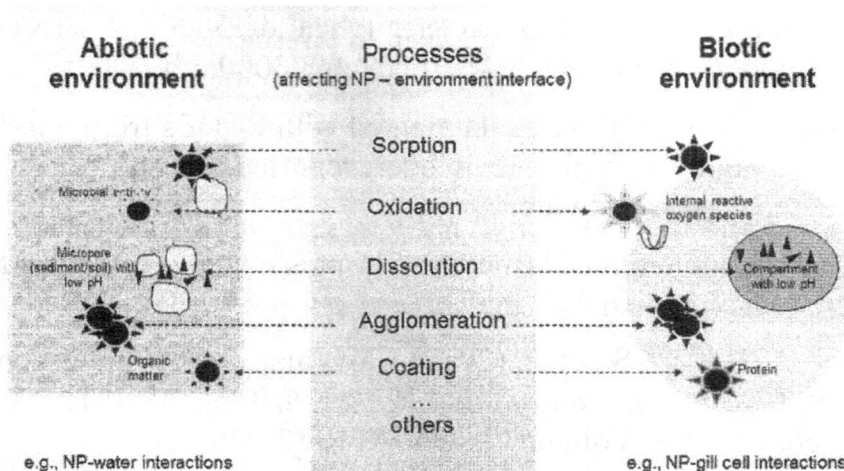

Abiotic environment	Processes (affecting NP – environment interface)	Biotic environment
	Sorption	
	Oxidation	
	Dissolution	
	Agglomeration	
	Coating	
	...	
	others	
e.g., NP-water interactions		e.g., NP-gill cell interactions

- "Because NP display a high diversity in chemical composition and physico-chemical characteristics and the fact that ecological experiments are time-consuming and complex, research efforts should put priority on those NP predicted to occur in the environment in larger quantities."

Schlining, K., von Thun, S., Kuhnz, L., et al. (2013) Debris in the deep: Using a 22-year video annotation database to survey marine litter in Monterey Canyon, central California, USA. *Deep-Sea Research*. 79. pg. 96-105. http://ac.els-cdn.com/S0967063713001039/1-s2.0-S0967063713001039-main.pdf?_tid=41c6d1fe-99ca-11e4-8d11-00000aacb35d&acdnat=1421005649_3c1f510319be8c07d6d4d4a0221e19da

- "Marine debris was technically defined by UNEP (2009) as 'any persistent manufactured or processed solid material discarded, disposed of or abandoned in the marine environment', and described perhaps more acutely by Moore and Allen (2000) as 'a visible expression of human impact on the marine environment.'"

- "Debris was most abundant within Monterey Canyon where aggregation and downslope transport of debris from the continental shelf are enhanced by natural canyon dynamics. The majority of debris was plastic (33%) and metal (23%). The highest relative frequencies of plastic and metal observations occurred below 2000 m, indicating that previous studies may greatly underestimate the extent of anthropogenic marine debris on the seafloor due to limitations in observing deeper regions."

- "Toxic chemicals from debris may leach into the surrounding environment."

Schmitt, C., Belliveau, M., Donahue, R. and Sears, A. (2007). *Body of evidence - a study of pollution in Maine people*. Alliance for a Clean and Healthy Maine, Portland, ME. http://www.cleanandhealthyme.org/bodyofevidencereport/tabid/55/default.aspx

- See Volume 3 of the *Phenomenology of Biocatastrophe* publication series (Brack 2010c) for a summary of the data in this study.

Schoeters, G. and Hoogenboom, R. (2006). Contamination of free-range chicken eggs with dioxins and dioxin-like polychlorinated biphenyls. *Mol. Nutr. Food. Res.* 50(10). pg. 908-14. http://www.ncbi.nlm.nih.gov/pubmed/16676378

Schonfelder, G., Flick, B., Mayr, E., et al. (2002). In utero exposure to low doses of bisphenol A lead to long-term deleterious effects in the vagina. *Neoplasia.* 4. pg. 258-63. http://www.ncbi.nlm.nih.gov/pmc/articles/PMC1550317/pdf/neo0402_0098.pdf

- See Volume 3 of the *Phenomenology of Biocatastrophe* publication series (Brack 2010c) for a summary of the data in this study.

Schultz, M. M. and Furlong, E. T. (2008). Trace analysis of antidepressant pharmaceuticals and their select degradates in aquatic matrixes by LC/ESI/MS/MS. *Analytical Chemistry.* 80(5). pg. 1756-62. http://www.ncbi.nlm.nih.gov/pubmed/18229944

- "Treated wastewater effluent is a potential environmental point source for antidepressant pharmaceuticals."

- "Venlafaxine was the predominant antidepressant observed in wastewater and river water samples. Individual antidepressant concentrations found in the wastewater effluent ranged from 3 (duloxetine) to 2190 ng/L (venlafaxine)."

Schultz, M., Furlong, E., Kolpin, D., et al (2010). Antidepressant pharmaceuticals in two U.S. effluent-impacted streams: Occurrence and fate in water and sediment, and selective uptake in fish neural tissue. *Environmental Science and Technology.* 44(6). pg. 1918-25. http://pubs.acs.org/doi/abs/10.1021/es9022706

- Included in this study were "antidepressants, including fluoxetine, norfluoxetine (degradate), sertraline, norsertraline (degradate), paroxetine, citalopram, fluvoxamine, duloxetine, venlafaxine, and bupropion."

Schwartz, John. (February 13, 2015). Study finds rising levels of plastics in oceans. *The New York Times.* pg. A4. http://www.nytimes.com/2015/02/13/science/earth/plastic-ocean-waste-levels-going-up-study-says.html

- "The problem is more than an aesthetic one: Exposed to saltwater and sun, and the jostling of the surf, the debris shreds into tiny pieces that become coated with toxic substances like PCBs and other pollutants."

Science for Environment Policy. (2011). *Plastic waste: Ecological and human health impacts*. European Commission: Science for Environment Policy. http://ec.europa.eu/environment/integration/research/newsalert/pdf/IR1_en.pdf

- "Plastic waste also has the ability to attract contaminants, such as persistent organic pollutants (POPs)… Plastic could potentially transport these chemicals to otherwise clean environments and, when ingested by wildlife, plastic could cause the transfer of chemicals into the organism's system."

- "With their large surface area-to-volume ratio, microplastics may have the capacity to make chemicals more available to wildlife and the environment in comparison to larger sized plastics."

- "However, once ingested, microplastics may pass through the digestive system more quickly than larger plastics, potentially providing less opportunity for chemicals to be absorbed into the circulatory system."

- No mention is made of the evolution of microplastics into plastic nanoparticles (PNP), their increasing sorbtion of ecotoxins as their mass decreases, and their tendency to be ingested by microzooplankton, which then transfer and biomagnify these environmental chemicals to higher trophic levels.

- Also not noted is that thousands of environmental chemicals are now being sorbed and translocated by plastic nanoparticles. Since their quantities and the nano and pico particles that sorb them are too small to measure, they must not exist.

- Where are the "otherwise clean environments" noted in the report, which provides no information about the ecological behavior of ecotoxins associated with invisible PNP.

- Many important research articles are not cited in this relatively recent report.

- This report contains an extensive bibliography on visible plastic pollution.

Scribner, E., Battaglin, W., Dietze, J. and Thurman, E. (2003). *Reconnaissance data for glyphosate, other selected herbicides, their degradation products, and antibiotics in 51 streams in nine Midwestern States, 2002*. U.S. Geological Survey Open-File Report 03-21. http://ks.water.usgs.gov/pubs/reports/ofr.03-217.html

Scrinis, G. and Lyons, K. (2007). The emerging nano-corporate paradigm: Nanotechnology and the transformation of nature, food and agri-food systems. *International Journal of Sociology, Agriculture and Food*. 15(2).

SCCS. (2010). *Opinion on triclosan (antimicrobial resistance)*. European Commission Scientific Committee on Consumer Safety. http://ec.europa.eu/health/scientific_committees/consumer_safety/docs/sccs_o_023.pdf

- "According to the information provided by COLIPA, the quantity of triclosan used within the EU reached approximately 450 tons (as 100% active) in the year 2006. Dye et al. (2007) estimated triclosan production in the EU to be 10-1,000 tonnes per year."

Seabra, A., Paula, A., Lima, R., et al. (2014). Nanotoxicity of graphene and graphene oxide. *Chemical Research in Toxicology*. 27(2). pg. 159-68. http://pubs.acs.org/doi/abs/10.1021/tx400385x

- "The toxicity of graphene is dependent on the graphene surface…the generation of reactive oxygen species in target cells is the most important cytotoxicity mechanism of graphene."

Seaton, A. and Donaldson, K. (2005). Nanoscience, nanotoxicology, and the need to think small. *The Lancet*. 365. pg. 923-4. http://www.ncbi.nlm.nih.gov/pubmed/15766981

Seneviratne, C., Leung, K., Wong, C., et al. (2014). Nanoparticle-encapsulated chlorhexidine against oral bacterial biofilms. *PLoS ONE*. 9(8). http://www.plosone.org/article/fetchObject.action?uri=info:doi/10.1371/journal.pone.0103234&representation=PDF

Senjen, Rye. (2009). Nanotechnologies in the 21st century. Nanomaterials: Health and environment concerns. 2. http://www.eeb.org/EEB/?LinkServID=540E4DA2-D449-3BEB-90855B4AE64E8CE6

- "It is estimated that in 2007/2008 about 350 tons of CNTs were produced worldwide."

- "Little or no detailed information is available on nanomaterials safety assessment either in research phase or in production, distribution and use in consumer products and hence no full exposure assessment can yet be performed."

- "Environmental and biological exposure pathways for many nanomaterials are still largely unknown as they have not been observed."

- "There is a lack of information about commercial products, including which contain nanomaterials, their precise nano content and therefore the public's potential level of exposure."

Service, R. (1998). Chemistry: Nanotubes: The next asbestos? *Science*. 281(5379). pg. 941. http://www.sciencemag.org/content/281/5379/941.summary

Setälä, O., Fleming-Lehtinen, V. and Lehtiniemi, M. (2014). Ingestion and transfer of microplastics in the planktonic food web. *Environmental Pollution*. 185. pg. 77-83. http://www.sciencedirect.com/science/article/pii/S0269749113005411

- "Experiments were carried out with different Baltic Sea zooplankton taxa to scan their potential to ingest plastics. Mysid shrimps, copepods, cladocerans, rotifers, polychaete larvae and ciliates were exposed to 10 um fluorescent polystyrene microspheres. These experiments showed ingestion of microspheres in all taxa studied…this study shows for the first time the potential of plastic microparticle transfer via planktonic organisms from one trophic level (mesozooplankton) to a higher level (macrozooplankton). The impacts of plastic transfer and possible accumulation in the food web need further investigations."

Shah, A., Hasan, F., Hameed, A. and Ahmed, S. (2008). Biological degradation of plastics: A comprehensive review. *Biotechnology Advances*. 26(3). pg. 246-65. http://www.sciencedirect.com/science/article/pii/S0734975008000141

Sharpe, R. M. and Skakkebaek, N. E. (2008). Testicular dysgenesis syndrome: Mechanistic insights and potential new downstream effects. *Fertility and Sterility*. 89. pg. 33-8. http://www.researchgate.net/profile/Richard_Sharpe2/publication/5544488_Testicular_dysgenesis_syndrome_mechanistic_insights_and_potential_new_downstream_effects/links/02bfe50f830cd2c46e000000.pdf

- Aside from phthalates, what other ecotoxins cause these disorders and what are their sources and pathways?

Shatkin, Jo Anne. (2008). Nanotechnology: Health and environmental risks. CRC Press, Taylor & Francis Group, Boca Raton, FL. http://handyfellow.com/downer/nano_ebooks/Nanotechnology._Health_and_Environmental_Risks,_2008,_p.194.pdf

- "All aspects of government, business and academia are subject to the influence of nanotechnology. All vertical industrial sectors will be impacted by nanotechnology—aerospace, health care, transportation, electronics and computing, telecommunications, biotechnology, agriculture, construction and energy."

- "The health (and environmental) consequences of nanomaterials are mostly unknown."

- This publication is part of the CRC Press publication series on nanotechnology. It contains 9 chapters (±150 pages) exploring the life cycle risk assessments, toxicology, and health and environmental risks of nanotechnology.

Shaw, Benjamin and Handy, Richard. (2011). Physiological effects of nanoparticles on fish: A comparison of nanometals versus metal ions. *Environment International*. 37. pg. 1083-97. http://www.ncbi.nlm.nih.gov/pubmed/21474182

- "The use of nanoscale materials is growing exponentially, but there are also concerns about the environmental hazard to aquatic biota. Metal-containing engineered nanoparticles (NPs) are an important group of these new materials."

- "Emerging studies on the acute toxicity of nanometals have so far shown that these materials can be lethal to fish in the mg-µgl(-1) range."

- "Nano-forms of some metals (Cu-NPs and ZnO NPs) may be more toxic to embryos or juveniles, than the equivalent metal salt."

- "We conclude that nanometals do have adverse physiological effects on fish, and the hazard for some metal NPs will be different… [from] the traditional dissolved forms of metals."

Shaw, David G. and Day, Robert. (1994). Colour and form- dependent loss of plastic microdebris from the North Pacific Ocean. *Marine Pollution Bulletin*. 28(1). pg. 39-43. http://www.sciencedirect.com/science/article/pii/0025326X94901848

- "Comparison of the size distribution of plastic observed with that predicted by a simple physical fragmentation model indicated that some forms, colours, and size fractions were significantly under-represented. We consider four possible explanations of these results and conclude that it is likely that marine organisms selectively remove plastic particles whose size, shape, and colour allow them to be mistaken for prey items. We further conclude that ingestion of small plastic objects by marine organisms occurs in substantial quantities."

Shaw, Susan D. (2002). *An investigation of persistent organic pollutants (POPs) and heavy metals in tissues of harbor seals and gray seals in the Gulf of Maine*. Final Report to the Maine Department of Environmental Protection, Augusta, ME.

http://www.meriresearch.org/Portals/0/Documents/Shaw%202002,%20Final%20Report%20to%20ME%20DEP,%20POPs%20and%20Metals%20in%20Seals.pdf

- "Seal pups in this study had much higher levels of PCBs and OC pesticides (mean 35.1 and 19.8 µg/g, lipid basis, for PCBs and p,p'-DDE) compared with other age groups, reflecting the importance of maternal transfer of lipophilic OCs to the OC burden of the young seal."

Shaw, Susan D. (2003). *Summary report: Gulf of Maine Forum 2002: Seals as sentinels for the Gulf of Maine ecosystem*. Marine Environmental Research Institute, Blue Hill, ME.
http://www.meriresearch.org/Portals/0/Documents/Shaw%202003,%20GoM%20Forum,%20Seals%20as%20Sentinels.pdf

- "Harbor seals (Phoca vitulina concolor) are widely distributed in the temperate coastal waters of the Gulf of Maine and are useful sentinels of food chain contamination because they occupy a high trophic level, are long-lived, and tissue samples can be obtained with relative ease."

- "Concern has focused on the polyhalogenated aromatic hydrocarbons (PHAHs) including the polychlorinated biphenyls (PCBs), dioxins and furans (PCDD/Fs), DDT and its metabolites because of their chemical and toxicological properties and evidence of their immune- and endocrine-disrupting effects in animals and humans."

Shaw, Susan D., Bourakovsky, A., Brenner, D., et al. (2005a). Polybrominated diphenyl ethers (PBDEs) in farmed salmon (Salmo salar) from Maine and eastern Canada. *Organohalogen Compounds*. 67. pg. 644-6.
http://www.meriresearch.org/Portals/0/Documents/Shaw%20et%20al%202005,%20Org%20Cmpds,%20PBDEs%20in%20Salmon.pdf

- "A recent survey found that PBDE levels in US foods are higher than levels in food from other countries, and the highest PBDE levels are in fish such as salmon as compared with other farm-raised animals (chicken, beef, pork) destined for human consumption."

Shaw, Susan D., Brenner, D., Bourakovsky, A., et al. (2005b). PCBs, dioxin-like PCBs and organochlorine pesticides in farmed salmon (*Salmo salar*) from Maine and eastern Canada. *Organohalogen Compounds*. 67. pg. 1571-6.
http://www.meriresearch.org/Portals/0/Documents/Shaw%20et%20al%202005,%20Org%20Cmpds,%20PCBs%20in%20salmon.pdf

- "In Maine, the industry tripled production between 1990 and 2003, and now supplies ~18% of US domestic consumption of farmed salmon."

- "The PCB WHO-TEQs in the organically farmed salmon from Norway in this study were an order of magnitude higher than those of the conventional farmed salmon from Maine and eastern Canada."

Shaw, Susan D., Brenner, D., Bourakovsky, et. al. (2005c). Polychlorinated biphenyls and chlorinated pesticides in harbor seals (*Phoca vitulina concolor*) from the northwestern Atlantic coast. *Marine Pollution Bulletin*. 50. pg. 1069-84. http://www.ncsu.edu/project/bio183de/Black/science/science_reading/lab1article.pdf

Shaw, Susan D., Brenner, D., Berger, M., et al. (2006a). PCBs, PCDD/Fs, and organochlorine pesticides in farmed Atlantic salmon from Maine, eastern Canada, and Norway, and wild salmon from Alaska. *Environmental Science and Technology*. 40. pg. 5347-54. http://www.meriresearch.org/Portals/0/Documents/Shaw%20et%20al%202005,%20Org%20Cmpds,%20PCBs%20in%20salmon.pdf

Shaw, Susan D., Berger, M., Brenner, D., et. al. (2006b). Polybrominated diphenyl ethers (PBDEs) in harbor seals (*Phoca vitulina concolor*) from the northwestern Atlantic. *Organohalogen Compounds*. 68. pg. 600-3. http://www.meriresearch.org/Portals/0/Documents/Shaw%20et%20al%202006,%20Org%20Cmpds,%20PBDEs%20in%20NW%20Atlantic%20Harbor%20Seal.pdf

- "The Gulf of Maine is a shallow, semi-enclosed sea receiving significant riverine, urban, agricultural, and industrial pollutant inputs from large urban centers in the Northeast as well as via long-range atmospheric transport."

- "PBDE concentrations in northwestern Atlantic harbor seals are relatively high, with mean concentrations of 13174, 7943, and 1540 ng/g, lw in the pups, yearlings, and adult males, respectively."

Shaw, Susan D. (2007a). *Chapter 7. How are seals, as top predators, impacted by toxic contaminants in Casco Bay and the Gulf of Maine?* Toxic Pollution in Casco Bay: Sources and Impacts, Casco Bay Estuary Partnership. http://muskie.usm.maine.edu/cascobay/pdfs/Toxics%20Chapter%207.pdf

- "Dr. Susan Shaw and co-workers at the Marine Environmental Research Institute (MERI), Center for Marine Studies, in Blue Hill, Maine, have been studying the impacts of environmental pollutants on seals in the Gulf of Maine and along the mid-Atlantic coast since 2001 as part of the *Seals as Sentinels* project."

- "Harbor seals (Phoca vitulina) are widely distributed in the temperate near-shore waters of the northern hemisphere and are important indicators of coastal contamination because they occupy a high trophic level, are long-lived (35-40 years), and accumulate high concentrations of persistent organic pollutants (POPs) and mercury through the food chain."

- "Harbor seals feed on a variety of fish including hake, herring, alewife, haddock, redfish, and winter flounder in coastal and estuarine environments and are exposed to contaminated habitats and prey across their range."

- "Lipophilic (fat soluble) POPs including PCBs, dioxins, and DDT build up in fatty tissues such as blubber and have been shown to cause immune- and endocrine-disrupting effects in seals and other marine wildlife."

- "PCB burdens in harbor seals from the northwestern Atlantic exceed the estimated threshold level of 17μg PCB/g, lw in blubber for adverse effects on immune function (Kannan et al. 2000), and fall within the estimated threshold level of 25-77 μg PCB/g, lw for reproductive effects in marine mammals."

Shaw, Susan D., Brenner, D., Berger, M., et al. (2007b). Polybrominated diphenyl ethers (PBDEs) in harbor seals from the northwestern Atlantic: Are seals debrominating DecaBDE? *Organohalogen Compounds*. 69. 1752-6.
http://www.meriresearch.org/Portals/0/Documents/Shaw%20et%20al%202007,%20Org%20Cmpds,%20PBDEs%20in%20NW%20Atlantic%20HS,%20Debromi.pdf

- "Harbor seals (Phoca vitulina concolor) inhabiting the northwestern Atlantic are closely associated with polluted near-shore environments and are highly contaminated by PBDEs and other persistent organic pollutants (POPs)."

Shaw, Susan D., Brenner, D., Berger, M., et al. (2007c). Patterns and trends of PCBs and PCDD/Fs in northwestern Atlantic harbor seals: Revisiting threshold levels using the new TEFs. *Organohalogen Compounds*. 69. pg. 1752-6.
http://www.meriresearch.org/Portals/0/Documents/Shaw%20et%20al%202007,%20Org%20Cmpds,%20PCBs%20Dioxins%20in%20HS,%20Toxic%20Thresh.pdf

- "Here we report, for the first time, the levels, patterns, and trends of PCBs, dioxins and furans (PCDD/Fs) in harbor seals from this region."

- "No temporal trend was found in concentrations between 1991 and 2005, suggesting a continuous input of PCBs in the northwestern Atlantic."

Shaw, Susan D., Brenner, D., Berger, M., et al. (2007d). PCBs, PCDD/Fs, and organochlorine pesticides in farmed Atlantic salmon from Maine, eastern Canada, and Norway, and wild salmon from Alaska (comment/correction). *Environmental Science*

and Technology. 41. pg. 4180.
http://www.meriresearch.org/RESEARCH/ToxicContaminantsinbrFarmedFish/PCBsDi
oxinsandPesticidesinFarmedSalmonfro/tabid/178/Default.aspx

Shaw, Susan D., Brenner, D., Berger, M., et al. (2008a). Bioaccumulation of
polybrominated diphenyl ethers in harbor seals from the northwest Atlantic.
Chemosphere. 73. pg. 1773-80. http://www.ncbi.nlm.nih.gov/pubmed/18950831

- "Unlike the trend for PCBs, no decreasing gradient from urban to rural/remote
  areas was observed for PBDEs...No significant temporal trend was observed for
  PBDEs in harbor seals between 1991 and 2005."

Shaw, Susan D., Berger, M., Brenner, D., et al. (2008b). Polybrominated diphenyl
ethers (PBDEs) in farmed and wild salmon marketed in the Northeastern United States.
Chemosphere. 71. pg. 1422-31.
http://www.meriresearch.org/Portals/0/Documents/Shaw%20et%20al%202005,%20Org
%20Cmpds,%20PBDEs%20in%20Salmon.pdf

Shaw, Susan D., Berger, M. L., Brenner, D., et al. (2008c). Bioaccumulation of
polybrominated diphenyl ethers and hexabromocyclododecane in the northwest Atlantic
marine food web. *Organohalogen Compounds.* 70. pg. 841-5.
http://www.meriresearch.org/RESEARCH/SealsasSentinels/BrominatedFlameRetardan
tsintheFoodWeb/tabid/272/Default.aspx

Shaw, Susan D., Berger, M., Brenner, D., et al. (2008d). Specific accumulation of
perfluorochemicals in harbor seals (*Phoca vitulina concolor*) from the northwest
Atlantic. *Chemosphere.* 74. pg. 1037-43.
http://www.ncbi.nlm.nih.gov/pubmed/19101009

- "Perfluorooctane sulffonate (PFOS) concentrations were the highest in liver (8-
  1388 ng/g, ww), followed by perfluoroundecanoic acid (PFUnDA) (<1-30.7
  ng/g, ww)."

Shaw, Susan D. and Kannan, Kurunthachalam. (2009a). Polybrominated diphenyl
ethers in marine ecosystems of the American continents: Foresight from current
knowledge. *Reviews on Environmental Health.* 24(3). pg. 157-229.
http://www.ncbi.nlm.nih.gov/pubmed/19891120

- "The oceans are considered global sinks for PBDEs, as higher levels are found in
  marine organisms than in terrestrial biota. For the past three decades, North
  America has dominated the world market demand for PBDEs, consuming 95%
  of the penta-BDE formulation. Accordingly, the PBDE concentrations in marine
  biota and people from North America are the highest in the world and are

increasing. Despite recent restrictions on penta- and octa-BDE commercial formulations, penta-BDE containing products will remain a reservoir for PBDE release for years to come, and the deca-BDE formulation is still in high-volume use."

- "We outline here our concerns about the potential future impacts of large existing stores of banned PBDEs in consumer products, and the vast and growing reservoirs of deca-BDE as well as new and naturally occurring brominated compounds on marine ecosystems."

Shaw, Susan D., Berger, M., Brener, D., et al. (2009b). Bioaccumulation of polyrominated diphenyl ethers and hexabromocyclododecane in the northwest Atlantic marine food web. *Science of the Total Environment.* 407. 3323-9. http://www.gulfofmaine.org/2/wp-content/uploads/2014/03/Shaw-et-al-2009-Sci-Tot-Env-article-in-press.pdf

Shaw, Susan D., Berger, M., Brenner, D., et al. (2009c). Specific accumulation of perfluorochemicals in harbor seals (*Phoca vintulina concolor*) from the northwest Atlantic. *Chemosphere.* 74. pg. 1037-43. http://www.researchgate.net/publication/23682529_Specific_accumulation_of_perfluorochemicals_in_harbor_seals_(Phoca_vitulina_concolor)_from_the_northwest_Atlantic

Shaw, Susan D., Covaci, A., Weijs, L., et al. (2009d). Accumulation of hexabromocyclododecanes (HBCDs) and their metabolites in pup and adult harbour seals from the northwest Atlantic. *Organohalogen Compounds.* 71. pg. 1486-90. http://www.meriresearch.org/RESEARCH/SealsasSentinels/tabid/85/Default.aspx

Shaw, Susan D., Berer, M., Weijs, L. and Covaci, A. (2012). Tissue-specific accumulation of polybrominated diphenyl ethers (PBDEs) including Deca-BDE and hexabromcyclododecaes (HBCDs) in harbor seals from the northwest Atlantic. *Environment International.* 44. pg. 1-6. http://www.ncbi.nlm.nih.gov/pubmed/22321537

- "BDE-209 concentrations in liver were up to five times higher than those in blubber."

- "Although detection frequency was low, BDE-209 levels in seal liver were up to ten times higher than those in their prey fish, suggesting that the accumulation/biomagnification of Deca-BDE in marine food webs is tissue-specific."

Sheavly, S. and Register, K. (2007. Marine debris and plastics: Environmental concerns, sources, impacts and solutions. *Journal of Polymers and the Environment*. http://link.springer.com/article/10.1007%2Fs10924-007-0074-3#page-1

Sherman, L. (2012). *Sub-micron additives make strides (just don't say 'nano'). Plastics Technology*. Gardner Business Media, Cincinnati, OH. http://www.ptonline.com/articles/sub-micron-additives-make-strides-just-dont-say-nano

- "High-surface-area, sub-micron-size additives have been making significant commercial strides in thermoplastic applications, albeit generally later than had been expected just a decade ago. Moreover, they are playing functional roles such as biocides, barrier additives, flame retardants, and electrostatic dissipation (ESD) agents, in addition to fulfilling their initially expected role as reinforcing fillers."

- "One type of sub-micron additive is gaining popularity in formulating electrically conductive thermoplastics: carbon nanotubes (CNTs), especially the more affordable and more widely available multi-walled variety, or MWCNTs."

- "Nanoadditives and nanoadditive combinations, plus hybrid combinations of nano-scale and micro-scale additives, will add design flexibility for engineering specialty compounds and composites."

Sherr, B. F., Sherr, E. B. and Newell, S. Y. (1984). Abundance and productivity of heterotrophic nanoplankton in Georgia coastal waters. *Journal of Plankton Research*. 6. pg. 195-202. http://plankt.oxfordjournals.org/content/6/1/195.full.pdf

Sherr, B. F., Sherr, E. B., Andrew, T. L., et al. (1986a). Trophic interactions between heterotrophic Protozoa and bacterioplankton in estuarine water analyzed with selective metabolic inhibitors. *Marine Ecological Progress*. 32. pg. 169-79. http://www.int-res.com/articles/meps/32/m032p169.pdf

Sherr, E. B., Sherr, B. F., Fallon, R. D. and Newell, S. Y. (1986b). Small, aloricate ciliates as a major component of the marine heterotrophic nanoplankton. *Limnology and Oceanography*. 31. pg. 177-83. https://ir.library.oregonstate.edu/xmlui/handle/1957/12790

Shi, J., Evans, D., Khan, A. and Harrison, R. (2001). Sources and concentrations of nanoparticles <10 nm diameter in the urban atmosphere. *Atmospheric Environment*. 35(2001). pg. 1193-202. http://www.researchgate.net/profile/Douglas_Evans4/publication/223472216_Sources_and_concentration_of_nanoparticles_%2810nm_diameter%29_in_the_urban_atmosphe re/links/00463527177f4594c1000000.pdf

- "Measurements in this study show that road traffic [both a diesel- and a petrol-fuelled vehicle] and stationary combustion sources generate a significant number of nanoparticles of diameter <10 nm."

Shi, G., Zhou, X., Jiang, S., et al. (2014). Further insights into the composition, source, and toxicity of PAHs in size-resolved particulate matter in a megacity in China. *Environmental Toxicological Chemistry*. http://www.ncbi.nlm.nih.gov/pubmed/25400005

- "The average concentrations of PM_{10} and $PM^{2.5}$ reached 209.75 $\mu g/m^3$ and 141.87 $\mu g/m^3$, respectively, and those of $\Sigma PAHs$ were 41.46 ng/m^3 for PM_{10} and 36.77 ng/m^3 for $PM_{2.5}$."

- "Diesel exhaust contributed 46.77% (PM_{10}) and 41.12% ($PM_{2.5}$)… gasoline exhaust contributed 31.02% (PM_{10}) and 39.47% ($PM_{2.5}$)… and coal combustion contributed 22.22% (PM_{10}) and 19.41% ($PM_{2.5}$)."

- PM concentration of other sorbed environmental chemicals were not noted.

Shiohara, A., Hoshino, A., Hanaki, K., et al. (2004). On the cyto-toxicity caused by quantum dots. *Microbiology and Immunology*. 48. pg. 669-75. http://www.ncbi.nlm.nih.gov/pubmed/15383704

Silicon Valley Toxics Coalition. (2014). *2014 solar scorecard*. Silicon Valley Toxics Coalition. http://www.solarscorecard.com/2014/2014-SVTC-Solar-Scorecard.pdf

- SVTC 2014 Solar Scorecard Key includes:
  - "Emissions Transparency—10 points A sunny score means that the company reports all categories of emissions through its annual report, its website, and/or third-party auditing or government agencies. Points are awarded for reporting: chemical emissions, including chemical waste, hazardous waste disposal, and/ or heavy metals; air pollutants, including NOx, SOx, volatile organic compounds (VOCs) and particulate matter (PC); emissions of ozone depleting substances; and information regarding landfill disposal."
  - "Module Toxicity—10 points For a sunny score, a company's PV modules do not contain toxic heavy metals (no more lead or cadmium than allowed under RoHS)."

Silicon Valley Toxics Coalition. (2015). *Silicon Valley Toxics Coalition website.* SVTC. http://www.svtc.org

- "The U.S Environmental Protection Agency (EPA) estimates that the stream of e-waste is growing two to three times the rate of any other source of waste. Only 15-20 percent of e-waste is recycled, and, according to the EPA, the "vast majority" of that waste is exported. California alone exported an estimated 20 million pounds of e-waste in 2006."
- "Be sure to ask where the e-waste is going before you drop-off any of your electronics."

Silva, M., Barr, D., Reidy, J., et al. (2004). Urinary levels of seven phthalate metabolites in the U.S. population from the National Health and Nutrition Examination Survey (NHANES) 1999–2000. *Environmental Health Perspectives*. 112. pg. 331-8. http://www.ncbi.nlm.nih.gov/pmc/articles/PMC1241863/pdf/ehp0112-000331.pdf

- "Phthalates are a class of widely used industrial compounds…[with] various toxicologic and chemical characteristics."
- "We saw significant differences in metabolite concentrations across the demographic groups."
- Aside from exposure due to the everyday use of common consumer products, what are the pathways of phthalates used in industrial applications?

Simonich, S. and Hites, R. (1995). Global distribution of persistent organochloride compounds. *Science*. 269(5232). pg. 1851-4. http://www.sciencemag.org/content/269/5232/1851

Simoneit, B., Medeiros, P. and Didyk, B. (2005). Combustion products of plastics as indicators for refuse burning in the atmosphere. *Environmental Science & Technology*. 39(18). pg. 6961-70. http://www.researchgate.net/publication/7562860_Combustion_products_of_plastics_as_indicators_for_refuse_burning_in_the_atmosphere/links/02bfe50c99d5bbba2f000000.pdf

Sing, N., Manshian, B., and Jenkins, G. (2009). "NanoGenotoxicology: The DNA damaging potential of engineered nanomaterials. *Biomaterials*. 30(23-24). pg. 3891-914. http://ac.els-cdn.com/S0142961209004062/1-s2.0-S0142961209004062-main.pdf?_tid=f83f18c8-afd8-11e4-a966-00000aacb35e&acdnat=1423430894_124bdfa171fe3e07ed3bf19729b0b688

Sivan, Alex. (2011). New perspectives in plastic biodegradation. *Current Opinion in Biotechnology*. 22(3). pg. 422-6. https://wiki.umn.edu/pub/ESPM3241W/S12TopicSummaryTeamFour/New_Perspectives_on_Plastic_Biodegradation.pdf

Sjödin, A., Jones, R. S., Focant, J.-F., et al. (2004). Retrospective time-trend study of polybrominated diphenyl ether and polybrominated and polychlorinated biphenyl levels in human serum from the United States. *Environ Health Perspect.* 112. pg. 655-8.

Sjodin, A., Wong, L., Jones, R., et al. (2008). *Serum concentrations of polybrominated diphenyl ethers (PBDEs) and polybrominated biphenyl (PBB) in the United States population: 2003-2004.* Centers for Disease Control and Prevention, Atlanta, GA. http://www.cdc.gov/exposurerePort/pdf/brominated_flame_retardants_1.pdf

- "2,062 serum samples, from participants in the National Health and Nutrition Examination Survey (NHANES) 2003-2004 aged 12 years and older, were analyzed for PBDEs and BB-153 [2,2',4,4',5,5'-hexabromobiphenyl (BB-153)]."

- "The highest sum of PBDE levels observed in the NHANES 2003-2004 participants was 3,680 ng/g lipid, which included BDE-47 at 2,350 ng/g lipid and concentrations of BDE-99, BDE-100 and BDE-153 at 692, 339 and 152 ng/g lipid, respectively."

- "A recent time trend study of U.S. residents, although non-representative, showed increasing serum PBDE concentrations from the mid-1980s to 2002."

- For additional data and updates since 2004 see the *Fourth National Report on Human Exposure to Chemicals* (CDC 2009).

Skaare, J. U., Larsen, H. J., Lie, E., et al. (2002). Ecological risk assessment of persistent organic pollutants in the arctic. *Toxicology.* 181. pg. 193-7.

Skakkebaek, N., Toppari, J., Soder, O., et al. (2011). The exposure of fetuses and children to endocrine disrupting chemicals: A European Society for Pediatric Endocrinology (ESPE) and Pediatrics Endocrine Society call to action statement. *Journal of Clinical Endocrinology and Metabolism.* 96(10). pg. 3056-8.

Slack, R. J., Gronow, J. R. and Voulvoulis, N. (2005). Household hazardous waste in municipal landfills: Contaminants in leachate. *Science of the Total Environment.* 337. pg. 119-37. http://www.sciencedirect.com/science/article/pii/S0048969704005017

- "Household hazardous waste (HHW) includes waste from a number of household products such as paint, garden pesticides, pharmaceuticals, photographic chemicals, certain detergents, personal care products, fluorescent tubes, waste oil, heavy metal-containing batteries, wood treated with dangerous substances, waste electronic and electrical equipment and discarded CFC-containing equipment."

- "Data on the amounts of HHW discarded are very limited and are hampered by insufficient definitions of what constitutes HHW."

Sloan, Jeff. (2011). Carbon fiber market: Cautious optimism. *High Performance Composites*. http://www.compositesworld.com/articles/carbon-fiber-market-cautious-optimism

Smedile, F., Messina, E., La Cono, V., et al. (2013). Metagenomic analysis of hadopelagic microbial assemblages thriving at the deepest part of Mediterranean Sea, Matapan-Vavilov Deep. *Environmental Microbiology*. 15(1). pg. 167-82. http://www.ncbi.nlm.nih.gov/pubmed/22827264

Smiley, Robert. (2002). *Ullmann's encyclopedia of industrial chemistry: Phenylene- and toluenediamines*. Wiley-VCH, Weinheim.

Smink, A., Ribas-Fito, N., Garcia, R., et. al. (2008). Exposure to hexachlorobenzene during pregnancy increases the risk of overweight in children aged 6 years. *Acta Paediatrics*. 97(10). pg. 1465-9. http://www.ncbi.nlm.nih.gov/pubmed/18665907

- "Prenatal exposure to HCB is associated with an increase in BMI and weight at age 6.5 years."

Smith, C., Shaw, B. and Handy, R. (2007). Toxicity of single walled carbon nanotubes to rainbow trout, (*Oncorhynchus mykiss*): Respiratory toxicity, organ pathologies, and other physiological effects. *Aquatic Toxicology*. 82. pg. 94-109. http://www.ncbi.nlm.nih.gov/pubmed/17343929

Soares, A., Guieysse, B., Jefferson, B., et al. (2008). Nonylphenol in the environment: A critical review on occurrence, fate, toxicity and treatment in wastewaters. http://ac.els-cdn.com/S0160412008000081/1-s2.0-S0160412008000081-main.pdf?_tid=7725fec6-d000-11e4-b736-00000aab0f6c&acdnat=1426966294_1444d76dac6a1118a10c60f853ac00e3

- "The occurrence of nonylphenol in the environment is clearly correlated with anthropogenic activities such as wastewater treatment, landfilling and sewage sludge recycling."

- "The impacts of nonylphenol in the environment include feminization of aquatic organisms, [and] decrease in male fertility."

Society of the Plastics Industry. (2014). *SPI resin identification code – guide to correct use*. The Plastics Industry Trade Association, Washington, DC. http://www.plasticsindustry.org/AboutPlastics/content.cfm?ItemNumber=823&navItem Number=2144

Soenen, S., Rivera-Gil, P., Montenegro, J., et al. (2011). Cellular toxicity of inorganic nanoparticles: Common aspects and guidelines for improved nanotoxicity evaluation. *Nano Today*. 6. pg. 446-65. http://nanosaude.ensp.fiocruz.br/sites/default/files/Cellular%20toxicity%20of%20inorganic%20nanoparticles-Common.pdf

- "The safe use of inorganic nanoparticles (NPs) in biomedical applications remains an unresolved issue. The present review presents an overview of the cytotoxic effects of commonly used inorganic NPs: quantum dots, gold and iron oxide nanoparticles."

Solenthaler, Balz and Bunge, Rainer. (2003). *Waste incineration in China*. Institut für angewandte Umwelttechnik. Switzerland. http://www.seas.columbia.edu/earth/wtert/sofos/Waste_Incineration_China.pdf

Song, Y., Li, X. and Du, X. (2009). Exposure to nanoparticles is related to pleural effusion, pulmonary fibrosis, and granuloma. *European Respiratory Journal*. 34. pg. 559-67. http://erj.ersjournals.com/content/34/3/559.full.pdf+html

- "Using transmission electron microscopy, nanoparticles were observed to lodge in the cytoplasm and caryoplasm of pulmonary epithelial and mesothelial cells, but are also located in chest fluid."

Sørmo, E. G., Salmer, M. P., Jenssen, B. M., et al. (2006). Biomagnification of polybrominated diphenyl ether and hexabromocyclododecane flame retardants in herring gulls in the polar bear food chain in Svalbard, Norway. *Environ Toxicol Chem*. 25. pg. 2502-11.

Soto, A., Vandenberg, L., Maffini, M. and Sonnenschein, C. (2008). Does breast cancer start in the womb? *Basic Clinical Pharmacological Toxicology*. 102. pg. 125-33. http://www.ncbi.nlm.nih.gov/pmc/articles/PMC2817934/pdf/nihms173627.pdf

- "Foetal exposure to xenooestrogens may be an underlying cause of the increased incidence of breast cancer observed over the last 50 years."

- "The contamination of our environment with endocrine disrupting chemicals is providing evidence that mammalian development is far more malleable than previously thought, as both natural and synthetic oestrogen exposure during development results in morphological and functional effects that persist into adulthood."

- What other endocrine disrupting chemicals are transmitted in cord blood serum and what are their sources and pathways?

Southwood, J. M., Muir, D. C. G. and Mackay, D. (1999). Modeling agrochemical dissipation in surface microlayers following aerial deposition. *Chemosphere*. 38. pg. 121-41. http://www.ncbi.nlm.nih.gov/pubmed/10903096

- "Conventional chemical fate models which treat the water compartment as being well-mixed may not adequately describe the fate of chemicals deposited onto lakes and ponds from aerial applications or spray drift."

- "The model was applied to four pesticides: bromoxynil octanoate, cypermethrin, deltamethrin and fenitrothion."

Spear, L., Ainley, D. and Ribic, C. (1995). Incidence of plastic in seabirds from the tropical pacific, 1984-1991: Relation with distribution of species, sex, age, season, year and body-weight. *Marine Environmental Research*. 40. pg. 123-46.

Spigoni, Valentina, Cito, Monia, Alinovi, Rossella, et al. (2015). Effects of TiO_2 and Co_3O_4 nanoparticles on circulating angiogenic cells. *PLoS ONE*. 10(3). http://journals.plos.org/plosone/article?id=10.1371/journal.pone.0119310

- "Sparse evidence suggests a possible link between exposure to airborne nanoparticles (NPs) and cardiovascular (CV) risk, perhaps through mechanisms involving oxidative stress and inflammation."

Srivastava, D. (2004). Computational nanotechnology of carbon nanotubes. In: *Carbon nanotubes: Science and applications*. CRC, Boca Raton, FL.

Stahlhut, R., van Wijngaarden, E., Dye, T., Cook, S. and Swan, S. (2007). Concentrations of urinary phthalate metabolites are associated with increased waist circumference and insulin resistance in adult U.S. males. *Environmental Health Perspectives*. 115(6). pg. 876-82. http://www.ncbi.nlm.nih.gov/pmc/articles/PMC1892109/pdf/ehp0115-000876.pdf

- "In this national cross-section of U.S. men, concentrations of several prevalent phthalate metabolites showed statistically significant correlations with abdominal obesity and insulin resistance."

Staples, C. A., Woodburn, K. Caspers, N. et al. (2002). A weight of evidence approach to the aquatic hazard assessment of bisphenol A. *Hum. Ecol. Risk Assess*. 8. pg. 1083-105.

Stapleton, H. M., Letcher, R. J. and Baker, J. E. (2004). Debromination of polybrominated diphenyl ether congeners BDE 99 and BDE 183 in the intestinal tract of the common carp (Cyprinus carpio). *Environmental Science and Technology*. 38. pg. 1054-61.

Stapleton, H., Dodder, N., Offenberg, J., et. al. (2005). Polybrominated diphenyl ethers in house dust and clothes dryer lint. *Environmental Science and Technology*. 39(4). pg. 925-31. http://pubs.acs.org/doi/abs/10.1021/es0486824

- "Dust samples were analyzed for 22 individual PBDE congeners and our results found PBDEs present in every sample."

- "Using estimates of inadvertent dust ingestion (0.02–0.2 g/day) by young children (ages 1–4), we estimate ingestion of total PBDEs to range from 120 to 6000 ng/day."

Stapleton, H. M., Dodder, N. G., Kucklick, J. R., et al. (2006). Determination of HBCD, PBDEs and MeO-BDEs in California sea lions (*Zalophus californianus*) stranded between 1993 and 2003. *Mar Pollut Bull*. 52. pg. 522-31.

Stapleton, H., Eagle, S., Sjödin, A. and Webster, T. (2012). Serum PBDEs in a North Carolina toddler cohort: Associations with handwipes, house dust, and socioeconomic variables. *Environmental Health Perspectives*. 120(7). pg. 1049-54. http://www.ncbi.nlm.nih.gov/pmc/articles/PMC3404669/pdf/ehp.1104802.pdf

- "Polybrominated diphenyl ethers (PBDEs) are flame retardant chemicals that have been used to reduce the flammability of polymers and resins found in commercial products such as furniture, electronics."

- "The high use of pentaBDE in North America may explain why these populations have some of the highest PBDE levels ever recorded in non-occupationally exposed populations."

- "Our study suggests that hand-to-mouth activity may be a significant source of exposure to PBDEs. Furthermore, age, socioeconomic status, and breast-feeding were significant predictors of exposure, but associations varied by congener. Specifically, serum $\sum BDE3$ was inversely associated with socioeconomic status, whereas serum BDE-153 was positively associated with duration of breast-feeding and mother's education."

Stavreva, D., George, A., Klausmeyer, P., et al. (2012). Prevalent glucocorticoid and androgen activity in US water sources. *Scientific Reports*. 2(937). pg. 1-8. http://www.nature.com/srep/2012/121206/srep00937/pdf/srep00937.pdf

- "We report previously unrecognized glucocorticoid activity in 27%, and androgen activity in 35% of tested water sources from 14 states in the US. Steroids of both classes impact body development, metabolism, and interfere with reproductive, endocrine, and immune systems."

Steele, J. H. and Henderson, E. W. (1992). The role of predation in plankton models. *Journal of Plankton Research*. 14. pg. 157-72. http://plankt.oxfordjournals.org/content/14/1/157.abstract

Steenland, K., Piacitelli, L., Deddens, J., et al. (1999). Cancer, heart disease, and diabetes in workers exposed to 2,3,7,8-tetrachlorodibenzo-p-dioxin. *J Natl Cancer Inst.* 91. pg. 779-86.

Stefatos, A. and Charalampakis, M. (1999). Marine debris on the seafloor of the Mediteranean Sea: Examples from two enclosed gulfs in Western Greece. *Marine Pollution Bulletin*. 38(9). pg. 389-93. http://www.sciencedirect.com/science/article/pii/S0025326X98001416

Steffen, W. et al. (2004). *Global change and the earth system: A planet under pressure*. Springer Verlag, Germany.

Stevenson, C. (2011). *Plastic debris in the California marine ecosystem: A summary of current research, solution efforts and data gaps*. University of Southern California Sea Grant. California Ocean Science Trust, Oakland, CA. http://calost.org/pdf/science-initiatives/marine%20debris/Plastic%20Report_10-4-11.pdf

- "Because solar radiation and thermal oxidation are factors in the breakdown of plastic into smaller and smaller pieces, and both factors are absent in deep ocean environments, it is unlikely that any plastic breaks down on the seafloor."

- "Recent studies now focus on the fact that plastic particles floating in the ocean also serve as concentrating and transport devices for environmental pollutants; some studies, in fact, indicate that plastics may be better concentrators than natural sediment."

- "Plastic marine debris is acting as a concentrating and transport mechanism for pollutants of concern."

- "BPA is a common additive in hard polycarbonate plastics or coatings (e.g., food and beverage cans and containers), CDs, DVDs, printer ink, medical equipment."

- "Phthalates, a class of chemicals, are used as softening additives in an array of products including clothing, toys, hoses, personal care products, insulation, flooring, inflatable structures, health-care products (catheters, blood bags), pesticides, construction materials in the form of PVC, and even in the pharmaceutical field as the coatings for some medications."

- "Plastic can be a significant transporter of contaminants to marine sediments, and therefore, sediment dwelling organisms can be contaminated via directly ingesting the plastic or simply ingesting the sediment."

Stevenson, L., Dickson, H., Klanjscek, T., et al. (2013). Mitigation of silver nanoparticle toxicity to *Chlamydomonas reinhardtii* by algal-produced organic compounds. *PLoS ONE*. 8(9). http://journals.plos.org/plosone/article?id=10.1371/journal.pone.0074456

Stokstad, Erik. (2012). Field research on bees raises concern about low-dose pesticides. *Science*. 335. http://www.whaleofatime.org/forms/Editorial_science-bees.pdf

Stone, Wesley W., Gilliom, Robert J. and Ryberg, Karen R. (2014). Pesticides in U. S. streams and rivers: Occurrence and trends during 1992-2011. *Environmental Science & Technology*. 48(19). pg. 11025-30. http://pubs.acs.org/doi/pdf/10.1021/es5025367

Stoodley, L., Costerton, J. and Stoodley, P. (2004). Bacterial biofilms: From the natural environment to infectious diseases. *Nature Reviews*. 2. pg. 95-108. http://www.nature.com/nrmicro/journal/v2/n2/abs/nrmicro821.html

- "Biofilms – matrix-enclosed microbial accretions that adhere to biological or non-biological surfaces—represent a significant and incompletely understood mode of growth for bacteria."

Streets, S., Ferrey, M., Solem, L., et. al. (2008). *Endocrine disrupting compounds: A report to the Minnesota legislature*. Minnesota Pollution Control Agency. http://www.pca.state.mn.us/index.php/view-document.html?gid=3943

- "Endocrine disruption is a means by which a chemical exerts an adverse effect or endpoint; it is not a discrete toxic effect of the chemical itself."

- "A wide array of effects has been attributed to EDCs including impacts on growth, development, reproduction, changes in behavior, immune suppression, and cancer. Effects may occur at multiple levels of biological organization, including the molecular, cellular, tissue, individual organism, and population levels."

- "Known and potential EDCs exist among many classes of chemicals including pharmaceuticals and personal care products, general anthropogenic (man-made) compounds, pesticides, biogenic (naturally occurring) compounds, and inorganics and organometallic compounds."

- "Currently [in 2008], there are more than 87,000 chemicals produced and used worldwide and more are being produced all the time. Many of these chemicals may have endocrine-disrupting potential."

Strempel, S., Scheringer, M., Ng, C., et al. (2012). Screening for PBT chemicals among the "existing" and "new" chemicals of the EU. *Environmental Science and Technology*. 46. pg. 5680-7.

Su, Y., He, Y., Lu, H., et al. (2009). The cytotoxicity of cadmium based, aqueous phase-synthesized, quantum dots and its modulation by surface coating. *Biomaterials*. 30. pg. 19-25. http://www.sinap.ac.cn/english/text/The%20cytotoxicity%20of%20cadmium%20based.pdf

- "CdTe QDs are highly toxic for cells due to the release of cadmium ions...the cytotoxicity of QDs can be modulated through elaborate surface coatings."

Su, Y., Peng, F., Jiang, Z., et al. (2011). In vivo distribution, pharmacokinetics, and toxicity of aqueous synthesized cadmium-containing quantum dots. *Biomaterials*. 32. pg. 5855-862. http://cheng.sinano.ac.cn/admin/sdcms/2012718104158109944.pdf

- "There exist no comprehensive studies concerning in vivo behavior of aqueous synthesized QDs (aqQDs) up to present."

- "Previous studies have suggested that cytotoxicity of QDs is ascribed to release of toxic metals and production of reactive oxygen species, which could be largely alleviated by surface modification (e.g., epitaxial growth of ZnS shell)."

- "We have systematically studied short- and long-term in vivo biodistribution, pharmacokinetic, and toxicity of the aqQDs with extremely small hydrodynamic diameters (2.9e4.5 nm)."

- "Mice intravenous injected with the aqQDs survived for 80 days without evident toxic effects."

Sudaryanto, A., Isobe, T., Suzuki, G., et al. (2009). Characterization of brominated flame retardants in house dust and their role as non-dietary source for humans in Indonesia. *Interdisciplinary Studies on Environmental Chemistry*. pg. 133-41. http://www.terrapub.co.jp/onlineproceedings/ec/02/pdf/ERA15.pdf

- "Concentrations of PBDEs in house dust (range: 20–1500 ng/g dust, mean 200 ng/g dust and median 120 ng/g dust) were higher than HBCDs (range: 1.5–75 ng/g dust."

- "PCBs levels were the lowest (ranged 1.5–78 ng/g dust, mean 14 ng/g dust, median 10 ng/g dust). Levels of PBDEs and HBCDs in house dust from Indonesia were among the lowest when compared globally."

Sutherland, W., Clout, M., Côté, I., et al. (2011). A horizon scan of global conservation issues for 2010. *Trends in Ecology and Evolution*. 26(1). http://www.conservation.cam.ac.uk/sites/default/files/file-attachments/Horizon%20Scanning%20TREE%202011.pdf

Swan, S., Kruse, R., Liu, E., et al. (2003). Semen quality in relation to biomarkers of pesticide exposure. *Environmental Health Perspectives*. 111. pg. 1478-84. http://www.healthandenvironment.org/docs/Swan_2003_Pesticides_and_semen_quality.pdf

- "These associations between current-use pesticides and reduced semen quality suggest that agricultural chemicals may have contributed to the reduction in semen quality in fertile men from mid-Missouri we reported previously."

Swan, S., Main, K., Liu, F., et al. (2005). Decrease in anogenital distance among male infants with prenatal phthalate exposure. *Environmental Health Perspectives*. 113(8). pg. 1056-61. http://www.ncbi.nlm.nih.gov/pmc/articles/PMC1280349/pdf/ehp0113-001056.pdf

- "The median concentrations of phthalate metabolites that are associated with short AGI and incomplete testicular descent are below those found in one-quarter of the female population of the United States, based on a nationwide sample."

- "These data support the hypothesis that prenatal phthalate exposure at environmental levels can adversely affect male reproductive development in humans."

- "These changes in male infants, associated with prenatal exposure to some of the same phthalate metabolites that cause similar alterations in male rodents, suggest that commonly used phthalates may undervirilize humans as well as rodents."

Tabb, M. and Blumberg, B. (2006). New modes of action for endocrine-disrupting chemicals. *Molecular Endocrinology*. 20. pg. 475-82. http://press.endocrine.org/doi/full/10.1210/me.2004-0513

- "The concept of endocrine disruption [is] the inappropriate modulation of the endocrine system by dietary and environmental changes."

- "Xenobiotics and environmental contaminants can act as hormone sensitizers by inhibiting histone deacetylase activity and stimulating mitogen-activated protein kinase activity."

- "Some endocrine disrupters can have genome-wide effects on DNA methylation status. Others can modulate lipid metabolism and adipogenesis, perhaps contributing to the current epidemic of obesity."

Takada, S. (2013). International pellet watch: Studies of the magnitude and spatial variation of chemical risks associated with environmental plastics. In: *Accumulation: The material politics of plastic*. Gabrys, J., Hawkins, G. and Michael, M. eds. Routledge, Oxon, UK.

Takada, Hideshige. (2006). Call for pellets! International Pellet Watch global monitoring of POPs using beached plastic resin pellets. *Marine Pollution Bulletin*. 52. pg. 1547-8. http://www.ncbi.nlm.nih.gov/pubmed/17113110

Takada, Hideshige. (2013). *Microplastics and the threat to our seafood*. Ocean Health Index. http://www.oceanhealthindex.org/News/Microplastics

- "POPs are hazardous human-made chemicals that are resistant to degradation in the environment. Polychlorinated biphenyls (PCBs), different sorts of organochlorine pesticides (e.g. DDTs and HCHs) and brominated flame-retardants are all POPs. Because they are basically lipophilic (i.e. have a high affinity for oils and fats), POPs accumulate in fatty tissues of marine organisms. They have the potential to cause many adverse effects in wildlife and humans (e.g. cancer, malformation, decrease in the immune response, impaired reproductive ability). Plastic pellets are also lipophilic and have an extremely high affinity for POPs. The concentration of POPs in plastic resin pellets is a million times higher than in the surrounding seawater."

- "These spatial patterns are consistent with those found by traditional monitoring methods (e.g., mussel watch), indicating the reliability of IPW as a monitoring tool. The spatial pattern of POPs in pellets was also concordant with those in plastic fragments (4). This means that similar accumulation of POPs occurs on plastic fragments as on pellets, confirming that pellets can be considered surrogates for plastic fragments and microplastics in general."

- "In addition to the absorption of POPs, marine plastics contain additives such as plasticizers, antioxidants, anti-static agents and flame retardants. Some additives and additive-derived chemicals (e.g., nonylphenol, bisphenol A) cause endocrine disruption—that is they interfere with body processes mediated by hormones."

- "Our latest study demonstrated that endocrine disrupter nonylphenols are present even in water bottle caps."

- "Scientists are concerned that chemicals associated with microscopic plastic could be transferred to the internal system of lower-tropic-level organisms such as mussels, oysters or copepods, then biomagnified to animals at higher tropic levels. This would mean not only persistence but increases in levels of toxic chemicals."

- "Reduction of plastic waste is essential for a sustainable society. Reduction of single-use plastic is a fundamental and effective way to decrease the risks associated with plastics."

Takagi, M., Ukibe, D., Murota, K. and Minami, T. (2013). *Effect of lipids on the direct absorption of methyl mercury by lymph and blood*. Science and Engineering Research Laboratory of Kinki University. http://www.rist.kindai.ac.jp/no.26/minami.pdf

- "The quantity of lymph and blood absorption of mercury does not differ even if methyl mercury is given with various lipids."

Taleb, Nassim Nicholas. (2010). *The black swan: The impact of the highly improbable*. Penguin Press, London, UK.

- The ménage à trois of nanoparticles, environmental chemicals, and microbiota is a black swan event.

Talsness, C., Andrade, A., Kuriyama, S., Taylor, J. and vom Saal, F. (2009). Components of plastic: Experimental studies in animals and relevance for human health. *Philosophical Transcripts of the Royal Society*. 364(1526). pg. 2079-96.

- "Phthalates function as anti-androgens while the main action attributed to BPA is oestrogen-like activity. PBDE and TBBPA have been shown to disrupt thyroid hormone homeostasis while PBDEs also exhibit anti-androgen action."

- "The spectrum of effects following perinatal exposure of male rats to phthalates has remarkable similarities to the testicular dysgenesis syndrome in humans."

- "In general, EDCs may disrupt the endocrine system by competing with endogenous steroid hormone binding to receptors and hormone transport proteins or by altering the metabolism or synthesis of endogenous hormones, eventually influencing recruitment of transcription factors and altering gene expression in cells."

- "Identification of the chemicals used in products, such as baby toys, food and beverage containers or paper products, is not required."

Table 1. Prenatal–neonatal exposure of mice and rats to BPA at human exposure levels in relation to human health trends.

effects in mice and rats	human health trends
prostate hyperplasia and cancer	prostate cancer increase
mammary hyperplasia and cancer	breast cancer increase
abnormal urethra/ obstruction	hypospadias
sperm count decrease	sperm count decrease
early puberty in females	early sexual maturation
ovarian cysts/uterine fibroids	polycystic ovary syndrome/ uterine fibroids
abnormal oocyte chromosomes	miscarriage
body weight increase	obesity increase
insulin resistance	type 2 diabetes
hyperactivity/impaired learning	attention deficit hyperactivity disorder

Tammemagi, Hans. (1999). *The waste crisis*. Oxford University Press, Oxford, UK.

Tanabe, S. (2002). Contamination and toxic effects of persistent endocrine disrupters in marine mammals and birds. *Marine Pollution Bulletin*. 45. pg. 69–77. http://www.ncbi.nlm.nih.gov/pubmed/12398369

Tanaka, K., Takada, H., Yamashita, R., et al. (2013). Accumulation of plastic-derived chemicals in tissues of seabirds ingesting marine plastics. *Marine Pollution Bulletin*. 69(1-2). pg. 219-22. http://www.ncbi.nlm.nih.gov/pubmed/23298431

Tangahu, B., Abdullah, S., Basri, H., et al. (2011). A review on heavy metals (As, Pb, and Hg) uptake by plants through phytoremediation. *International Journal of Chemical Engineering*. http://www.hindawi.com/journals/ijce/2011/939161/

Taylor, R., Coulombe, S., Otanicar, T., et al. (2013). Small particles, big impacts: A review of the diverse applications of nanofluids. *Journal of Applied Physics*. 113.

Teixeira, E., Migliavacca, D., Pereira, S., Machado, A. and Dallarosa, J. (2008). Study of wet precipitation and its chemical composition in South of Brazil. *Anais de*

Academia Brasileira de Ciencias. 80(2). pg. 381-95.
http://www.scielo.br/pdf/aabc/v80n2/a16v80n2.pdf

Teuten, E. L., Rowland, S. J., Galloway, T. S. and Thompson, R. C. (2007). Potential for plastics to transport hydrophobic contaminants. *Environmental Science and Technology*. 41(22). pg. 7759-64. http://pubs.acs.org/doi/abs/10.1021/es071737s

- "We estimate that the addition of as little as 1 μg of contaminated polyethylene to a gram of sediment would give a significant increase in phenanthrene accumulation by *A. marina*. Thus, plastics may be important agents in the transport of hydrophobic contaminants to sediment-dwelling organisms."

- "Weathering of polymers increases their surface area through cracking and fragmentation, while photo-oxidation functionalizes the surface."

Teuten, E. L., Saquing, J. M., Knappe, D., et al. (2009). Transport and release of chemicals from plastics to the environment and to wildlife. *Philosophical Transcripts of the Royal Society*. 364(1526). pg. 2027-45. http://vnu.edu.vn/upload/scopus/330.pdf

- "Model calculations and experimental observations consistently show that polyethylene accumulates more organic contaminants than other plastics such as polypropylene and polyvinyl chloride."

- "Plastics debris in the marine environment, including resin pellets, fragments and microscopic plastic fragments, contain organic contaminants, including polychlorinated biphenyls (PCBs), polycyclic aromatic hydrocarbons, petroleum hydrocarbons, organochlorine pesticides (2,2′-bis(p- chlorophenyl)-1,1,1-trichloroethane, hexachlorinated hexanes), polybrominated diphenylethers, alkylphenols and bisphenol A, at concentrations from sub ng g^{-1} to μg g^{-1}."

- "Concentrations of hydrophobic contaminants adsorbed on plastics showed distinct spatial variations reflecting global pollution patterns."

- "Plasticizers, other plastics additives and constitutional monomers also present potential threats in terrestrial environments because they can leach from waste disposal sites into groundwater and/or surface waters."

Thayer, K. A., Heindel, J. J., Bucher, J. R., et al. (2012). Role of environmental chemicals in diabetes and obesity: A National Toxicology Program Workshop report. *Environ Health Perspect*. 120(6). pg. 779-89.

Thiel, M. and Gutow, L. (2005). The ecology of rafting in the marine environment: The rafting organisms and community. *Oceanography and Marine Biology: An Annual Review*. 43. pg. 279-418. http://epic.awi.de/11613/1/Thi2005a.pdf

- "Rafting organisms comprised cyanobacteria, algae, protists, invertebrates from most marine but also terrestrial phyla, and even a few terrestrial vertebrates. Marine hydrozoans, bryozoans, crustaceans and gastropods were the most common taxa that had been observed rafting."

- "Floating plastic shows simple surface structure offering primarily settlement substratum for sessile organisms."

Thiel, M., Hijonosa, I., Miranda, L., et al. (2013). Anthropogenic marine debris in the coastal environment: A multi-year comparison between coastal waters and local shores. *Marine Pollution Bulletin*. 71. pg. 307-16. http://www.pubfacts.com/fulltext_frame.php?PMID=23507233&title=Anthropogenic%20marine%20debris%20in%20the%20coastal%20environment:%20a%20multi-year%20comparison%20between%20coastal%20waters%20and%20local%20shores

Thomas, G. O., Moss, S. E. W., Asplund, L. and Hall A. J. (2005). Absorption of decabromodiphenyl ether and other organohalogen chemicals by grey seals (*Halichoerus grypus*). *Environ Pollut*. 133. pg. 581-6.

Thompson, E., Sayers, B., Glista-Baker, E., et al. (2013). Innate immune responses to nanoparticle exposure in the lung. *Journal of Environmental Immunology and Toxicology*. 1(3). pg. 150-6. http://bonnerlab.wordpress.ncsu.edu/files/2013/10/Thompson-et-al.-2013-JEIT.pdf

- "Biopersistent engineered nanomaterials (ENMs) stimulate immune, inflammatory, and fibroproliferative responses in the lung...Due to their nanoscale dimensions and increased surface area per unit mass, ENMs have a much greater potential to reach the distal regions of the lung and generate ROS."

- "ENMs also migrate from the lungs across epithelial, endothelial, or mesothelial barriers to stimulate or suppress systemic immune responses."

Thompson, R. C., Olsen, Y., Mitchell, R., et al (2004). Lost at sea: Where is all the plastic? *Science*. 304(5672). pg. 838. http://www.researchgate.net/publication/8575062_Lost_at_Sea_Where_Is_All_the_Plastic

Thompson, R. C. (2006). Plastic debris in the marine environment: Consequences and solutions. In: *Marine nature conservation in Europe*. Federal Agency for Nature Conservation, Stralsund, Germany. pg. 107-15. http://dev1.bfn.eu/fileadmin/MDB/documents/themen/meeresundkuestenschutz/downloads/Fachtagungen/Marine-Nature-Conservation-2006/Proceedings-Marine_Nature_Conservation_in_Europe_2006.pdf#page=111

Thompson, R. C., Moore, C., Saal, F. and Swan, S. (2009a). Plastics, the environment, and human health: Current consensus and future trends. *Philosophical Transactions of the Royal Society*. 364(1526).
http://rstb.royalsocietypublishing.org/content/royptb/364/1526/2153.full.pdf

- "Annual production is likely to exceed 300 million tonnes by 2010."

Thompson, R. C., Swan, S., Moore, C. and vom Saal, F. (2009b). Our plastic age. *Philosophical Transactions of the Royal Society*. 364. pg. 1973-6.
http://rstb.royalsocietypublishing.org/content/royptb/364/1526/1973.full.pdf

Thompson, R. C. (2013). *Additional written evidence document WQ17*. Written Evidence of Water Quality: Priority Substances - Science and Technology Committee.
http://www.publications.parliament.uk/pa/cm201213/cmselect/cmsctech/writev/932/wq17.pdf

Thrall, L. (2006a). Are red blood cells defenseless against smaller nanoparticles? *Environmental Science & Technology*. 40(14). pg. 4327-8.
http://www.ncbi.nlm.nih.gov/pubmed/16903263

Thrall, L. (2006b). Study links TiO2 nanoparticles with potential for brain-cell damage. *Environmental Science & Technology*. 40(14). pg. 4326-7.

Tidd, Michael. (2008). The big idea: Polonium, radon and cigarettes. *Journal of the Royal Society of Medicine*. 101(3). pg. 156-7.
http://www.ncbi.nlm.nih.gov/pmc/articles/PMC2270238/

- "Smoking increases the number of particles in the air of a room by several orders of magnitude."
- "Submicroscopic electrically charged solids…become attached to the smoke particles, some of which are so small that they may remain suspended almost indefinitely."
- Po-210 in smoke nanoparticles may be transported by evaporation, washout or dry deposition to abiotic media where they may adhere to hydrophobic plastic nanoparticles.

Tokiwa, Y., Calabia, B., Ugwu, C. and Aiba, S. (2009). Biodegradability of plastics. *Int. J. Mol. Sci.* 10. pg. 3722-42.

Tomy, G., Budakowski, W., Halldorson, T., et al. (2004). Biomagnification of alpha- and gamma-hexabrom;ocyclododecane isomers in a lake Ontario food web. *Environmental Science and Technology*. 38(8). pg. 2298-303.
http://www.ncbi.nlm.nih.gov/pubmed/15116833

- "Whole body concentrations (ng/g, wet wt) of alpha- and gamma-HBCD were highest in the top predator lake trout samples ranging from 0.4 to 3.8 ng/g for the alpha-isomer and 0.1 to 0.8 ng/g for the gamma-isomer."

Tong, Z., Bischoff, M., Nies, L., Applegate, B. and Turco, F. Impact of fullerenes on a soil microbial community. *Environmental Science and Technology*. 441. pg. 2985-91. http://www.ph.ucla.edu/ehs/EHS280/articles/C60%20fullerenes%20%28Holden%29.pdf

- "The impact of manufactured nanomaterials on key soil processes must be addressed so that an unbiased discussion concerning the environmental consequences of nanotechnology can take place."
- "The introduction of fullerene, as either C60 or nC60, has little impact on the structure and function of the soil microbial community and microbial processes."

Topcu, E., Tonay, A., Dede, A., et. al. (2013). Origin and abundance of marine litter along sandy beaches of the Turkish Western Black Sea Coast. *Marine Environmental Research*. 85. pg. 21-8. http://www.sciencedirect.com/science/article/pii/S0141113612002243

Torkelson, A. R., Lanza, G. M., Birmingham, B. K., et al. (1988). Concentrations of insulin-like growth factor 1 (IGF-1) in bovine milk: Effect of herd, stage of lactations, and sometribove. *Journal of Dairy Science.*71. pg. 52.

Tovar-Sanchez, A., Sanchez-Quiles, D., Basterretxea, G., et al. (2013). Sunscreen products as emerging pollutants to coastal waters. *PLoS ONE*. 8(6). http://www.plosone.org/article/fetchObject.action?uri=info:doi/10.1371/journal.pone.0065451&representation=PDF

- "Sunscreen products are a significant source of organic and inorganic chemicals that reach the sea with potential ecological consequences on the coastal marine system."

Tratnyek, P. and Johnson, R. (2006). Nanotechnologies for environmental cleanup. *Nanotoday*. 1(2). pg. 44-8. http://www.denix.osd.mil/cmrmd/upload/NANORESTORATIONMAY06.PDF.

- "Among the many applications of nanotechnology that have environmental implications, remediation of contaminated groundwater using nanoparticles containing zero-valent iron (nZVI) is one of the most prominent examples of a rapidly emerging technology with considerable potential benefits. There are, however, many uncertainties regarding the fundamental features of this technology, which have made it difficult to engineer applications for optimal performance or to assess the risk to human or ecological health."

- "There are, however, many characteristics of this technology about which very little is known: e.g. how quickly nZVI will be transformed and to what products, whether this residue will be detectable in the environment, and how surface modifications of nZVI will alter its long-term environmental fate and effectiveness for remediation."

- "Particles have properties that are unique, or at least qualitatively different than those of larger particles. The most compelling examples of such properties arise only for particles smaller than ~10 nm, where particle size approaches the length-scale of certain molecular properties."

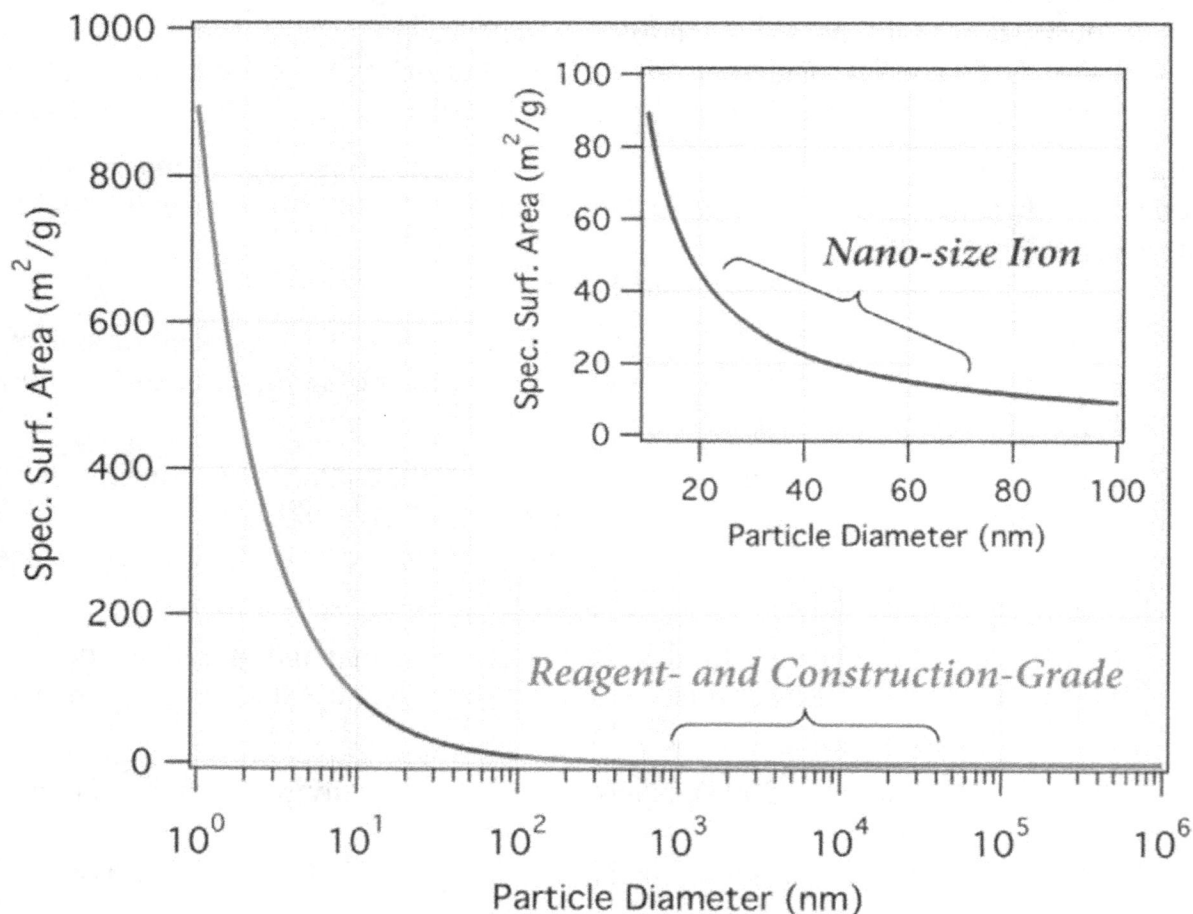

Fig. 2 *Particle surface area calculated from diameter assuming spherical geometry and density 6.7 g/cm³ (based on the average of densities for pure Fe⁰ and Fe₃O₄).*

- Fig. 2 is reprinted from *Nanotoday*, 1(2), Tratnyek, P. and Johnson, R. Nanotechnologies for environmental cleanup, pg. 46, Copyright 2006, with permission from Elsevier.

Trouiller, B., Reliene, R., Westbrook, A., Solaimani, P. and Schiestl, R. (2009). Titanium dioxide nanoparticles induce DNA damage and genetic instability in vivo in mice. *Cancer Research*. 69(22). http://www.ncbi.nlm.nih.gov/pmc/articles/PMC3873219/pdf/nihms-146780.pdf

Tsutumi, T., Amakura, Y., Nakamura, M., et al. (2003). Validation of the CALUX bioassay for the screening of PCDD/Fs and dioxin-like PCBs in retail fish. *Analyst*. 128(5). pg. 486-92. http://www.ncbi.nlm.nih.gov/pubmed/12790202

- "The chemical-activated luciferase expression (CALUX) assay is a reporter gene assay that detects dioxin-like compounds based on their ability to activate the aryl hydrocarbon receptor (AhR) and thus expression of the reporter gene. In this paper, the CALUX assay was examined for its application in the screening of polychlorinated dibenzo-p-dioxins (PCDDs), dibenzofurans (PCDFs) and dioxin-like polychlorinated biphenyls (dioxin-like PCBs) in retail fish."

Turner, J. T. (2002). Zooplankton fecal pellets, marine snow, and sinking phytoplankton blooms. *Aquatic Microbiology Ecology*. 27. pg. 57-102. http://www.int-res.com/articles/ame/27/a027p057.pdf

- "The sedimentary flux of fecal pellets, marine snow and sinking phytoplankton is an important component of the biological pump that not only transports and recycles materials in the sea but also may help scrub greenhouse gases from the atmosphere."

- "Some fecal pellets attach to marine snow, some forms of marine snow are fecal pellets, and some phytoplankton detritus forms marine snow."

- "Portions of this flux reach the benthos, but much, if not most, is repackaged or recycled in the water column."

- "The sedimentary flux of fecal pellets, marine snow and phytodetritus is important not only to communities on the sea bottom but also to those in the water column."

- Fecal pellets and marine snow are noted as a component of sedimentary flux.

- Comprehensive bibliography.

Turner, A. and Holmes, L. (2011). Occurrence, distribution and characteristics of beached plastic production pellets on the island of Malta (central Mediterranean).

Marine Pollution Bulletin. 62 (2). pg. 377-81.
http://www.ncbi.nlm.nih.gov/pubmed/21030052

Ueno, D., Takahashi, S., Tanaka, H., et al. (2003). Global pollution monitoring of PCBs and organochlorine pesticides using skipjack tuna as a bioindicator. *Archives of Environmental Contaminants Toxicology.* 45. pg. 378-89.
http://link.springer.com/article/10.1007/s00244-002-0131-9#page-1

- "Concentrations of organochlorines (OCs) representing persistent organic pollutants (POPs), such as polychlorinated biphenyls (PCBs), dichlorodiphenyl trichloroethane and its metabolites (DDTs), chlordane compounds (CHLs), hexachlorocyclohexane isomers (HCHs), and hexachlorobenzene (HCB), were determined in the liver of skipjack tuna (*Katsuwonus pelamis*) collected from the offshore waters of various regions in the world (offshore waters around Japan, Taiwan, Philippines, Indonesia, Seychelles, and Brazil, and the Japan Sea, the East China Sea, the South China Sea, the Bay of Bengal, and the North Pacific Ocean). OCs were detected in livers of all of the skipjack tuna collected from the locations surveyed."

- "OC residue levels were rather uniform among the individuals… this species is a suitable bioindicator for monitoring the global distribution of OCs in offshore waters and the open ocean."

Ugolini, A., Ungherese, G., Ciofini, M., et al. (2013). Microplastic debris in sandhoppers. *Estuarine, Coastal and Shelf Science.* 129. pg. 19-22.
https://www.deepdyve.com/lp/elsevier/microplastic-debris-in-sandhoppers-c0Q1DCGRXz

UK DEFRA. (2006). *UK voluntary reporting scheme for engineered nanoscale materials.* Chemicals and Nanotechnologies Division Defra, London, UK.
http://archive.defra.gov.uk/environment/quality/nanotech/documents/vrs-nanoscale.pdf

- "The nature and extent of risks to consumer health from ingestion of engineered free nanoparticles via food and drinks are currently unknown."

- "The likelihood of consumer exposure from the use of nanotechnology derived FCMs is intrinsically linked to the migration of nanoparticles into food and drinks."

- "Migration of nanoparticles is likely to be dependent on the type and composition of the polymer."

- "The study has highlighted a number of major knowledge gaps that require further research; for example, to understand the behavior, fate and toxicology of nanoparticles."

- No mention is made of the life cycle assessments of food contact materials after waste disposal and translocation to biogeochemical trophic level cycling.

UN-ECE. (1998). *Protocol to the 1979 convention on long range transboundary air pollution on persistent organic pollutants and executive body decision 1998/2 on information to be submitted and the procedure for adding substances to annexes I, II or III to the protocol on persistent organic pollutants*. ECE/EB.AIR/60. United Nations, New York and Geneva.
http://www.unece.org/fileadmin/DAM/env/lrtap/ExecutiveBureau/Handbbok.E.pdf

UNEP. (2001). *Stockholm Convention on Persistent Organic Pollutants*. United Nations Environment Programme.
http://www.chem.unep.ch/pops/POPs_Inc/dipcon/meetingdocs/25june2001/conf4_final act/en/FINALACT-English.PDF

- "In its decision 19/13 C of 7 February 1997, the Governing Council of the United Nations Environment Programme (UNEP) requested the Executive Director of UNEP, together with relevant international organizations, to prepare for and convene, by early 1998, an intergovernmental negotiating committee with a mandate to prepare an international legally binding instrument for implementing international action on certain persistent organic pollutants (POPs), initially beginning with 12 specified POPs: aldrin, chlordane, dieldrin, DDT, endrin, heptachlor, hexachlorobenzene, mirex, toxaphene, PCBs, dioxins and furans."

UNEP. (2005). *Marine litter: An analytical overview*. United Nations Environment Programme, Nairobi, Kenya.
http://www.unep.org/regionalseas/marinelitter/publications/docs/anl_oview.pdf

UNEP. (2011a). *Marine debris as a global environmental problem*. Scientific and Technical Advisory Panel, United Nations Environment Programme.
http://www.thegef.org/gef/sites/thegef.org/files/publication/STAP%20MarineDebris%2 0-%20website.pdf

UNEP. (2011b). *UNEP year book 2011: Emerging issues in our global environment*. United Nations Environment Programme, United Nations Publications, Nairobi, Kenya.

UNEP. (2013). *Microplastics brochure*. United Nations Environment Programme, United Nations Publications, Nairobi, Kenya.
http://www.unep.org/yearbook/2013/pdf/Microplastic_english.pdf

UNEP. (2014). *UNEP year book 2014: Emerging issues in our global environment*. United Nations Environment Programme, United Nations Publications, Nairobi, Kenya. http://www.unep.org/yearbook/2014/PDF/UNEP_YearBook_2014.pdf

UNEP Chemicals. (1999). *Guidelines for the identification of PCBs and materials containing PCBs*. United Nations Environment Programme (UNEP), Geneva, Switzerland. http://www.chem.unep.ch/pops/pdf/PCBident/pcbid1.pdf

UNEP and WHO. (2012). *State of the science of endocrine disrupting chemicals-2012*. http://www.unep.org/chemicalsandwaste/UNEPsWork/EndocrineDisruptingChemicals/tabid/130226/Default.aspx

Underwood, J. C., Harvey, R. W., Metge, D. W., et al. (2011). Effects of the antimicrobial sulfamethoxazole on groundwater bacterial enrichment. *Environmental Science and Technology*. 45(7). pg. 3096-101. http://toxics.usgs.gov/highlights/antibiotics_gw/

- "SMX is a sulfonamide antibiotic that is commonly used to treat a variety of bacterial infections. Previous studies have documented that SMX is a contaminant in both U. S. streams and groundwater, and that wastewater treatment plants and septic tanks are sources of antibiotics to the environment."

Unger, M., Harvey, E., Vadas, G. and Vecchione, M. (2008). Persistent pollutants in nine species of deep-sea cephalopods. *Marine Pollution Bulletin*. 56. pg. 1486-512. http://www.sciencedirect.com/science/article/pii/S0025326X0800218X

- "A key to understanding the mechanisms governing the fate of these compounds is to identify the source of POP's to predatory species."

- "Cephalopods are important in the diet of cetaceans; they form the primary food for 28 odontocete species. To determine if cephalopods were a potential source of contaminants to pelagic predators, nine species were collected in May 2003 from 1000-2000 m depths…Species were selected for chemical analysis based on their importance as prey."

- "Surrogate recoveries averaged 87% (SD=19) for the TBT analysis and 64% (SD=24) for the organic contaminants in the cephalopod samples."

University of Wisconsin Madison. (2014). *How small are nanotubes?* Materials Research Science and Engineering Center and the Internships in Public Science Education Project of the University of Wisconsin – Madison. http://education.mrsec.wisc.edu/IPSE/educators/activities/supplements/nanotube-Handout.pdf

United Nations General Assembly. (2006). *The impacts of fishing on vulnerable marine ecosystems: Actions taken by States and regional fisheries management organizations and arrangements to give effect to paragraphs 66 to 69 of General Assembly resolution 59/25 on sustainable fisheries, regarding the impacts of fishing on vulnerable marine ecosystems.* UN. http://www.un.org/Depts/los/general_assembly/documents/impact_of_fishing.pdf

US CDC. (2009). *Fourth national report on human exposure to environmental chemicals.* Centers for Disease Control. http://www.cdc.gov/exposurereport/

- "Polychlorinated dibenzo-*p*-dioxins and dibenzofurans are two similar classes of chlorinated aromatic chemicals that are produced as contaminants or by-products. They have no know commercial or natural use."

- "Both the synthesis and heat-related degradation of polychlorinated biphenyls (PCBs) will produce dibenzofuran byproducts. Releases from industry sources have decreased approximately 80% since the 1980s."

- "Today, the largest release of these chemicals occurs as a result of the open burning of house-hold and municipal trash, landfill fires, and agricultural and forest fires. When advanced analytical techniques are used, most soil and water samples will reveal trace amounts of polychlorinated dibenzo-*p*-dioxins and dibenzofurans."

- 1,2,3,6,7,8-Hexachlorodibenzo-p-dioxin (HxDDD) (whole weight) is reported in serum concentrations (in fg/g of serum or parts per quadrillion) for the U.S. population from the National Health and Nutrition Examination Survey. This CDC table analyzes data from 1999 to 2004 in two year increments. The highest levels of contamination were reported in 2001-02. For example, female sample sizes of 997, 670, and 951 showed levels of 400, 644, and 363 fg/g of serum in years 1999-2000, 2001-02, and 2003-04.

- Most dioxin cogeners are reported in parts per trillion pg/g in serum or lipids.

- The CDC measures dioxins in parts per quadrillion (fg/g). Why is National Security Agency (NSA) survey data on environmental chemicals in biotic media, such as CAFO meats or high fructose corn syrup, not made publically available? Where is Edward Snowden when we need him?

US CDC. (2010). *Prevalence of overweight among children and adolescents: United States, 2003-2004.* Centers for Disease Control. http://www.cdc.gov/nchs/data/hestat/overweight/overweight_child_03.htm

US CDC. (2011). *Diabetes: Successes and opportunities for population-based prevention and control at a glance, 2011*. Centers for Disease Control. http://www.cdc.gov/chronicdisease/resources/publications/AAG/ddt.htm

US CDC. (2013). *Biomonitoring summary: Polybrominated diphenyl ethers and 2,2',4,4',5,5'-Hexabromodiphenyl(BB-153)*. National Biomonitoring Program. http://www.cdc.gov/biomonitoring/PBDEs_BiomonitoringSummary.html

US CDC. (2014). *Fourth national report on human exposure to environmental chemicals: Updated tables*. Centers for Disease Control. http://www.cdc.gov/exposurereport/

US CDC. (2015). *National biomonitoring program: Chemical fact sheets*. Centers for Disease Control and Prevention. http://www.cdc.gov/biomonitoring/chemical_factsheets.html

US Department of Agriculture. (2012). *Colony collapse disorder progress report*. USDA. http://www.ars.usda.gov/is/br/ccd/ccdprogressreport2012.pdf

US EIA. (2015a). *International energy statistics database*. US Energy Information Administration. http://www.eia.gov/cfapps/ipdbproject/IEDIndex3.cfm?tid=1&pid=7&aid=1

US EIA. (2015b). *U.S. field production of crude oil*. US Energy Information Administration. http://www.eia.gov/dnav/pet/hist/LeafHandler.ashx?n=PET&s=MCRFPUS1&f=M

US Environmental Protection Agency. (1973). *Combustion products from the incineration of plastics*. US EPA, Washington, DC. http://deepblue.lib.umich.edu/handle/2027.42/3702

US Environmental Protection Agency. (1997a). *TRI: Toxics release inventory*. *Chemicals in the Environment*. 3(2). http://www2.epa.gov/toxics-release-inventory-tri-program

- "From 1987 to 1990, TRI data focused on release and transfer data for approximately 2300 chemicals and 28,000 facilities. With the passage of the Pollution Prevention Act of 1990 (PPA), information on the amount of toxic materials leaving a facility in waste was added."

- "The rules implementing the PPA were subjected to rigorous review and EPA was unable to finalize working definitions that would assure consistency in the data."

US Environmental Protection Agency. (1997b). *Evaluation of emissions from the burning of household waste in barrels*. US EPA, Washington, DC. http://www.epa.gov/ttn/catc/dir1/barlbrn2.pdf

US Environmental Protection Agency. (2003). *Basic questions regarding acute exposure guideline levels (AEGLs) in emergency planning and response*. US EPA, Washington, DC. https://www.osha.gov/SLTC/emergencypreparedness/chemical/pdf/tier_2-aegls_basic_usachppm1_03.pdf

US Environmental Protection Agency. (2004). *Chemical hazard classification and labeling: Comparison of OPP requirements and the GHS*. US EPA, Washington, DC. http://www.epa.gov/oppfead1/international/global/ghscriteria-summary.pdf

US Environmental Protection Agency. (2004). *Polychlorinated biphenyl inspection manual*. US EPA, Washington, DC. http://www2.epa.gov/sites/production/files/2013-09/documents/pcbinspectmanual.pdf

US Environmental Protection Agency. (2005). *Water quality study of bays in coastal Mississippi water quality report*. Science and Ecosystem Support Division, US EPA, Washington, DC. http://www.epa.gov/region4/sesd/reports/2005-0926/05msbay-report.WebFinal.pdf

US Environmental Protection Agency. (2006a). *Response to 2005 hurricanes: Test results*. US EPA, Washington, DC. http://www.epa.gov/katrina/testresults/index.html

US Environmental Protection Agency. (2006b). *An inventory of sources and environmental releases of dioxin-like compounds in the US for the years 1987, 1995, and 2000 (final, Nov 2006)*. US EPA, Washington, DC. http://cfpub.epa.gov/ncea/CFM/recordisplay.cfm?deid=159286

- "The major identified sources of environmental releases of dioxin-like compounds are grouped into six broad categories: combustion sources, metals smelting, refining and process sources, chemical manufacturing sources, natural sources, and environmental reservoirs."

- "The quantitative results are expressed in terms of the toxicity equivalent (TEQ) of the mixture of polychlorinated dibenzo-p-dioxin (CDD) and polychlorinated dibenzofuran (CDF) compounds present in environmental releases using a procedure sanctioned by the World Health Organization (WHO) in 1998."

- "This analysis indicates that between reference years 1987 and 2000, there was approximately a 90% reduction in the releases of dioxin-like compounds to the circulating environment of the United States from all known sources combined."

- "In 1987 and 1995, the leading source of dioxin emissions to the U.S. environment was municipal waste combustion; however, because of reductions in dioxin emissions from municipal waste combustors, it dropped to the fourth ranked source in 2000."

- "Burning of domestic refuse in backyard burn barrels remained fairly constant over the years, but in 2000, it emerged as the largest source of dioxin emissions to the U.S. environment."

US Environmental Protection Agency. (2007). *Nanotechnology white paper*. US EPA, Washington, DC. http://www.epa.gov/osa/pdfs/nanotech/epa-nanotechnology-whitepaper-0207.pdf

US Environmental Protection Agency. (2008). *Toxicological review of 2,2',4,4'-tetrabromodiphenyl ether (BDE-47)*. US EPA, Washington, DC. http://www.epa.gov/iris/toxreviews/1010tr.pdf

US Environmental Protection Agency. (2009a). Consumer factsheet on polychlorinated biphenyls. In: *National primary drinking water regulations*. US EPA, Washington, DC. http://www.epa.gov/ogwdw/pdfs/factsheets/soc/pcbs.pdf

US Environmental Protection Agency. (2009b). *Targeted national Sewage Sludge Survey*. US EPA, Washington, DC. http://water.epa.gov/scitech/wastetech/biosolids/upload/2009_04_23_biosolids_tnsss-overview.pdf

- "Section 405(d) of the Clean Water Act (CWA) requires the U.S. Environmental Protection Agency (EPA) to identify and regulate toxic pollutants that may be present in biosolids at levels of concern for public health and the environment."

- Briefly, the survey found:
  - The four anions were found in every sample.
  - 27 metals were found in virtually every sample, with one metal (antimony) found in no less than 72 samples.
  - Of the six semivolatile organics and polycyclic aromatic hydrocarbons, four were found in at least 72 samples, one was found in 63 samples, and one was found in 39 samples.
  - Of the 72 pharmaceuticals, three (i.e., cyprofloxacin, diphenhydramine, and triclocarban) were found in all 84 samples and nine were found in at least 80 of the samples. However, 15 pharmaceuticals were not found in any sample and 29 were found in fewer than three samples.
  - Of the 25 steroids and hormones, three steroids (i.e., campesterol, cholestanol,

and coprostanol) were found in all 84 samples and six steroids were found in at least 80 of the samples. One hormone (i.e., 17α-ethynyl estradiol) was not found in any sample and five hormones were found in fewer than six samples.

- All of the flame retardants except one (BDE-138) were essentially found in every sample; BDE-138 was found in 54 out of 84 samples.

- "It is not appropriate to speculate on the significance of the results until a proper evaluation has been completed and reviewed."

US Environmental Protection Agency. (2010a). *An exposure assessment of Polybrominated Diphenyl Ethers (PBDE) (FINAL)*. EPA/600/R-08/086F. US EPA, Washington, DC. http://cfpub.epa.gov/si/si_public_record_Report.cfm?dirEntryId=210404

- "Unlike other POPs, however, the key routes of human exposure are thought to be from their use in household consumer products, and their presence in house dust, and not from dietary routes."

- "The estimated adult intake dose of 7.1 ng/kg-day was predicted to result in a body burden of 31.0 ng/g lipid weight (lwt). This compared to body burdens of 36.3 ng/g lwt found in blood and 44.1 ng/g lwt found in mother's milk."

- "Children's estimated intakes were higher at 47.2 ng/kg/day for ages 1–5, 13.0 ng/kg/day for 6–11, and 8.3 ng/kg/day for 12–19. Infant intakes due to ingestion of mother's milk were the highest at 141 ng/kg/day."

- This is a comprehensive survey of PBDEs in abiotic and biotic environments, including humans. It has 378 pages with 29 tables and 9 figures.

- Translocation of PBDE by PNP to humans would be a tertiary route, but may be a primary vector of exposure for fish and marine mammals such as seals.

US Environmental Protection Agency. (2010b). *Persistent organic pollutants (POPs) in humans and wildlife*. US EPA, Washington, DC. http://www.who.int/ipcs/publications/en/ch_3a.pdf

- "Options to reduce the risk of exposure to 'dioxin-like' compounds by humans and wildlife include the destruction of existing stockpiles of PCBs and related compounds (e.g. >800°C incineration); changes to the manufacturing and combustion processes that lead to the formation of such compounds (e.g. dioxins and furans as by-products of pesticide and herbicide manufacture, as by-products of the pulp bleaching process, or as by-products of low temperature combustion); and the remediation of contaminated sites."

US Environmental Protection Agency. (2011a). *Marine debris in the North Pacific: A summary of existing information and identification of data gaps*. Weston Solutions. http://www.epa.gov/region9/marine-debris/pdf/MarineDebris-NPacFinalAprvd.pdf

- "Further data is also needed to assess trophic transfer dynamics of POPs via plastic throughout the marine food web. Currently, there is only a limited amount of data on the transfer of POPs from plastic marine debris to conspicuous marine organisms, such as sea birds. By studying the effects of plastics on the planktonic and benthic invertebrate communities, as well as top predators, a better understanding of potentially important and currently understudied impacts of plastics can be gained. Trophic transfer potential also has important human health risk ramifications. Impacts to humans from consumption of fish and invertebrates that ingest plastics are in large part currently unknown. Therefore, collection of tissues from fish that comprise part of the pelagic food web will be important for understanding the potential for transfer of plastic particles and POPS to humans."

US Environmental Protection Agency. (2011b). *Exposure factors handbook*. National Center for Environmental Assessment. Washington, DC. http://www.epa.gov/ncea/efh/pdfs/efh-frontmatter.pdf

- "The Exposure Factors Handbook provides information on various physiological and behavioral factors commonly used in assessing exposure to environmental chemicals."

US Environmental Protection Agency. (2013a). *Outdoor air: Industry, business, and home: Backyard trash burning – additional information*. US EPA, Washington, DC. http://www.epa.gov/oaqps001/community/details/barrelburn_addl_info.html

US Environmental Protection Agency. (2013b). *1,3-Butadiene*. US EPA, Washington, DC. http://www.epa.gov/airtoxics/hlthef/butadien.html

US Environmental Protection Agency. (2013c). *Phthalic anhydride*. US EPA, Washington, DC. http://www.epa.gov/airtoxics/hlthef/phthalic.html

US Environmental Protection Agency. (2014a). *Combustion*. U.S. EPA, Washington, DC. http://epa.gov/climatechange/wycd/waste/downloads/combustion-chapter10-28-10.pdf

US Environmental Protection Agency. (2014b). *Municipal solid waste generation, recycling, and disposal in the United States: Tables and figures for 2012*. US Environmental Protection Agency. http://www.epa.gov/osw/nonhaz/municipal/pubs/2012_msw_dat_tbls.pdf and http://www.epa.gov/osw/nonhaz/municipal/pubs/2012_msw_fs.pdf

- "In 2012, Americans generated about 251 million tons of trash and recycled and composted almost 87 million tons of this material, equivalent to a 34.5 percent recycling rate... On average, Americans recycled and composted 1.51 pounds out of our individual waste generation rate of 4.38 pounds per person per day."

- "EPA is thinking beyond waste and seeking a systematic approach that provides a transition from waste management to sustainable materials management (SMM)."

- Table 7 shows that 31,750 thousand tons of plastics were found in MSW; only 2,800 thousand tons were recovered (8.8%).

- Approximately 5,000,000 tons per year are destined for marine environments (6 months to 10 years translocation time).

- No mention is made of any household or workshop chemicals in this report.

- The chemical fallout cover-up continues.

US Environmental Protection Agency. (2014c). *Plastics: Introduction to WARM and plastics*. US EPA, Washington, DC. http://www.epa.gov/climatechange/wycd/waste/downloads/plastics-chapter10-28-10.pdf

US Environmental Protection Agency. (2015). *TRI National Analysis 2013 Updated January 2015*. http://www2.epa.gov/sites/production/files/2015-01/documents/2013-tri-national-analysis-complete_1.pdf

US Geological Survey. (2000). *Atmospheric deposition program of the U.S. Geological Survey*. USGS. https://bqs.usgs.gov/acidrain/program.pdf

US Geological Survey. (2006). *Pesticides in the nation's streams and groundwater, 1992-2001: A summary*. USGS. http://pubs.usgs.gov/fs/2006/3028/pdf/fs2006-3028.pdf

- "Pesticides are frequently present in streams and ground water, are seldom at concentrations likely to affect humans, but occur in many streams at concentrations that may have effects on aquatic life or fish-eating wildlife."

US Geological Survey. (2013). *Deficit in nation's aquifers accelerating*. USGS. http://www.usgs.gov/newsroom/article.asp?ID=3595&from=rss#.VTK0XSHBzGc

- "*Groundwater Depletion in the United States (1900-2008)* comprehensively evaluates long-term cumulative depletion volumes in 40 separate aquifers (distinct underground water storage areas) in the United States."

- "From 1900 to 2008, the Nation's aquifers, the natural stocks of water found under the land, decreased (were depleted) by more than twice the volume of water found in Lake Erie. Second, groundwater depletion in the U.S. in the years 2000-2008 can explain more than 2 percent of the observed global sea-level rise during that period."

- "During the most recent period of the study (2000–2008)…the depletion rate averaged almost 25 cubic kilometers per year."

- "Substantial pumping of the High Plains aquifer for irrigation since the 1940s has resulted in large water-table declines that exceed 160 feet in places."

US Geological Survey. (2014a). *Complex response to decline in atmospheric deposition of mercury*. USGS. http://toxics.usgs.gov/highlights/2014-06-12-mercury_deposition.html

US Geological Survey. (2014b). *Contaminants found in groundwater*. USGS. https://water.usgs.gov/edu/groundwater-contaminants.html

- "Industrial discharges, urban activities, agriculture, ground-water pumpage, and disposal of waste all can affect ground-water quality. Contaminants can be human-induced, as from leaking fuel tanks or toxic chemical spills. Pesticides and fertilizers applied to lawns and crops can accumulate and migrate to the water table."

US Geological Survey. (2014c). *Neonicotinoid insecticides documented in Midwestern U.S. streams*. USGS. http://toxics.usgs.gov/highlights/2014-07-21-neonics.html

US Geological Survey. (2014d). *New knowledge on the fate and transport of emerging contaminants in rivers*. USGS. http://toxics.usgs.gov/highlights/redwood_river.html

- "A detergent degradation product (4-nonylphenol) and a biogenic hormone (17β-estradiol) added to the Redwood River, Minnesota, were attenuated by biodegradation and other natural processes."

- "Because sorption occurs more rapidly than biodegradation, these compounds will bioaccumulate in biofilms, which are an important source of food for stream food webs."

US Government Accountability Office. (1988). *Toxic substances: PCB spill at the Guam naval power generating plant*. US GAO, Washington, DC. http://www.gao.gov/assets/150/146926.pdf

US Government Accountability Office. (2010). *GAO Report GAO-11-63, Afghanistan and Iraq: DOD should improve adherence to its guidance on open pit burning and*

solid waste management. US GAO, Washington, DC.
http://www.gao.gov/htext/d1163.html

US National Park Service. (2005). *Wet deposition monitoring protocol*. National Park Service, Air Resources Division Research and Monitoring Branch, Lakewood, CO.
http://www.nature.nps.gov/air/Monitoring/docs/200508FinalWetDepProtocol.pdf

- The National Atmospheric Deposition Monitoring Network (NADP) measurement parameters include mercury, wet and dry deposition of sulfate, nitrate, ammonia, wet deposition of calcium, magnesium, potassium, sodium, chloride, and dry deposition of nitric acid and sulfur dioxide.

- No mention is made of chemical fallout (±10,000 chemicals) often in quantities below the limit of detection, but destined to be biomagnified in the biotic environment.

US National Park Service. (2007). *Final preliminary assessment report for potential radiological contamination at Great Kills Park*. National Park Service, Washington Office, Boulder, CO.
http://www.nps.gov/gate/parkmgmt/upload/AR0000305_PA_Report-2.pdf

US NRC. (1999). *Hormonally active agents in the environment*. National Research Council. National Academies Press. Washington, DC.

Usenko, C., Harper, S. and Tanguay, R. (2008). Fullerene C60 exposure elicits an oxidative stress response in embryonic zebrafish. *Toxicological Applied Pharmacology*. 229. pg. 44-55.
http://www.ncbi.nlm.nih.gov/pmc/articles/PMC2421009/pdf/nihms50228.pdf

- "C60 can act as a pro-oxidant and elicit a toxic response via oxidative stress."

Uskokovic, Vuk. (2007). Nanotechnologies: What we do not know. *Technology in Society*. 29. pg. 43-61.

Uskokovic, V. and Bertassoni, L. (2010). Nanotechnology in dental sciences: Moving towards a finer way of doing dentistry. *Materials*. 3. pg. 1674-91.
http://www.mdpi.com/1996-1944/3/3/1674/pdf

Vajda, A., Barber, L., Gray, J., et al. (2011). Demasculinization of male fish by wastewater treatment plant effluent. *Aquatic Toxicology*. 103. pg. 213-21.
http://web.stcloudstate.edu/aquatictox/projects/Publications/Vajda11_FEM.pdf

- "These results support the hypothesis that the reproductive disruption observed in this watershed is due to endocrine-active chemicals in the WWTP effluent."

Valavanidis, A., Iliopoulos, N., Gotsis, G. and Fiotakis, K. (2008). Persistent free radicals, heavy metals and PAHs generated in particulate soot emissions and residue ash from controlled combustion of common types of plastic. *Journal of Hazardous Materials*. 156(1-3). pg. 277-84. http://www.ncbi.nlm.nih.gov/pubmed/18249066

- "Low molecular weight PAHs were at higher concentrations in the airborne particulate soot than in the residue solid ash."

Valavanidis, A. and Vlachogianni, T. (2010). *Nanomaterials and nanoparticles in the aquatic environment: Toxicological and ecotoxicological risks*. Department of Chemistry, University of Athens. http://chem-tox-ecotox.org/wp/wp-content/uploads/2010/09/05-Nanoparticles-27_09_20101.pdf

- "The small size of nanoparticles and their properties can become easily a vehicle for binding and transport of toxic chemical pollutants."

- "NPs and nanotubes can be released in the environment and cause harmful effects to humans and/or living organisms."

- "Uptake by endocytotic routes are identified as probable major mechanisms of entry cellular compartments leading to various types of toxic cell injury through free radical reactions."

VanBriesen, J. and Bridges, T. (2011). *Final report: Modeling and decision support tools based on the effects of sediment geochemistry and microbial populations on contaminant reactions in sediments*. Strategic Environmental Research and Development Program. https://www.serdp-estcp.org/Program-Areas/Environmental-Restoration/Contaminated-Sediments/ER-1495

- "PCBs are a primary contaminant driving risk at many Department of Defense facilities."

- "Lack of time and resources for extensive characterization at all sites introduces significant uncertainty in assessing biodegradation potential."

Van den Berg, M., Birnbaum, L. S., Denison, M., et al. (2006). The 2005 World Health Organization reevaluation of human and mammalian toxic equivalency factors for dioxins and dioxin-like com-pounds. *Toxicological Science*. 93. pg. 223-41.

Vandenberg, L., Hauser, R., Marcus, M., Olea, N. and Welshons, W. (2007). Human exposure to bisphenol A. *Reproductive Toxicology*. 24(2). pg. 137-77. http://www.loe.org/images/content/070803/Vandenberg%20Exposure%20Rep%20Tox%20resubmission.pdf

Vandenberg, L., Maffini, M., Sonnenschein, C., et. al. (2009). Bisphenol-A and the great divide: A review of controversies in the field of endocrine disruption. *Endocrinology Review*. 30. pg. 75-95. http://press.endocrine.org/doi/pdf/10.1210/er.2008-0021

- "The term 'endocrine disruptor' was agreed upon to describe a class of chemicals including those that act as agonists and antagonists of the estrogen receptors (ERs), androgen receptor, thyroid hormone receptor, and others."

- "Six additional issues that have divided scientists in the field of BPA research, namely: 1) mechanisms of BPA action; 2) levels of human exposure; 3) routes of human exposure; 4) pharmacokinetic models of BPA metabolism; 5) effects of BPA on exposed animals; and 6) links between BPA and cancer."

Van der Wiel, S., De Franceschi, J., Elzerman, T., et al. (2003). Electron transport through double quantum dots. *Revolutionary Modern Physics*. 75(1). pg. 1-22. http://journals.aps.org/rmp/abstract/10.1103/RevModPhys.75.1

Vasiliu, O., Cameron, L., Gardiner, J., Deguire, P. and Karmaus, W. (2006). Polybrominated biphenyls, polychlorinated biphenyls, body weight, and incidence of adult-onset diabetes mellitus. *Epidemiology*. 17. pg. 352-9. http://journals.lww.com/epidem/pages/articleviewer.aspx?year=2006&issue=07000&article=00003&type=abstract

Veltman, K., Hendriks, J., Huijbregts, M., et al. (2005). Accumulation of organochlorines and brominated flame retardants in estuarine and marine food chains: Field measurements and model calculations. *Mar Pollut Bull*. 50. pg. 1085-102.

Velzeboer, I., Kwadijk, C. and Koelmans, A. (2014). Strong sorption of PCBs to nanoplastics, microplastics, carbon nanotubes, and fullerenes. *Environmental Science and Technology*. 48(9). pg. 4869-76. http://pubs.acs.org/doi/abs/10.1021/es405721v

- "PCB Sorption to MWCNT and nano-PS was nonlinear. PCB sorption to MWCNT and C_{60} was 3–4 orders of magnitude stronger than to OM and micro-PE. Sorption to nano-PS was 1–2 orders of magnitude stronger than to micro-PE, which was attributed to the higher aromaticity and surface–volume ratio of nano-PS. Organic matter effects varied among sorbents, with the largest OM fouling effect observed for the high surface sorbents MWCNT and nano-PS."

- "The exceptionally strong sorption of (planar) PCBs to C_{60}, MWCNT, and nano-PS may imply increased hazards upon membrane transfer of these particles."

Venkatesan, R., Dwarakadasa, E. and Ravindran, M. (2003). Biofilm formation on structural materials in deep sea environments. *Indian Journal of Engineering &*

Materials Sciences. 10. pg. 486-91.
http://nopr.niscair.res.in/bitstream/123456789/24315/1/IJEMS%2010%286%29%20486-491.pdf

Vennila, A., Jayasiri, H. and Pandey, P. (2014). Plastic debris in the coastal and marine ecosystem: A menace that needs concerted efforts. *International Journal of Fisheries and Aquatic Studies*. 2(1). pg. 24-9. http://fisheriesjournal.com/vol2issue1/Pdf/16.1.pdf

Verplanck, P., Taylor, H., Nordstrom, D. and Barber, L. (2005). Aqueous stability of gadolinium in surface waters receiving sewage treatment plant effluent, Boulder Creek, Colorado. *Environmental Science and Technology*. 39(18). pg. 6923-9.
http://pubs.acs.org/doi/abs/10.1021/es048456u

Versar. (2010). *Review of exposure data and assessments for select dialkyl ortho-phthalates*. US Consumer Product Safety Commission, Bethesda, MD.
http://www.cpsc.gov/PageFiles/126552/pthalexp.pdf

Vethaak, A., Winther-Nielsen, M. and Reifferscheid, G. (2014). Microplastics in freshwater ecosystems: What we know and what we need to know. *Environmental Sciences Europe*. 26(12). http://www.enveurope.com/content/pdf/s12302-014-0012-7.pdf

- "To assess the environmental risk associated with MP, comprehensive data on their abundance, fate, sources, and biological effects in freshwater ecosystems are needed."

Vijgen, John. (2006). *The legacy of Lindane HCH isomer production*. International HCH & Pesticides Association.
http://www.ihpa.info/docs/library/reports/Lindane%20Main%20Report%20DEF20JAN06.pdf

vom Saal, F. S. and Meyers, J. P. (2008). Bisphenol A and risk of metabolic disorders. *JAMA*. 300. pg. 1353-5.
http://endocrinedisruptors.missouri.edu/pdfarticles/vomsaal/2008/vomsaal%20JAMA.%20NHANES%20BPA%20Editorial%202008.pdf

- "Based on their analysis of data from the National Health and Nutrition Examination Survey 2003-2004, Lang et al report a significant relationship between urine concentrations of BPA and cardiovascular disease, type 2 diabetes, and liver-enzyme abnormalities in a representative sample of the adult US population."

Vom Saal, F. S., Nagel, S. C., Coe, B. L., et al. (2012). The estrogenic endocrine disrupting chemical bisphenol A (BPA) and obesity. *Mol. Cell. Endocrinol.* 354(1-2). pg. 74-84.

Voparil, I. and Mayer, L. (2000). Dissolution of sedimentary polycyclic aromatic hydrocarbons into the lugworm's (*Arenicola marina*) digestive fluids. *Environmental Science & Technology.* 34. pg. 1221-8. http://pubs.acs.org/doi/abs/10.1021/es990885i

- "Gut fluid concentrations of high molecular weight PAHs are greater than those predicted from equilibrium partitioning theory, indicating the importance of the digestive pathway for hydrophobic organic contaminant exposure and bioaccumulation."

Vorkamp, K., Dam, M., Riget, F., et al. (2004). *Screening of "new" contaminants in the marine environment of Greenland and the Faroe Islands*. NERI Technical Report No. 525. National Environmental Research Institute, Denmark.

Votavova, L. Dobias, J., Voldrich, M. and Cizkova, H. (2009). Migration of nonylphenols from polymer packaging materials into food stimulants. *Czech Journal of Food Science.* 27(4). pg. 293-9.http://www.agriculturejournals.cz/publicFiles/10123.pdf

Wagner, M., Scherer, C., Alvarez-Muñoz, D., et al. (2005). Characterization of trophic transfer for polychlorinated dibenzo-p-dioxins, dibenzofurans, non and mono ortho polychlorinated biphenyls in the marine food web of Bohai Bay, North China. *Environmental Science and Technology.* 39(8). pg. 2417-25. http://pubs.acs.org/doi/pdf/10.1021/es048657y

Wagner, M. and Oehlmann, J. (2009). Endocrine disruptors in bottled mineral water: Estrogenic activity in the E-Screen. *Environmental Science Pollution Research.* 16(3). pg. 278-86. http://www.researchgate.net/profile/Joerg_Oehlmann/publication/47678469_Endocrine _disruptors_in_bottled_mineral_water_estrogenic_activity_in_the_E-Screen/links/02e7e51dc223ca705b000000.pdf

- "When comparing water of the same spring that is packed in glass or plastic bottles made of polyethylene terephthalate (PET), estrogenic activity is three times higher in water from plastic bottles."

- "These data support the hypothesis that PET packaging materials are a source of estrogen-like compounds…the contamination of bottled water with endocrine disruptors is a transnational phenomenon."

Wang, D., Atkinson, S., Hoover-Miller, A. and Li, Q. X. (2007). Polychlorinated naphthalenes and coplanar polychlorinated biphenyls in tissues of harbour seals (*Phoca vitulina*) from the northern Gulf of Alaska. *Chemosphere*. 67. pg. 2044-57.

Wang, Q., Zhao, L., Fang, X., et al. (2013). Gridded usage inventories of chlordane in China. *Frontiers of Environmental Science & Engineering*. 7(1). pg. 10-8. http://link.springer.com/article/10.1007/s11783-012-0458-z

Wang, X., Hong, H., Zhao, D. and Hong, L. (2008). Environmental behavior of organotin compounds in the coastal environment of Xiamen, China. *Marine Pollution Bulletin*. 57. pg. 419-24. http://mel.xmu.edu.cn/upload_eg/20094231154889143.pdf

- "OTs can be enriched in the microlayer water and that the potential toxicity of these pollutants to larval organisms should be of some concern."

Wang, X., Li, Q., Xie, J., Jin, Z., Wang, J., Li, Y., Jiang, K. and Fan, S. (2009). Fabrication of ultralong and electrically uniform single-walled carbon nanotubes on clean substrates. http://www.chem.pku.edu.cn/page/liy/labhomepage/publications/2009/2009NL.pdf

Wang, X., Guo, Y., Yang, L., et al. (2012). Nanomaterials as sorbents to remove heavy metal ions in wastewater treatment. *Evnironmental and Analytical Toxicology*. 2(7). http://omicsonline.org/nanomaterials-as-sorbents-to-remove-heavy-metal-ions-in-wastewater-treatment-2161-0525.1000154.pdf

- "With the development of nanotechnology, nanomaterials are used as the sorbents in wastewater treatment...due to their unique structure properties. Three kinds of nanomaterials are presented in this paper, including nanocarbon materials, nanometal particles, and polymer-supported nanoparticles. For heavy metal ions, all these nanomaterials show high selectivities and adsorption capacities."

Wang, Ying-Ying, Lai, Samuel K., So, Conan, et al. (2011). Mucoadhesive nanoparticles may disrupt the protective human mucus barrier by altering its microstructure. *PLoS ONE*. 6(6). http://journals.plos.org/plosone/article?id=10.1371/journal.pone.0021547

Wania, F. and Mackay, D. (1996). Tracking the distribution of POPs. *Environ Sci Technol*. 30. pg. 390-6.

Ward, J. and Kach, D. (2009). Marine aggregates facilitate ingestion of nanoparticles by suspension-feeding bivalves. *Marine Environment Research*. 68(3). pg. 137-42. http://www.ncbi.nlm.nih.gov/pubmed/19525006

- "Aggregates significantly enhance the uptake of 100-nm particles. Nanoparticles had a longer gut retention time than 10-micron polystyrene beads."

Ward, J. and Shumway, S. (2004). Separating the grain from the chaff: Particle selection in suspension- and deposit-feeding bivalves. *Journal of Experimental Marine Biology and Ecology*. 300(1-2). pg. 83-130. http://sandy.heupel.com/pubs/journal_of_experimental_marine_biology_ecology/ward_shumway_2004.pdf

- "Particle feeding on suspended and deposited material is a common mode of food collection among many groups of the Metazoa."

- "Introduction of electronic particle counters led to a better understanding of the rates and efficiencies at which particle removal occurred."

- "The introduction of more advanced technologies (e.g., flow cytometry, video endoscopy, confocal microscopy)…allowed more detailed studies of the mechanisms associated with particle uptake and selection by these animals."

Warheit, D., Laurence, B., Reed, K., et al. (2004). Comparative pulmonary toxicity assessment of single-wall carbon nanotubes in rats. *Toxicological Sciences*. 77(1). pg 117-25. http://toxsci.oxfordjournals.org/content/77/1/117.full.pdf

- "The pulmonary effects of carbon nanotubes must be evaluated by generating aerosols of SWCNT and, thus, conducting an inhalation toxicity study of carbon nanotubes in rats."

Warheit, D. (2006). What is currently known about the health risks of carbon nanotube exposures? *Carbon*. 44(6). pg. 1064-9. http://www.researchgate.net/publication/223916967_What_is_currently_known_about_the_health_risks_related_to_carbon_nanotube_exposures

- "The potential hazards of inhalation exposure to carbon nanotubes have not been sufficiently evaluated."

- "SWCNT have a strong tendency to agglomerate following intratracheal exposures."

Waring, R. and Harris, R. (2011). Endocrine disrupters: A threat to women's health? *Maturitas*. 68. pg. 111-5. http://www.maturitas.org/article/S0378-5122(10)00386-5/pdf

- "Endocrine disrupters (EDs) are compounds which may be of industrial or natural origin and which act to dysregulate steroid function and metabolism."

- "They not only affect reproductive function but also affect a range of tissues which are steroid sensitive such as the central nervous system and thyroid."

- "EDs may also affect the immune system, glucose homeostasis and can act as epigenetic modulators resulting in transgenerational effects."

- "Despite many years of effort, the effects on human health of long-term environmental exposure to EDs, whether singly or as mixtures, remain unknown."

Watkins, D., McClean, M., Fraser, A., et. al. (2012). Exposure to PBDEs in the office environment: Evaluating the relationship between dust, handwipes and serum. *Environmental Science and Technology*. 46(2). pg. 1192-200. http://www.ncbi.nlm.nih.gov/pmc/articles/PMC3230398/

- "Exposure to pentaBDE in the office environment contributes to pentaBDE body burden, with exposure likely linked to PBDE residues on hands."

Watras, C. J. and Huckabee, J. W. eds. (1994). *Mercury pollution: Integration and synthesis*. Lewis Publishers, Chelsea, MI.

Watson, R. and Pauly, D. (2001). Systematic distortions in world fisheries catch trends. *Nature*. 414. pg. 534-6. http://www.seaaroundus.org/researcher/dpauly/PDF/2001/JournalArticles/SystematicDistortionsWorldFisheriesCatchTrends.pdf

Watts, A., Lewis, C., Goodhead, R., et al. Uptake and retention of microplastics by the shore crab *Carcinus maenas*. *Environmental Science Technology*. 48(15). pg. 8823-30. http://pubs.acs.org/doi/abs/10.1021/es501090e

Wegner, A., Besseling, E., Foekema, E., et al. (2012). Effects of nanopolystyrene on the feeding behavior of the blue mussel (*Mytilus edulis*). *Environmental Toxicological Chemistry*. 31. pg. 2490-7. http://www.ncbi.nlm.nih.gov/pubmed/22893562

- "As the industrial production of nanoplastic and the degradation of microplastic into smaller particles at sea increase, the potential amount of nanoplastics in the marine environment rises."

- "The state of nano PS aggregation in the exposure medium was assessed using dynamic light scattering."

- "*M. edulis* reduced its filtering activity when nano PS was present."

- "The presence of nano PS around the foot of *M. edulis* after the bioassay confirmed that the organism removed nano PS from the water."

Weigelt, D., Mehinto, A., Ankley, G., et al. (2014). Transcriptomic effects-based monitoring for endocrine active chemicals: Assessing relative contribution of treated wastewater to downstream pollution. *Environmental Science and Technology*. 48(4). pg. 2385-94. http://pubs.acs.org/doi/abs/10.1021/es404027n

Weight-control Information Network. (2014). *Overweight and obesity statistics*. US Department of Health and Human Services. Washington, DC. http://win.niddk.nih.gov/publications/PDFs/stat904z.pdf

Weijs, L., Shaw, S., Berger, M., et al. (2014). Methoxylated PBDEs (MeO-PBDEs) and hydroxylated PBDEs (HO-PBDEs) and PCBs (HO-PCBs) in liver smaples of harbor seals from the northwest Atlantic. 493. pg. 606-14. http://www.ncbi.nlm.nih.gov/pubmed/24982026

- "Metabolites of PCBs and PBDEs are shown to influence the thyroid hormone homeostasis and therefore, could have an influence on the growth of newborn or young animals."

Weisse, T. (1990). Trophic interactions among heterotrophic microplankton, nanoplankton, and bacteria in Lake Constance. *Hydrobiologia*. 191. pg. 111-22. http://link.springer.com/article/10.1007%2FBF00026045#page-1

Weisse, T. and Scheffel-Moser, U. (1991). Uncoupling the microbial loop: Growth and grazing loss rates of bacteria and heterotrophic nanoflagellates in the North Atlantic. *Marine Ecology Progress*. 71. pg. 195-205. http://www.int-res.com/articles/meps/71/m071p195.pdf

Weldy, C., Liu, Y., Liggit, H. and Chin, M. (2014). *In Utero* exposure to diesel exhaust air pollution promotes adverse intrauterine conditions, resulting in weight gain, altered blood pressure, and increased susceptibility to heart failure in adult mice. *PLoS ONE*. 9(2). http://www.plosone.org/article/fetchObject.action?uri=info:doi/10.1371/journal.pone.0088582&representation=PDF

Weschler, C. and Nazaroff, W. (2008). Semivolatile organic compounds in indoor environments. *Atmospheric Environment*. 42(40). pg. 9018-40. http://www.researchgate.net/publication/223485264_Semivolatile_organic_compounds_in_indoor_environments

Weston, Donald P. and Lydy, Michael J. (2014). Toxicity of the insecticide fipronil and its degradates to benthic macroinvertebrates of urban streams. *Environmental Science & Technology*. 48(2). pg. 1290-7. http://pubs.acs.org/doi/abs/10.1021/es4045874

White, S. S., Fenton, S. E. and Hines E. P. (2011). Endocrine disrupting properties of perfluorooctanoic acid. *J. Steroid Biochem. Mol. Biol.* 127(1-2). pg. 16-26. http://www.ncbi.nlm.nih.gov/pmc/articles/PMC3335904

Wiesner, M., Lowry, G., Alvarez, P., et. al. (2006). Assessing the risks of manufactured nanomaterials. *Environmental Science and Technology*. 39. pg. 4307-45. http://alvarez.blogs.rice.edu/files/2012/02/73.pdf

- "Commercial applications of nanomaterials currently available or soon to appear include nanoengineered titania particles for sunscreens and paints; fullerene nanotube composites in tires, tennis rackets, and video screens; fullerene cages in cosmetics; silica nanoparticles as solid lubricants; metal nanoparticles for groundwater remediation; and protein-based nanomaterials in soaps, shampoos, and detergents."

- "The production, use, and disposal of nanomaterials will inevitably lead to their appearance in air, water, soils, or organisms."

Williams, Timothy. (January 10th, 2015). Garbage incinerators make comeback, kindling both garbage and debate. *The New York Times*. http://www.nytimes.com/2015/01/11/us/garbage-incinerators-make-comeback-kindling-both-garbage-and-debate.html

Windham, G., Pinney, S., Sjodin, A., et. al. (2010). Body burdens of brominated flame retardants and other persistent organo-halogenated compounds and their descriptors in US girls. *Environmental Research*. 110(3). pg. 251-7. http://www.ncbi.nlm.nih.gov/pmc/articles/PMC2844779/pdf/nihms176214.pdf

- "Several of these potential HAAs were detected in nearly all of these young girls, some at relatively high levels, with variation by geographic location and other demographic factors that may reflect exposure pathways. The higher PBDE levels in California likely reflect differences in fire regulation and safety codes, with potential policy implications."

Winters, Nancy and Graunke, Kyle. (2014). *Roofing materials assessment: Investigation of toxic chemicals in roof runoff*. Washington State Department of Ecology. https://fortress.wa.gov/ecy/publications/publications/1403003.pdf

- "From February through April 2013, the Washington State Department of Ecology collected runoff from 18 constructed roofing panels following 10 rain events for contaminant analysis. Analysis of the runoff included total and dissolved metals (arsenic, cadmium, copper, lead, and zinc) and organic compounds [polycyclic aromatic hydrocarbons (PAHs), phthalates, and polybrominated diphenyl ethers (PBDEs)]."

- "Testing of metals' release from aged roofing panels also has shown that the potential for pollutant release still exists after 60 years."

Wirgin, I., Roy, N., Loftus, M., et. al. (2011). Mechanistic basis of resistance to PCBs in Atlantic tomcod from the Hudson River. *Science*. 331. pg. 1322-5. http://www.sciencemag.org/content/331/6022/1322

Wolkers, H., van Bavel, B., Derocher, A. E., et al. (2004). Congener-specific accumulation and food chain transfer of polybrominated diphenyl ethers in two Arctic food chains. *Environmental Science and Technology*. 38. pg. 1667-74.

Woodall, L., Sanchez-Vidal, A., Canals, M., et al. (2015). The deep sea is a major sink for microplastic debris. *Royal Society Open Science*. http://rsos.royalsocietypublishing.org/content/royopensci/1/4/140317.full.pdf

- "Microplastic, in the form of fibres, was up to four orders of magnitude more abundant (per unit volume) in deep-sea sediments from the Atlantic Ocean, Mediterranean Sea and Indian Ocean than in contaminated seasurface waters."

- "Given the vastness of the deep sea and the prevalence of microplastics at all sites we investigated, the deepsea floor appears to provide an answer to the question—*where is all the plastic*?"

- In fact, the presence of plastic fibers in deep sea sediments does not explain the current location of plastic nanoparticles (PNP) which are derived from microplastic debris—a topic not mentioned in this article.

Woodard, Colin. (Oct. 25 through Oct. 30, 2015). Mayday: Gulf of Maine in distress, parts 1 through 6. *Portland Press Herald*. pg. A1. http://www.pressherald.com/2015/10/25/climate-change-imperils-gulf-maine-people-plants-species-rely/

- "The administration, he said, left a number of key goals unaddressed, including identifying and reducing acidification-causing nutrient pollution from unpermitted sources, stepping up ocean chemistry monitoring, or creating a body that can coordinate the necessary work."

Woodruff, T., Zota, A. and Schwartz, J. (2011). Environmental chemicals in pregnant women in the United States: NHANES 2003-2004. *Environmental Health Perspectives*. 119(6). pg. 878-85. http://www.ncbi.nlm.nih.gov/pmc/articles/PMC3114826/pdf/ehp-119-878.pdf

Woods Hole Oceanographic Institution (WHOI). (2014). *Current research projects (2014)*. Hahn Lab at WHOI. http://www.whoi.edu/sbl/liteSite.do?litesiteid=48972&articleId=87369

- "The overall objective of the proposed research is to elucidate the molecular mechanisms by which short-term exposure to harmful algal bloom (HAB) toxins and marine toxicants during development causes physiological and neurological abnormalities later in life."

- "It is now well known that the early life environment can have a profound effect on the health of adults (*the developmental origins of health and disease*). However, the mechanisms by which developmental exposure elicits effects later in life are not understood."

World Bank. (2013). *Fish to 2030: Prospects for fisheries and aquaculture.* The World Bank.

World Bank. (2014). *Press release: GEF grant to enhance the environmental performance of municipal solid waste incinerators in Chinese cities.* The World Bank. http://www.worldbank.org/en/news/press-release/2014/11/14/gef-grant-to-enhance-the-environmental-performance-of-municipal-solid-waste-incinerators-in-chinese-cities

World Health Organization. (2003). *Concise international chemical assessment document 55: Polychlorinated biphenyls: Human health aspects.* World Health Organization. http://www.who.int/ipcs/publications/cicad/en/cicad55.pdf

- "They have been used in plasticizers, surface coatings, inks, adhesives, flame retardants, pesticide extenders, paints, and microencapsulation of dyes for carbonless duplicating paper…[and] in dielectric fluids in transformers and capacitors."

- "The pyrolysis of PCB mixtures produces hydrogen chloride and polychlorinated dibenzofurans (PCDFs), and pyrolysis of mixtures containing chlorobenzenes also produces polychlorinated dibenzodioxins (PCDDs)."

- "In 1978, the estimated dietary intake of PCBs by adults in the USA was 0.027 µg/kg body weight per day, but it declined to 0.0005 µg/kg body weight per day in 1982-1984 and <0.001 µg/kg body weight per day for the period 1986-1991."

- This report contains a comprehensive bibliography of over 400 citations.

- An unanswered question in 2015: though production of PCBs has been prohibited for ±4 decades, why have body burdens of PCBs remained relatively constant in the 21st century?

- What is the difference between "estimated intake" and actual intake given the proliferation of ecotoxin-laden nanoparticles in pathways to human consumption?

World Health Organization. (2011). *Safe drinking-water from desalination.* WHO Press, Geneva, Switzerland.

http://www.who.int/water_sanitation_health/publications/2011/desalination_guidance_en.pdf

World Health Organization. (2012). *State of the science of endocrine disrupting chemicals, 2012*. Inter-organization Programme for the Sound Management of Chemicals (IPCS). United Nations Environment Programme. http://www.who.int/ceh/publications/endocrine/en/

- "Close to 800 chemicals are known or suspected to be capable of interfering with hormone receptors, hormone synthesis or hormone conversion. However, only a small fraction of these chemicals have been investigated in tests capable of identifying overt endocrine effects in intact organisms."

- "Unlike 10 years ago, we now know that humans and wildlife are exposed to far more EDCs than just those that are POPs."

- "There is far more knowledge on exposure to EDCs and potential EDCs today compared with 10 years ago. This applies to the diversity of chemicals being implicated as EDCs and to the exposure routes and levels in humans and wildlife. As examples, brominated flame retardants were mentioned only briefly and perfluorinated compounds not at all when the IPCS document on EDCs was prepared 10 years ago. In addition to these, there are now many more EDCs being found in both humans and wildlife."

- "Unlike 10 years ago, it is now better understood that humans and wildlife are exposed to far more EDCs than just POPs. However, only a fraction of the potential EDCs in the environment are currently known."

- "The known EDCs may not be representative of the full range of relevant molecular structures and properties due to a far too narrow focus on halogenated chemicals for many exposure assessments and testing for endocrine disrupting effects. Thus, research is needed to identify other possible EDCs. Endocrine disruption is no longer limited to estrogenic, androgenic and thyroid pathways. Chemicals also interfere with metabolism, fat storage, bone development and the immune system, and this suggests that all endocrine systems can and will be affected by EDCs."

- "Over one hundred pharmaceuticals (not including their metabolites) used by humans have been detected in effluents and surface waters at concentrations ranging from low parts per trillion (ng/L) to parts per billion (mg/L) and include analgesics, anti-inflammatories, anti-depressants, anti-epileptics, lipid regulators, several classes of antibiotics, β-blockers, antineoplastics, and hormones."

World Public Union. (2013). *Plastic numbers to avoid: BPA numbers.* World Public Union. http://www.worldpublicunion.org/2013-06-01-NEWS-plastic-numbers-to-avoid-bpa-numbers.html

Worldwatch Institute. (2013). *State of the world 2013: Is sustainability still possible?* Worldwatch Institute, Washington, DC.

Worldwatch Institute. (2014). *State of the world 2014: Governing for sustainability.* Worldwatch Institute, Washington, DC.

Wright, P. and Adams, C. (1976). Toxicity of combustion products from burning polymers: Development and evaluation of methods. *Environmental Health Perspectives.* 17. pg. 75-83. http://www.ncbi.nlm.nih.gov/pmc/articles/PMC1475268/

Wright, R., Coffin, R. and Lebo, M. (1987). Dynamics of planktonic bacteria and heterotrophic flagellates in the Parker estuary, northern Massachusetts. *Continental Shelf Research.* 7. pg. 1383-97. http://adsabs.harvard.edu/abs/1987CSR.....7.1383W

Wright, S., Rowe, D., Thompson, R. and Galloway, T. (2013). Microplastic ingestion decreases energy reserves in marine worms. *Current Biology.* 23(23). pg. 1031-3. http://www.cell.com/current-biology/abstract/S0960-9822(13)01343-2

- "On highly impacted beaches, microplastic concentrations (<1mm) can reach 3% by weight, presenting a global conservation issue…Depleted energy reserves arise from a combination of reduced feeding activity, longer gut residence times of ingested material and inflammation."

Wright, S., Thompson, R. and Galloway, T. (2013). The physical impacts of microplastics on marine organisms: A review. *Environmental Pollution.* 178. pg. 483-92. http://www.resodema.org/publications/publication9.pdf

- "Plastic debris at the micro-, and potentially also the nano-scale, are widespread in the environment. Microplastics have accumulated in oceans and sediments worldwide in recent years, with maximum concentrations reaching 100,000 particles m^3. Due to their small size, microplastics may be ingested by low trophic fauna, with uncertain consequences for the health of the organism."

- "This review focuses on marine invertebrates and their susceptibility to the physical impacts of microplastic uptake. Some of the main points discussed are (1) an evaluation of the factors contributing to the bioavailability of microplastics including size and density; (2) an assessment of the relative susceptibility of different feeding guilds; (3) an overview of the factors most likely to influence the physical impacts of microplastics such as accumulation and translocation;

and (4) the trophic transfer of microplastics. These findings are important in guiding future marine litter research and management strategies."

- "Archived plastic samples from the west North Atlantic Ocean over the past 24 years have revealed a decrease in mean particle size from 10.66 mm in the 1990s to 5.05 mm in the 2000s. Sixty nine per cent of fragments were 2e6 mm, highlighting a prevalence of small plastic particles. Given the continual fragmentation of plastic items, particle concentrations are likely to increase with decreasing size."

- "Morét-Ferguson et al. (2010) report a shift in the abundance of plastic debris to smaller size categories in the western North Atlantic Ocean."

- "Continuous fragmentation of plastic into smaller particles will increase its availability."

- "At the base of the food web, the freshwater and freshwater/marine algal cells Chlorella and Scenedesmus respectively adsorbed charged nanoplastics (20 nm). A preference for positively charged particles was reported, probably due to the electrostatic attraction between the beads and cellulose constituent of the living cells."

- "Microplastics are unlikely to be digested or absorbed and can therefore be considered bio-inert. However, they may pass through cell membranes and become incorporated into body tissues following ingestion."

- "The rapid translocation of smaller particles is applicable to both invertebrates and vertebrates."

Writer, J., Barber, L., Taylor, H., et al. (2010). Anthropogenic tracers, endocrine disrupting chemicals, and endocrine disruption in Minnesota lakes. *Science of the Total Environment*. 409(1). pg. 100-11.
http://www.sciencedirect.com/science/article/pii/S0048969710007126

- "Endocrine disrupting chemicals, including bisphenol A, 17β-estradiol, estrone, and 4-nonylphenol were detected in 90% of the lakes at part per trillion concentrations. Endocrine disruption was observed in caged fathead minnows and resident fish in 90% of the lakes."

Writer, J., Barber, L., Ryan, J. and Bradley, P. (2011). Biodegradation and attenuation of steroidal hormones and alkylphenols by stream biofilms and sediments. *Environmental Science and Technology*. 45(10). pg. 4370-6.
http://pubs.acs.org/doi/abs/10.1021/es2000134

Writer, J., Ryan, J., Keefe, S. and Barber, L. (2012). Fate of 4-nonylphenol and 17β-estradiol in the Redwood River of Minnesota. *Environmental Science and Technology*. 46(2). pg. 860-8. http://pubs.acs.org/doi/abs/10.1021/es2031664

- "Attenuation of 17β-estradiol (k_{stream} = −3.2 ± 1.0 day^{-1}) was attributed primarily due to sorption and biodegradation by the stream biofilm and bed sediments."

- "Additive concentrations from multiple sources and transformation of parent compounds into degradates having estrogenic activity can explain their environmental persistence and widespread observations of biological disruption in surface waters."

Writer, J., Ferrer, I., Barber, L. and Thurman, E. (2013). Widespread occurrence of neuro-active pharmaceuticals and metabolites in 24 Minnesota rivers and wastewaters. *Science of the Total Environment*. 461-462. pg. 519-27. http://www.sciencedirect.com/science/article/pii/S0048969713005408

Wurl, O. and Obbard, J. (2004). A review of pollutants in the sea-surface microlayer (SML): A unique habitat for marine organisms. *Marine Pollution Bulletin*. 48(11-12). pg. 1016-30. http://www.ncbi.nlm.nih.gov/pubmed/15172807

- "The sea-surface microlayer [SML] (uppermost 1–1000 um layer) forms the boundary layer interface between the atmosphere and ocean."

- "The SML is highly contaminated in many urban and industrialized areas of the world, resulting in severe ecotoxicological impacts."

Xia, K., Luo, M. B., Lusk, C., et al. (2008). Polybrominated diphenyl ethers (PBDEs) in biota representing different trophic levels of the Hudson River, New York: From 1999 to 2005. *Environ Sci Technol*. 42. pg. 4331-7.

Xia, T., Kovochich, M., Brant, J., et al (2006). Comparison of the abilities of ambient and manufactured nanoparticles to induce cellular toxicity according to an oxidative stress paradigm. *Nano Letters*. 6(8). pg. 1794-807. http://www.ncbi.nlm.nih.gov/pubmed/16895376

- "Ambient ultrafine particles (UFPs) and cationic PS nanospheres were capable of inducing cellular ROS production, GSH depletion, and toxic oxidative stress. This toxicity involves mitochondrial injury through increased calcium uptake and structural organellar damage."

- "Cationic polystyrene nanospheres induced mitochondrial damage and cell death without inflammation."

- "The assessment of ROS production and generation of oxidative stress is a valid mechanism of comparing the toxicity of manufactured or ambient NPs."

Yager, T. J. B., Furlong, E. T., Kolpin, D. W., et al. (2014). Dissipation of contaminants of emerging concern in biosolids applied to nonirrigated farmland in eastern Colorado. *Journal of the American Water Resources Association*. 50(2). pg. 343-57.

Yamashita, N., Taniyasu, S., Hanari, N., et al. (2003). Polychlorinated naphthalene contamination of some recently manufactured industrial products and commercial goods in Japan. *Journal of Environmental Science and Health A*. 38. pg. 1745-59.

Yang, W., Peters, J. and Williams, R. (2008). Inhaled nanoparticles – A current review. *International Journal of Pharmaceutics*. 356(1-2). pg. 239-47. http://www.ncbi.nlm.nih.gov/pubmed/18358652

Yin, J., Falconer, R., Chen, Y. and Probert, S. (2000). Water and sediment movements in harbours. http://www.academia.edu/1484832/Water_and_sediment_movements_in_Harbours

Young, S., Balluz, L. and Malilay, J. (2004). Natural and technological hazardous material releases during and after natural disasters: A review. *Science of the Total Environment*. 322. pg. 3-20. http://digitalcommons.unl.edu/cgi/viewcontent.cgi?article=1089&context=publichealthr esources

Youngs, L., Bevan, R. and Ashdown, L. (2011). *Endocrine disrupting chemicals: Overview of recent published literature*. Institute of Environment and Health, Cranfield University, UK. http://endichem.defra.gov.uk/EDC%20Oct%202010%20-%20Dec%202010.pdf

- "The term EDCs is normally applied to describe chemicals that are synthetic agents that mimic or block hormones resulting in the potential disruption of critical endocrine processes including reproduction, growth and development."

- "The papers were selected because they address research areas that are considered of direct relevance to the health and environmental effects of EDCs."

- This English publication lists over 100 research articles published in 2010 (only).

Ye, L., Yong, K., Liu, L., et al. (2012). A pilot study in non-human primates shows no adverse response to intravenous injection of quantum dots. *Nature Nanotechnology*. 7. pg. 453-8. http://www.nature.com/nnano/journal/v7/n7/full/nnano.2012.74.html

Ye, S. and Andrady, A. (1991). Fouling of floating plastic debris under Biscayne Bay exposure conditions. *Marine Pollution Bulletin*. 22(12). pg. 608-13. http://www.sciencedirect.com/science/article/pii/0025326X9190249R

- "Most plastic samples undergo fouling to an extent to cause the sample to be negatively buoyant in sea water."

- "Free-floating plastics at sea may, under certain conditions, undergo fouling-induced sinking followed by resurfacing as floating debris."

Yonkos, L., Friedel, E., Perez-Reyes, A., Ghosal, S. and Arthur, C. (2014). Microplastics in four estuarine rivers in the Chesapeake Bay, U.S.A. *Environmental Science and Technology*. 48(24). pg. 14195-102. http://www.ncbi.nlm.nih.gov/pubmed/25389665

- "Plastics entering the environment are mechanically, photochemically, and/or biologically degraded to the extent that they become imperceptible to the naked eye yet are not significantly reduced in total mass."

- "Concentrations demonstrated statistically significant positive correlations with population density and proportion of urban/suburban development within watersheds."

Yu, Y., Li, Y., Wang, W., et al. (2013) Acute toxicity of amorphous silica nanoparticles in intravenously exposed ICR mice. *PLoS ONE*. 8(5). http://www.plosone.org/article/fetchObject.action?uri=info:doi/10.1371/journal.pone.0061346&representation=PDF

Zafar, U., Houlden, A. and Robson, G. (2013). Fungal communities associated with the biodegradation of polyester polyurethane buried under compost at different temperatures. *Applied and Environmental Microbiology*. pg. 7313-24. http://aem.asm.org/content/79/23/7313.full.pdf+html

Zahran, S., Mielke, H., Gonzales, C., et al. (2010). New Orleans before and after Hurricanes Katrina/Rita: A quasi-experiment of the association between soil lead and children's blood lead. *Environmental Science and Technology*. 44(12). pg. 4433-40. http://pubs.acs.org/doi/abs/10.1021/es100572s

- "Prior to Hurricanes Katrina and Rita (HKR), significant associations were noted between soil lead (SL) and blood lead (BL) in New Orleans."

- "Test results show that SL decreased from 328.54 to 203.33 mg/kg post-HKR (t = 3.296, p ≤ 0.01). Decreases in SL are associated with declines in children's BL response (r = 0.308, p ≤ 0.05). When SL decreased at least 1%, median children's BL declined 1.55 µg/dL. Declines in median BL are largest in census tracts with ≥50% decrease in SL."

- So, where did all the lead go?

Zarfl, C., Fleet, D., Fries, E., et al. Microplastics in oceans. *Marine Pollution Bulletin*. 62. pg. 1589-91. http://epic.awi.de/24543/1/Marine_Pollution_Bulletin_62_%282011%29_1589%E2%80%931591.pdf

Zarfl, C. and Matthies, M. (2010). Are marine plastic particles transport vectors for organic pollutants to the Arctic? *Marine Pollution Bulletin*. 60(10). pg. 1810-4. http://www.ncbi.nlm.nih.gov/pubmed/20579675

- "We have estimated mass fluxes of polychlorinated biphenyls (PCBs), polybrominated diphenyl ethers (PBDEs), and perfluorooctanoic acid (PFOA) to the Arctic via the main ocean currents and compared them to those in the dissolved state and in air."

Ze, Y., Sheng, L., Zhao, X., et al. (2014). TiO_2 nanoparticles induced hippocampal neuroinflammation in mice. *PLoS ONE*. 9(3). http://www.plosone.org/article/fetchObject.action?uri=info:doi/10.1371/journal.pone.0092230&representation=PDF

Zegers, B. N., Lewis, W. A., Booij, K., et al. (2003). Levels of polybrominated diphenyl ether flame retardants in sediment cores from Western Europe. *Environ Sci Technol*. 37. pg. 3803-7.

Zegers, B.N., Mets, A., Van Bommel, R., et al. (2005). Levels of Hexabromocyclododecane in harbor porpoises and common dolphins from western European seas, with evidence for stereoisomerspecific biotransformation by cytochrome P450. *Environ Sci Technol*. 39. pg. 2095-100.

Zettler E., Mincer T. and Amaral-Zettler, L. (2013). Life in the "Plastisphere": Microbial communities on plastic marine debris. *Environmental Science and Technology*. 47. pg. 7137-46. http://pubs.acs.org/doi/abs/10.1021/es401288x

- "Plastic marine debris (PMD) collected at multiple locations in the North Atlantic was analyzed with scanning electron microscopy (SEM)...[The characterization of] the attached microbial communities...[revealed] a diverse microbial community of heterotrophs, autotrophs, predators, and symbionts, a community we refer to as the 'Plastisphere'."

Zhan, B., Choi, J., Eum, S., et al. (2013). TLR4 signaling is involved in brain vascular toxicity of PCB153 bound to nanoparticles. *PLoS ONE*. 8(4).

http://www.plosone.org/article/fetchObject.action?uri=info:doi/10.1371/journal.pone.0061346&representation=PDF

Zhou, T., Taylor, M., DeVito, M. and Crofton, K. (2002). Developmental exposure to brominated diphenyl ethers results in thyroid hormone disruption. *Toxicological Science*. 66(1). pg. 105-16. http://toxsci.oxfordjournals.org/content/66/1/105.full.pdf

Zitko, V. and Hanlon, M. (1991). Another source of pollution by plastics: Skin cleansers with plastic scrubbers. *Marine Pollution Bulletin*. 22(1). pg. 41-42.

Yamashita, R., Takada, H., Fukuwaka, M. and Watanuki, Y. (2011). Physical and chemical effects of ingested plastic debris on short-tailed shearwaters, *Puffinus tenuirostris*, in the North Pacific ocean. *Marine Pollution Bulletin*. 62(12). pg. 2845-9. http://www.ncbi.nlm.nih.gov/pubmed/22047741

Zhang, B., Misak, H., Dhanasekaran, P., et al. (2011). Environmental impacts of nanotechnology and its products. *Proceedings of the 2011 Midwest Section Conference of the American Society for Engineering Education*. https://www.asee.org/documents/sections/midwest/2011/ASEE-MIDWEST_0030_c25dbf.pdf

- "Nanotechnology increases the strengths of many materials and devices, as well as enhances efficiencies of monitoring devices, remediation of environmental pollution, and renewable energy production. While these are considered to be the positive effect of nanotechnology, there are certain negative impacts of nanotechnology on environment in many ways, such as increased toxicological pollution on the environment due to the uncertain shape, size, and chemical compositions of some of the nanotechnology products (or nanomaterials)."

- "Nanoscaled particles have relatively larger surface area per unit mass which is the critical factor to increase mechanical modulus and other physical and chemical properties."

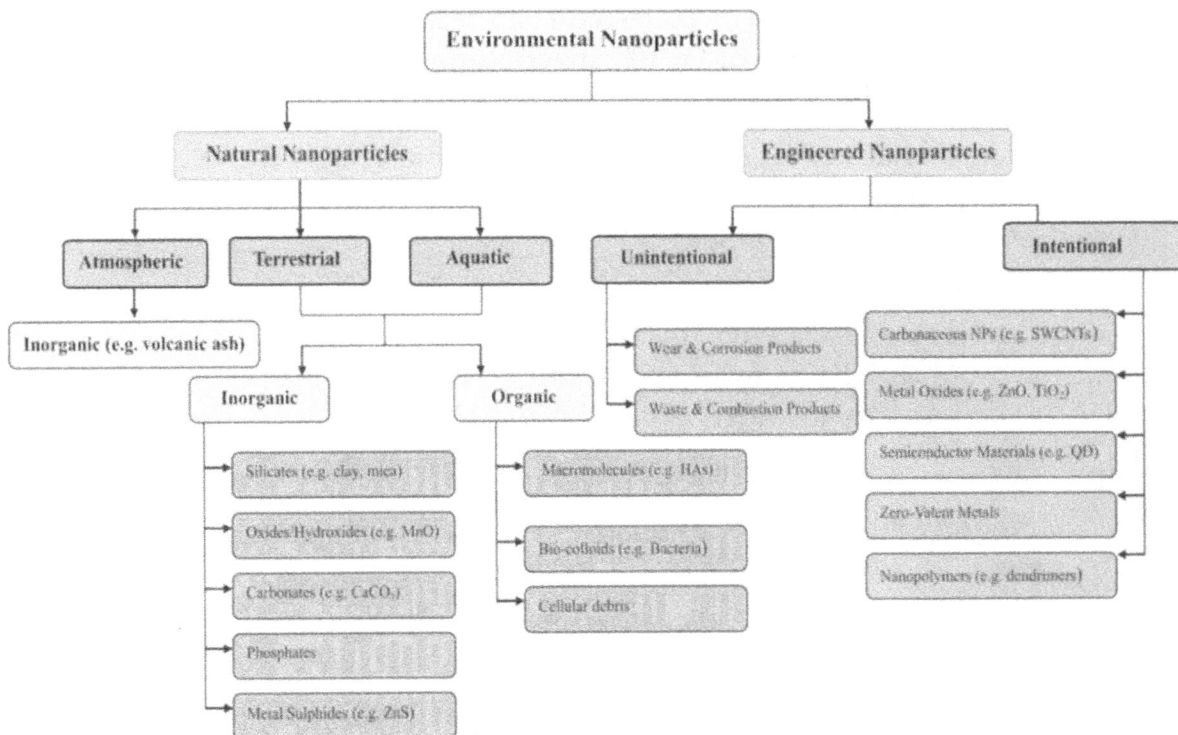

```
Environmental Nanoparticles
├── Natural Nanoparticles
│   ├── Atmospheric
│   │   └── Inorganic (e.g. volcanic ash)
│   ├── Terrestrial
│   │   └── Inorganic
│   │       ├── Silicates (e.g. clay, mica)
│   │       ├── Oxides/Hydroxides (e.g. MnO)
│   │       ├── Carbonates (e.g. CaCO₃)
│   │       ├── Phosphates
│   │       └── Metal Sulphides (e.g. ZnS)
│   └── Aquatic
│       └── Organic
│           ├── Macromolecules (e.g. HAs)
│           ├── Bio-colloids (e.g. Bacteria)
│           └── Cellular debris
└── Engineered Nanoparticles
    ├── Unintentional
    │   ├── Wear & Corrosion Products
    │   └── Waste & Combustion Products
    └── Intentional
        ├── Carbonaceous NPs (e.g. SWCNTs)
        ├── Metal Oxides (e.g. ZnO, TiO₂)
        ├── Semiconductor Materials (e.g. QD)
        ├── Zero-Valent Metals
        └── Nanopolymers (e.g. dendrimers)
```

Zhang, Y., Chen, W., Zhang, J., et al. (2007). In vitro and in vivo toxicity of CdTe nanoparticles. *Journal of Nanoscience and Nanotechnology*. 7. pg. 497-503. http://www.ncbi.nlm.nih.gov/pubmed/17450785

- "Cadmium telluride (CdTe) nanoparticles exhibit strong and stable fluorescence that is attractive for many applications such as biological probing and solid state lighting."

- "CdTe nanoparticles elicited cytotoxicity in a concentration- and size-dependent manner, with smaller-sized particles exhibiting somewhat higher potency."

- "The *in vitro* and *in vivo* toxic potential of different types of CdTe nanoparticles… suggest that the nervous system may be targeted by these nanoparticles under some conditions."

Zhao, Ling, Dong, Yuan-Hua and Wang, Hui. (2013). Residues of organochlorine pesticides and polycyclic aromatic hydrocarbons in farm-raised livestock feeds and manures in Jiangsu, China. *Science of the Total Environment*. 450-451. pg. 348-55. http://www.ncbi.nlm.nih.gov/pubmed/23058310

- "The residual levels of 8 organochlorine pesticides (OCPs) and 15 priority polycyclic aromatic hydrocarbons (PAHs) were determined in pig, chicken, and cow feed and manure samples collected from feedlots in Jiangsu province, China. The mean residuals of OCPs ranged from 25.35 to 65.62 ng g^{-1} in feeds and from 33.46 to 90.89 ng g^{-1} in manures."

- "The mean residuals of all of the PAHs varied from 128.94 to 389.66 ng g^{-1} in manures. The mean concentrations of seven carcinogenic PAHs in manures varied from 16.80 to 79.70 ng g^{-1}."

- "After the ban on OCPs usage in 1983, residuals of HCHs and DDTs in various environmental media have declined considerably, but large amounts still remain in the environment due to high persistency, possible illegal use, or occurrence of DDTs as an impurity in widely applied dicofol."

- Highlights:
  - "Slightly higher mean residuals of OCPs were found in manures than in feeds."
  - "α-HCH was the most abundant compound in all kinds of animal feeds and manures."
  - "The predominance of p.p'-DDE and p.p'-DDT of total DDTs was clearly observed."
  - "Phenanthrene was the most dominant PAH species in each kind of animal manure."
  - "PAHs with 3 rings were the primary components in the tested manures."

Zhao, Y., Wang, B., Feng, W. and Bai, C. (2007). Toxicological and biological effects of nanomaterials. *International Journal of Nanotechnology*. 4(1/2). pg. 179-96. http://www.eolss.net/Sample-Chapters/C05/E6-152-35-00.pdf

Zhao, Y., Wang, B., Feng, W. and Bai, C. (n.d.). Nanotoxicology: Toxicological and biological activities of nanomaterials. In: *Encyclopedia of life support systems* (EOLSS), Developed under the Auspices of the UNESCO, Eolss Publishers, Paris, France. http://www.eolss.net

- A comprehensive summary of the behavior and toxicity of nanoparticles.

- There is no mention of the life cycle assessment of nanoparticles and nanomaterials as they abrade and disintegrate and become components of the biotic environment, for example, in the piscivorous food web (algae – zooplankton – bivalve).

Zhu, L., Zhu, Y., Li, Z., et al. (2007). Developmental toxicity in zebrafish (*Danio rerio*) embryos after exposure to manufactured nanomaterials: Buckminsterfullerene aggregates (nC_{60}) and fullerol. *Environmental Toxicological Chemistry*. 26. pg. 976-9. http://www.ncbi.nlm.nih.gov/pubmed/17521145

- "nC60 at 1.5mg/L delayed zebrafish embryo and larval development, decreased survival and hatching rates, and caused pericardial edema."

Zhu, X., Tian, S. and Cai, Z. (2012). Toxicity assessment of iron oxide nanoparticles in zebra fish (*Danio rerio*) life stages. *PLoS ONE*. 7(9). http://www.plosone.org/article/fetchObject.action?uri=info%3Adoi%2F10.1371%2Fjournal.pone.0046286&representation=PDF

- "Iron oxide nanoparticles have been explored recently for their beneficial applications in many biomedical areas, in environmental remediation, and in various industrial applications. However, potential risks have also been identified with the release of nanoparticles into the environment. To study the ecological effects of iron oxide nanoparticles on aquatic organisms, we used early life stages of the zebrafish (Danio rerio) to examine such effects on embryonic development."

- "With the increasing use of metal oxide nanomaterials in catalysis, sensors, environmental remediation, and such commercial products as ones for personal care, there is a strong possibility that these nanomaterials will ultimately enter aquatic ecosystems through wastewater discharge and washing off during recreational activities, such as swimming and water skiing, thereby impacting on the environment and human health."

- Davenport, Coral. (October 28, 2015). A close-up look at Greenland, melting away. *The New York Times*. pg. A1, A10.

Zimmer, Carl. (October 29, 2015). Scientists urge national initiative to study microbiology. *The New York Times*. Pg. A24.

- "The trillions of microbes that live inside the human body, for example, play important roles in health, from fighting diseases to maintaining a balanced immune system. But the planet is home to a vast number of other microbiomes, from the microbial communities that live in undersea volcanoes to microbes that cling to existence in Antarctic deserts. These play an instrumental role in the environment: Ocean microbes, for example, produce half the oxygen we breathe."

Zolnik, B., Gonzalez-Fernandez, A., Sadrieh, N. and Dobrovolskaia, M. (2010). Minireview: Nanoparticles and the immune system. *Endocrinology*. 151. pg. 458-65. http://press.endocrine.org/doi/pdf/10.1210/en.2009-1082

- "The clear benefits of using nanosized products in various biological and medical applications are often challenged by concerns about the lack of adequate data regarding their toxicity. One area of interest involves the interactions between nanoparticles and the components of the immune system…[as well as] lower immunotoxicity…Lessons learned from previous studies include the importance of detection and prevention of potential particle contamination with such things as bacterial endotoxins and/or toxic synthesis byproducts."

Zota, A., Adamkiewicz, G. and Morello-Frosch, R. (2010). Are PBDEs an environmental equity concern? Exposure disparities by socioeconomic status. *Environmental Science and Technology*. 44(15). pg. 5691-2. http://pubs.acs.org/doi/pdf/10.1021/es101723d

Zubris, Kimberly and Richards, Brian. (2005). Synthetic fibers as indicator of land application of sludge. *Environmental Pollution*. 138. pg. 201-11. http://www.ncbi.nlm.nih.gov/pubmed/15967553

- "Synthetic fabric fibers have been proposed as indicators of past spreading of wastewater sludge…Fibers were detectable in field site soils up to 15 years after application."

Appendix 1: The Stockholm Convention: Overview, Definitions and Targeted Chemicals

Source: Secretariat of the Stockholm Convention. (2008). *Listing of POPs in the Stockholm Convention.*
http://chm.pops.int/TheConvention/ThePOPs/ListingofPOPs/tabid/2509/Default.aspx

"The Stockholm Convention on Persistent Organic Pollutants was adopted by the Conference of Plenipotentiaries on 22 May 2001 in Stockholm, Sweden. The Convention entered into force on 17 May 2004. [Chemicals that] remain intact in the environment for long periods, become widely distributed geographically, accumulate in the fatty tissue of humans and wildlife, and have harmful impacts on human health or on the environment. Exposure to Persistent Organic Pollutants (POPs) can lead to serious health effects including certain cancers, birth defects, dysfunctional immune and reproductive systems, greater susceptibility to disease and damages to the central and peripheral nervous systems. Given their long range transport, no one government acting alone can protect its citizens or its environment from POPs. In response to this global problem, the Stockholm Convention, which was adopted in 2001 and entered into force in 2004, requires its parties to take measures to eliminate or reduce the release of POPs into the environment... As set out in Article 1, the objective of the Stockholm Convention is to protect human health and the environment from persistent organic pollutants."

Numerous other articles and annexes describe the provisions and requirements of the convention, including the future listing of POPs.

Listing of POPs in the Stockholm Convention:

Annex A (Elimination)
Aldrin, Chlordane, Chlordecone, Dieldrin, Endrin, Heptachlor, Hexabromobiphenyl, Hexabromocyclododecane (HBCD), Hexabromodiphenyl ether and Heptabromodiphenyl ether, Hexachlorobenzene (HCB), Alpha hexachlorocyclohexane, Beta hexachlorocyclohexane, Lindane, Mirex, Pentachlorobenzene, Polychlorinated biphenyls (PCB), Technical endosulfan and its related isomers, Tetrabromodiphenyl ether and pentabromodiphenyl ether, Toxaphene

Annex B (Restriction)
DDT, Perfluorooctane sulfonic acid, its salts, and Perfluorooctane solfonyl fluoride

Annex C (Unintentional production)
Hexachlorobenzene (HCB), Pentachlorobenzene, Polychlorinated biphenyls (PCB), Polychlorinated dibenzo-*p*-dioxins (PCDD), Polychlorinated dibenzofurans (PCDF)

Appendix 2: EPA Drinking Water Contaminants List

Chemical Contaminants

Source: US Environmental Protection Agency. (2009). *Drinking Water Contaminants.* US EPA, Washington, DC. http://water.epa.gov/drink/contaminants/index.cfm

Substance Name	CASRN	Use
1,1,1,2-Tetrachloroethane	630-20-6	It is an industrial chemical used in the production of other substances.
1,1-Dichloroethane	75-34-3	It is an industrial chemical used as a solvent.
1,2,3-Trichloropropane	96-18-4	It is an industrial chemical used in paint manufacture.
1,3-Butadiene	106-99-0	It is an industrial chemical used in rubber production.
1,3-Dinitrobenzene	99-65-0	It is an industrial chemical and is used in the production of other substances.
1,4-Dioxane	123-91-1	It is used as a solvent or solvent stabilizer in the manufacture and processing of paper, cotton, textile products, automotive coolant, cosmetics and shampoos.
17alpha-estradiol	57-91-0	It is an estrogenic hormone and is used in pharmaceuticals.
1-Butanol	71-36-3	It is used in the production of other substances.
2-Methoxyethanol	109-86-4	It is used in consumer products, such as synthetic cosmetics, perfumes, fragrances, hair preparations, and skin lotions.
2-Propen-1-ol	107-18-6	It is used in the production of other substances, and in the manufacture of flavorings and perfumes.
3-Hydroxycarbofuran	16655-82-6	It is a carbamate and is a pesticide degradate. The parent, carbofuran, is used as an insecticide.
4,4'-Methylenedianiline	101-77-9	It is used in the production of other substances.
Acephate	30560-19-1	It is used as an insecticide.
Acetaldehyde	75-07-0	It is used in the production of other substances, and as a pesticide and food additive.
Acetamide	60-35-5	It is used as a solvent, solubilizer, plasticizer and stabilizer.

Substance Name	CASRN	Use
Acetochlor	34256-82-1	It is used as an herbicide for weed control on agricultural crops.
Acetochlor ethanesulfonic acid (ESA)	187022-11-3	Acetochlor ESA is an acetanilide pesticide degradate. The parent, acetochlor, is used as an herbicide for weed control on agricultural crops.
Acetochlor oxanilic acid (OA)	184992-44-4	Acetochlor OA is an acetanilide pesticide degradate. The parent, acetochlor, is used as an herbicide for weed control on agricultural crops.
Acrolein	107-02-8	It is used as an aquatic herbicide, rodenticide and industrial chemical.
Alachlor ethanesulfonic acid (ESA)	142363-53-9	Alachlor ESA is an acetanilide pesticide degradate. The parent, alachlor, is used as an herbicide for weed control on agricultural crops.
Alachlor oxanilic acid (OA)	171262-17-2	Alachlor OA is an acetanilide pesticide degradate. The parent, alachlor, is used as an herbicide for weed control on agricultural crops.
alpha-Hexachlorocyclohexane	319-84-6	It is a component of benzene hexachloride (BHC) and was formerly used as an insecticide.
Aniline	62-53-3	It is used as an industrial chemical, as a solvent, in the synthesis of explosives, rubber products and in isocyanates.
Bensulide	741-58-2	It is used as an herbicide.
Benzyl chloride	100-44-7	It is used in the production of other substances, such as plastics, dyes, lubricants, gasoline and pharmaceuticals.
Butylated hydroxyanisole	25013-16-5	It is used as a food additive (antioxidant).
Captan	133-06-2	It is used as a fungicide.
Chlorate	14866-68-3	Chlorate compounds are used in agriculture as defoliants or desiccants and may occur in drinking water related to use of disinfectants such as chlorine dioxide.
Chloromethane (Methyl chloride)	74-87-3	It is used as a foaming agent and in the production of other substances.
Clethodim	110429-62-4	It is used as an herbicide.

Substance Name	CASRN	Use
Cobalt	7440-48-4	It is a naturally-occurring element and was formerly used as cobaltus chloride in medicines and as a germicide.
Cumene hydroperoxide	80-15-9	It is used as an industrial chemical and is used in the production of other substances.
Cyanotoxins (3)*		Toxins naturally produced and released by cyanobacteria ("blue-green algae"). Various studies suggest three cyanotoxins for consideration: Anatoxin-a, Microcystin-LR, and Cylindrospermopsin.
Dicrotophos	141-66-2	It is used as an insecticide.
Dimethipin	55290-64-7	It is used as an herbicide and plant growth regulator.
Dimethoate	60-51-5	It is used as an insecticide on field crops, (such as cotton), orchard crops, vegetable crops, in forestry and for residential purposes.
Disulfoton	298-04-4	It is used as an insecticide.
Diuron	330-54-1	It is used as an herbicide.
equilenin	517-09-9	It is an estrogenic hormone and is used in pharmaceuticals.
equilin	474-86-2	It is an estrogenic hormone and is used in pharmaceuticals.
Erythromycin	114-07-8	It is used in pharmaceutical formulations as an antibiotic.
Estradiol (17-beta estradiol)	50-28-2	It is an estrogenic hormone and is used in pharmaceuticals.
estriol	50-27-1	It is an estrogenic hormone and is used in veterinary pharmaceuticals.
estrone	53-16-7	It is an estrogenic hormone and is used in veterinary and human pharmaceuticals.
Ethinyl Estradiol (17-alpha ethynyl estradiol)	57-63-6	It is an estrogenic hormone and is used in veterinary and human pharmaceuticals.
Ethoprop	13194-48-4	It is used as an insecticide.
Ethylene glycol	107-21-1	It is used as an antifreeze, in textile manufacture and is a cancelled pesticide.

Substance Name	CASRN	Use
Ethylene oxide	75-21-8	It is used as a fungicidal and insecticidal fumigant.
Ethylene thiourea	96-45-7	It is used in the production of other substances, such as for vulcanizing polychloroprene (neoprene) and polyacrylate rubbers, and as a pesticide.
Fenamiphos	22224-92-6	It is used as an insecticide.
Formaldehyde	50-00-0	It has been used as a fungicide, may be a disinfection byproduct and can occur naturally.
Germanium	7440-56-4	It is a naturally-occurring element and is commonly used as germanium dioxide in phosphors, transistors and diodes, and in electroplating.
Halon 1011 (bromochloromethane)	74-97-5	It is used as a fire-extinguishing fluid and to suppress explosions, as well as a solvent in the manufacturing of pesticides. May also occur as a disinfection by-product in drinking water.
HCFC-22	75-45-6	It is used as a refrigerant, as a low-temperature solvent, and in fluorocarbon resins, especially in tetrafluoroethylene polymers.
Hexane	110-54-3	It is used as a solvent and is a naturally-occurring alkane.
Hydrazine	302-01-2	It is used in the production of other substances, such as rocket propellants, and as an oxygen and chlorine scavenging compound.
Mestranol	72-33-3	It is an estrogenic hormone and is used in veterinary and human pharmaceuticals.
Methamidophos	10265-92-6	It is used as an insecticide.
Methanol	67-56-1	It is used as an industrial solvent, a gasoline additive and also as anti-freeze.
Methyl bromide (Bromomethane)	74-83-9	It has been used as a fumigant as a fungicide.
Methyl tert-butyl ether	1634-04-4	It is used as an octane booster in gasoline, in the manufacture of isobutene and as an extraction solvent.
Metolachlor	51218-45-2	It is used as an herbicide for weed control on agricultural crops.

Substance Name	CASRN	Use
Metolachlor ethanesulfonic acid (ESA)	171118-09-5	Metolachlor ESA is an acetanilide pesticide degradate. The parent, metolachlor, is used as an herbicide for weed control on agricultural crops.
Metolachlor oxanilic acid (OA)	152019-73-3	Metolachlor OA is an acetanilide pesticide degradate. The parent, metolachlor, is used as an herbicide for weed control on agricultural crops.
Molinate	2212-67-1	It is used as an herbicide.
Molybdenum	7439-98-7	It is a naturally-occurring element and is commonly used as molybdenum trioxide as a chemical reagent.
Nitrobenzene	98-95-3	It is used in the production of aniline, and also as a solvent in the manufacture of paints, shoe polishes, floor polishes, metal polishes, explosives, dyes, pesticides and drugs (such as acetaminophen).
Nitroglycerin	55-63-0	It is used in pharmaceuticals, in the production of explosives, and in rocket propellants.
N-Methyl-2-pyrrolidone	872-50-4	It is a solvent in the chemical industry, and is used for pesticide application and in food packaging materials.
N-nitrosodiethylamine (NDEA)	55-18-5	It is a nitrosamine used as an additive in gasoline and in lubricants, as an antioxidant, as a stabilizer in plastics and also may be a disinfection byproduct.
N-nitrosodimethylamine (NDMA)	62-75-9	It is a nitrosamine and has been formerly used in the production of rocket fuels, is used as an industrial solvent and an anti-oxidant, and also may be a disinfection byproduct.
N-nitroso-di-n-propylamine (NDPA)	621-64-7	It is a nitrosamine and may be a disinfection byproduct.
N-Nitrosodiphenylamine	86-30-6	It is a nitrosamine chemical reagent that is used as a rubber and polymer additive and may be a disinfection byproduct.
N-nitrosopyrrolidine (NPYR)	930-55-2	It is a nitrosamine used as a research chemical and may be a disinfection byproduct.
Norethindrone (19-Norethisterone)	68-22-4	It is a progresteronic hormone used in pharmaceuticals.

Substance Name	CASRN	Use
n-Propylbenzene	103-65-1	It is used in the manufacture of methylstyrene, in textile dyeing, and as a printing solvent, and is a constituent of asphalt and naptha.
o-Toluidine	95-53-4	It is used in the production of other substances, such as dyes, rubber, pharmaceuticals and pesticides.
Oxirane, methyl-	75-56-9	It is an industrial chemical used in the production of other substances.
Oxydemeton-methyl	301-12-2	It is used as an insecticide.
Oxyfluorfen	42874-03-3	It is used as an herbicide.
Perchlorate	14797-73-0	It is both a naturally occurring and human-made chemical. Perchlorate is used to manufacture fireworks, explosives, flares and rocket propellant.
Perfluorooctane sulfonic acid (PFOS)	1763-23-1	PFOS was used in fire fighting foams and various surfactant uses; few of which are still ongoing because no alternatives are available.
Perfluorooctanoic acid (PFOA)	335-67-1	PFOA is used in the manufacture of fluoropolymers, substances which provide non-stick surfaces on cookware and waterproof, breathable membranes for clothing
Permethrin	52645-53-1	It is used as an insecticide.
Profenofos	41198-08-7	It is used as an insecticide and an acaricide.
Quinoline	91-22-5	It is used in the production of other substances, as a pharmaceutical (anti-malarial) and as a flavoring agent.
RDX (Hexahydro-1,3,5-trinitro-1,3,5-triazine)	121-82-4	It is used as an explosive.
sec-Butylbenzene	135-98-8	It is used as a solvent for coating compositions, in organic synthesis, as a plasticizer and in surfactants.
Strontium	7440-24-6	It is naturally-occurring element and is used as strontium carbonate in pyrotechnics, in steel production, as a catalyst and as a lead scavenger.
Tebuconazole	107534-96-3	It is used as a fungicide.
Tebufenozide	112410-23-8	It is used as an insecticide.

Substance Name	CASRN	Use
Tellurium	13494-80-9	It is a naturally-occurring element and is commonly used as sodium tellurite in bacteriology and medicine.
Terbufos	13071-79-9	It is used as an insecticide.
Terbufos sulfone	56070-16-7	Terbufos sulfone is a phosphorodithioate pesticide degradate. The parent, terbufos, is used as an insecticide.
Thiodicarb	59669-26-0	It is used as an insecticide.
Thiophanate-methyl	23564-05-8	It is used as a fungicide.
Toluene diisocyanate	26471-62-5	It is used in the manufacture of plastics.
Tribufos	78-48-8	It is used as an insecticide and as a cotton defoliant.
Triethylamine	121-44-8	It is used in the production of other substances, as a stabilizer in herbicides and pesticides, in consumer products, in food additives, in photographic chemicals and in carpet cleaners.
Triphenyltin hydroxide (TPTH)	76-87-9	It is used as a pesticide.
Urethane	51-79-6	It is used as a paint ingredient.
Vanadium	7440-62-2	It is a naturally-occurring element and is commonly used as vanadium pentoxide in the production of other substances and as a catalyst.
Vinclozolin	50471-44-8	It is used as a fungicide.
Ziram	137-30-4	It is used as a fungicide.

Microbial Contaminants

Microbial Contaminant Name	Information
Adenovirus	Virus most commonly causing respiratory illness, and occasionally gastrointestinal illness
Caliciviruses	Virus (includes Norovirus) causing mild self-limiting gastrointestinal illness

Microbial Contaminant Name	Information
Campylobacter jejuni	Bacterium causing mild self-limiting gastroentestinal illness
Enterovirus	Group of viruses including polioviruses, coxsackieviruses and echoviruses that can cause mild respiratory illness
Escherichia coli (0157)	Toxin-producing bacterium causing gastrointestinal illness and kidney failure
Helicobacter pylori	Bacterium sometimes found in the environment capable of colonizing human gut that can cause ulcers and cancer
Hepatitis A virus	Virus that causes a liver disease and jaundice
Legionella pneumophila	Bacterium found in the environment including hot water systems causing lung diseases when inhaled
Mycobacterium avium	Bacterium causing lung infection in those with underlying lung disease, and disseminated infection in the severly immunocompromised
Naegleria fowleri	Protozoan parasite found in shallow, warm surface and ground water causing primary amebic meningoencephalitis
Salmonella enterica	Bacterium causing mild self-limiting gastrointestinal illness
Shigella sonnei	Bacterium causing mild self-limiting gastrointestinal illness and bloody diarrhea

Appendix 3: Chemicals in the CDC *Fourth Annual Report on Human Exposure to Environmental Chemicals Updated Tables*

Source: US CDC. (2014). *Fourth national report on human exposure to environmental chemicals: Updated tables.* Centers for Disease Control. http://www.cdc.gov/exposurereport/

Tobacco Smoke
Cotinine
NNAL*

Disinfection By-Products
Bromodichloromethane
Dibromochloromethane (Chlorodibromomethane)
Tribromomethane (Bromoform)
Trichloromethane (Chloroform)

Environmental Phenols
Benzophenone-3
Bisphenol A
4-*tert*-Octylphenol
Triclosan

Fungicides and Metabolites
ortho-Phenylphenol
Ethylene thiourea*
Pentachlorophenol
Propylene thiourea*

Herbicides and Metabolites
2,4-Dichlorophenoxyacetic acid
2,4,5-Trichlorophenoxyacetic acid

Sulfonylurea Herbicides
Bensulfuron-methyl*
Chlorsulfuron*
Ethametsulfuron-methyl*
Foramsulfuron*
Halosulfuron*
Mesosulfuron-methyl*
Metsulfuron-methyl*
Nicosulfuron*
Oxasulfuron*
Primisulfuron-methyl*
Prosulfuron*
Rimsulfuron*
Sulfometuron-methyl*
Sulfosulfuron*
Thifensulfuron-methyl*
Triasulfuron*
Triflusulfuron-methyl*

Carbamate Pesticide Metabolites

Carbofuranphenol
2-Isopropoxyphenol

Organochlorine Pesticides and Metabolites

Aldrin
Dieldrin
Endrin
Heptachlor epoxide
o,p'-Dichlorodiphenyltrichloroethane (DDT)
2,4,5-Trichlorophenol
2,4,6-Trichlorophenol

Other Pesticides and Metabolites

2,4-Dichlorophenol*
2,5-Dichlorophenol*

Organophosphorus Insecticides: Specific Metabolites

Acephate*
Dimethoate*
Methamidophos*
Omethoate*
Malathion dicarboxylic acid
2-Isopropyl-4-methyl-6-hydroxypyrimidine
para-Nitrophenol
3,5,6-Trichloro-2-pyridinol

Organophosphorus Insecticides: Dialkyl Phosphate Metabolites

Diethylphosphate (DEP)
Dimethylphosphate (DMP)
Diethylthiophosphate (DETP)
Dimethylthiophosphate (DMTP)
Diethyldithiophosphate (DEDTP)
Dimethyldithiophosphate (DMDTP)

Pyrethroid Metabolites

trans-3-(2,2-Dichlorovinyl)-2,2-dimethylcyclopropane carboxylic acid
cis-3-(2,2-Dibromovinyl)-2,2-dimethylcyclopropane carboxylic acid
4-Fluoro-3-phenoxybenzoic acid
3-Phenoxybenzoic acid

Metals and Metalloids

Antimony
Arsenic, Total
  Inorganic Arsenic-related Species*
  Arsenic (V) acid
  Arsenobetaine
  Arsenocholine
  Arsenous (III) acid
  Dimethylarsinic acid
  Monomethylarsonic acid
  Trimethylarsine oxide
Barium
Beryllium
Cadmium
Cesium
Cobalt
Copper*
Lead
Manganese*
Mercury (total; inorganic; ethyl* and methyl species*)
Molybdenum
Platinum
Selenium*
Strontium*
Thallium
Tin*
Tungsten
Uranium
Zinc*

Parabens

Butyl paraben*
Ethyl paraben*
Methyl paraben*
n-Propyl paraben*

Perchlorate and Other Anions

Nitrate*
Perchlorate
Thiocyanate*

Perfluorinated Compounds: Surfactants

Perfluorobutane sulfonic acid (PFBuS)
Perfluorodecanoic acid (PFDeA)
Perfluorododecanoic acid (PFDoA)
Perfluoroheptanoic acid (PFHpA)
Perfluorohexane sulfonic acid (PFHxS)
Perfluorononanoic acid (PFNA)
Perfluorooctanoic acid (PFOA)
Perfluorooctane sulfonic acid (PFOS)
Perfluorooctane sulfonamide (PFOSA)
2-(N-Ethyl-perfluorooctane sulfonamido) acetic acid (Et-PFOSA-AcOH)
2-(N-Methyl-perfluorooctane sulfonamido) acetic acid (Me-PFOSA-AcOH)
Perfluoroundecanoic acid (PFUA)

Phthalate and Phthalate Alternative Metabolites

Mono-benzyl phthalate (MBzP)
Mono-isobutyl phthalate (MiBP)
Mono-n-butyl phthalate (MnBP)
Mono-cyclohexyl phthalate (MCHP)
Mono-ethyl phthalate (MEP)
Mono-2-ethylhexyl phthalate (MEHP)
Mono-(2-ethyl-5-hydroxyhexyl) phthalate (MEHHP)
Mono-(2-ethyl-5-oxohexyl) phthalate (MEOHP)
Mono-(2-ethyl-5-carboxypentyl) phthalate (MECPP)
Mono-(carboxynonyl) phthalate (MCNP)*
Mono-isononyl phthalate (MiNP)
Mono-(carboxyoctyl) phthalate (MCOP)*
Mono-methyl phthalate (MMP)
Mono-(3-carboxypropyl) phthalate (MCPP)
Mono-n-octyl phthalate (MOP)
Cyclohexane-1,2-dicarboxylic acid mono(hydroxy-isononyl ester) (MHNCH)*

Phytoestrogens and Metabolites

Daidzein
Enterodiol
Enterolactone
Equol
Genistein
O-Desmethylangolensin

Polycyclic Aromatic Hydrocarbon Metabolites

2-Hydroxyfluorene
3-Hydroxyfluorene
9-Hydroxyfluorene
1-Hydroxyphenanthrene
2-Hydroxyphenanthrene
3-Hydroxyphenanthrene
4-Hydroxyphenanthrene
1-Hydroxypyrene
1-Hydroxynaphthalene (1-Naphthol)
2-Hydroxynaphthalene (2-Naphthol)

Volatile Organic Compounds (VOCs)

1,1,1-Trichloroethane (Methyl chloroform)
1,1,2,2-Tetrachloroethane
1,1,2-Trichloroethane
1,1-Dichloroethane
1,1-Dichloroethene (Vinylidene chloride)
1,2-Dibromo-3-chloropropane (DBCP)
1,2-Dichlorobenzene (o-Dichlorobenzene)
1,2-Dichloroethane (Ethylene dichloride)
cis-1,2-Dichloroethene
trans-1,2-Dichloroethene
1,2-Dichloropropane
1,3-Dichlorobenzene (m-Dichlorobenzene)
1,4-Dichlorobenzene (Paradichlorobenzene)
2,5-Dimethylfuran
Benzene
Chlorobenzene (Monochlorobenzene)
Dibromomethane
Dichloromethane (Methylene chloride)
Ethylbenzene
Hexachloroethane
Methyl-tert-butyl ether (MTBE)
Nitrobenzene
Styrene
Tetrachloroethene (Perchloroethylene)
Tetrachloromethane (Carbon tetrachloride)
Toluene
Trichloroethene (Trichloroethylene)
m-/p-Xylene
o-Xylene

Volatile Organic Compound (VOC) Metabolites

N-Acetyl-S-(benzyl)-L-cysteine*
N-Acetyl-S-(2-carbamoyl-2-hydroxyethyl)-L-cysteine*
N-Acetyl-S-(2-carbamoylethyl)-L-cysteine*
N-Acetyl-S-(2-carboxyethyl)-L-cysteine*
N-Acetyl-S-(3-hydroxypropyl)-L-cysteine*
N-Acetyl-S-(2-cyanoethyl)-L-cysteine*
N-Acetyl-S-(1,2-dichlorovinyl)-L-cysteine*
N-Acetyl-S-(2,2-dichlorovinyl)-L-cysteine*
N-Acetyl-S-(dimethylphenyl)-L-cysteine*
N-Acetyl-S-(N-methylcarbamoyl)-L-cysteine*
N-Acetyl-S-(3,4-dihydroxybutyl)-L-cysteine*
N-Acetyl-S-(2-hydroxy-3-butenyl)-L-cysteine*
N-Acetyl-S-(4-hydroxy-2-butenyl)-L-cysteine*
N-Acetyl-S-(1-hydroxymethyl-2-propenyl)-L-cysteine*
N-Acetyl-S-(2-hydroxyethyl)-L-cysteine*
N-Acetyl-S-(2-hydroxypropyl)-L-cysteine*
N-Acetyl-S-(3-hydroxypropyl-1-methyl)-L-cysteine*
N-Acetyl-S-(phenyl)-L-cysteine*
t,t-Muconic acid*
N-Acetyl-S-(phenyl-2-hydroxyethyl)-L-cysteine*
N-Acetyl-S-(n-propyl)-L-cysteine*
N-Acetyl-S-(trichlorovinyl)-L-cysteine*
2-Aminothiazoline-4-carboxylic acid*
Mandelic acid*
2-Methylhippuric acid*
3- and 4-Methylhippuric acid*
Phenylglyoxylic acid*
2-Thioxothiazolidine-4-carboxylic acid*

Perchlorate and Other Anions (Adult Cigarette Smokers and Nonsmokers: Special Sample)

Nitrate*
Perchlorate
Thiocyanate*

Polycyclic Aromatic Hydrocarbon Metabolites (Adult Cigarette Smokers and Nonsmokers: Special Sample)

2-Hydroxyfluorene
3-Hydroxyfluorene
9-Hydroxyfluorene
1-Hydroxyphenanthrene
2-Hydroxyphenanthrene
3-Hydroxyphenanthrene
4-Hydroxyphenanthrene
1-Hydroxypyrene
1-Hydroxynaphthalene (1-Naphthol)
2-Hydroxynaphthalene (2-Naphthol)

Metals and Metalloids (Adult Cigarette Smokers and Nonsmokers: Special Sample)

Antimony
Arsenic, Total
  Arsenic (V) acid
  Arsenobetaine
  Arsenocholine
  Arsenous (III) acid
  Dimethylarsinic acid
  Monomethylarsonic acid
  Trimethylarsine oxide
Barium
Cadmium
Cesium
Cobalt
Lead
Manganese*
Molybdenum
Strontium*
Thallium
Tin*
Tungsten
Uranium

Organochlorine Pesticides and Metabolites (Pooled Samples)

Oxychlordane
trans-Nonachlor
p,p'-DDT
p,p'-DDE
Hexachlorobenzene
beta-Hexachlorocyclohexane
gamma-Hexachlorocyclohexane
Mirex

Polybrominated Diphenyl Ethers and PBB 153 (Pooled Samples)

2,2',4'-Tribromodiphenyl ether (BDE 17)
2,4,4'-Tribromodiphenyl ether (BDE 28)
2,2',4,4'-Tetrabromodiphenyl ether (BDE 47)
2,3',4,4'-Tetrabromodiphenyl ether (BDE 66)
2,2',3,4,4'-Pentabromodiphenyl ether (BDE 85)
2,2',4,4',5-Pentabromodiphenyl ether (BDE 99)
2,2',4,4',6-Pentabromodiphenyl ether (BDE 100)
2,2',4,4',5,5'-Hexabromodiphenyl ether (BDE 153)
2,2',4,4',5,6'-Hexabromodiphenyl ether (BDE 154)
2,2',3,4,4',5',6-Heptabromodiphenyl ether (BDE 183)
2,2',3,3',4,4',5,5',6,6'-Decabromodiphenyl ether (BDE 209)*
2,2',4,4',5,5'-Hexabromobiphenyl (PBB 153)

398

Volatile Organic Compound (VOC) Metabolites
(Adult Cigarette Smokers and Nonsmokers: Special Sample)

N-Acetyl-S-(benzyl)-L-cysteine*
N-Acetyl-S-(2-carbamoyl-2-hydroxyethyl)-L-cysteine*
N-Acetyl-S-(2-carbamoylethyl)-L-cysteine*
N-Acetyl-S-(2-carboxyethyl)-L-cysteine*
N-Acetyl-S-(3-hydroxypropyl)-L-cysteine*
N-Acetyl-S-(2-cyanoethyl)-L-cysteine*
N-Acetyl-S-(1,2-dichlorovinyl)-L-cysteine*
N-Acetyl-S-(2,2-dichlorovinyl)-L-cysteine*
N-Acetyl-S-(dimethylphenyl)-L-cysteine*
N-Acetyl-S-(N-methylcarbamoyl)-L-cysteine*
N-Acetyl-S-(3,4-dihydroxybutyl)-L-cysteine*
N-Acetyl-S-(2-hydroxy-3-butenyl)-L-cysteine*
N-Acetyl-S-(4-hydroxy-2-butenyl)-L-cysteine*
N-Acetyl-S-(1-hydroxymethyl-2-propenyl)-L-cysteine*
N-Acetyl-S-(2-hydroxyethyl)-L-cysteine*
N-Acetyl-S-(2-hydroxypropyl)-L-cysteine*
N-Acetyl-S-(3-hydroxypropyl-1-methyl)-L-cysteine*
N-Acetyl-S-(phenyl)-L-cysteine*
t,t-Muconic acid*
N-Acetyl-S-(phenyl-2-hydroxyethyl)-L-cysteine*
N-Acetyl-S-(n-propyl)-L-cysteine*
N-Acetyl-S-(trichlorovinyl)-L-cysteine*
2-Aminothiazoline-4-carboxylic acid*
Mandelic acid*
2-Methylhippuric acid*
3- and 4-Methylhippuric acid*
Phenylglyoxylic acid*
2-Thioxothiazolidine-4-carboxylic acid*

Dioxin-like Polychlorinated Biphenyls: mono-*ortho*-substituted PCBs
(Pooled Samples)

2,3,3',4,4'-Pentachlorobiphenyl (PCB 105)
2,3,3',4,4'-Pentachlorobiphenyl (PCB 114)*
2,3',4,4',5-Pentachlorobiphenyl (PCB 118)
2',3,4,4',5-Pentachlorobiphenyl (PCB 123)*
2,3,3',4,4',5-Hexachlorobiphenyl (PCB 156)
2,3,3',4,4',5'-Hexachlorobiphenyl (PCB 157)
2,3',4,4',5,5'-Hexachlorobiphenyl (PCB 167)
2,3,3',4,4',5,5'-Heptachlorobiphenyl (PCB 189)

Non-Dioxin-Like Polychlorinated Biphenyls (Pooled Samples)

2,4,4'-Trichlorobiphenyl (PCB 28)

2,2'3,5'-Tetrachlorobiphenyl (PCB 44)

2,2',4,5'-Tetrachlorobiphenyl (PCB 49)

2,2',5,5'-Tetrachlorobiphenyl (PCB 52)

2,3',4,4'-Tetrachlorobiphenyl (PCB 66)

2,4,4',5-Tetrachlorobiphenyl (PCB 74)

2,2',3,4,5'-Pentachlorobiphenyl (PCB 87)

2,2',4,4',5-Pentachlorobiphenyl (PCB 99)

2,2',4,5,5'-Pentachlorobiphenyl (PCB 101)

2,3,3',4',6-Pentachlorobiphenyl (PCB 110)

2,2',3,3',4,4'-Hexachlorobiphenyl (PCB 128)

2,2',3,4,4',5' and 2,3,3',4,4',6-Hexachlorobiphenyl (PCB 138 & 158)

2,2',3,4',5,5'-Hexachlorobiphenyl (PCB 146)

2,2',3,4',5',6-Hexachlorobiphenyl (PCB 149)

2,2',3,5,5',6-Hexachlorobiphenyl (PCB 151)

2,2',4,4',5,5'-Hexachlorobiphenyl (PCB 153)

2,2',3,3',4,4',5-Heptachlorobiphenyl (PCB 170)

2,2',3,3',4,5,5'-Heptachlorobiphenyl (PCB 172)

2,2',3,3',4,5',6'-Heptachlorobiphenyl (PCB 177)

2,2',3,3',5,5',6-Heptachlorobiphenyl (PCB 178)

2,2',3,4,4',5,5'-Heptachlorobiphenyl (PCB 180)

2,2',3,4,4',5',6-Heptachlorobiphenyl (PCB 183)

2,2',3,4',5,5',6-Heptachlorobiphenyl (PCB 187)

2,2',3,3',4,4',5,5'-Octachlorobiphenyl (PCB 194)

2,2',3,3',4,4',5,6-Octachlorobiphenyl (PCB 195)

2,2',3,3',4,4',5,6' and 2,2',3,4,4',5,5',6-Octachlorobiphenyl (PCB 196 & 203)

2,2',3,3',4,5,5',6-Octachlorobiphenyl (PCB 199)

2,2',3,3',4,4',5,5',6-Nonachlorobiphenyl (PCB 206)

2,2',3,3',4,4',5,5',6,6'-Decachlorobiphenyl (PCB 209)

400

Appendix 4: World Health Organization List of Unique Groups of Endocrine-Disrupting Chemicals

Source: World Health Organization. (2012). *State of the science of endocrine disrupting chemicals, 2012*. Inter-organization Programme for the Sound Management of Chemicals (IPCS). United Nations Environment Programme. http://www.who.int/ceh/publications/endocrine/en/

Table 39: list of unique groups of chemicals

acrylamides	halogenated phenolic compounds	PFAS
alkaloids	heavy metals	pharmaceuticals
alkylphenols	heterocyclic amines	phosphinate flame retardants
antibiotics	HO-PCBs	phthalates
benzophenones	hydrocarbons	phycotoxins
biotoxins	isoflavones	phytoestrogens
bisphenols	mycotoxins	plasticisers
brominated flame retardants	nitrosamines	polycyclic musks
carbamates	non-dioxin like PCBs	polyfluoroalkyl phosphate
chlorinated compounds	HO-PCBs	surfactants
coccidiostats	organochlorine pesticides	POPs
dibenzofurans	organochlorines	soy
dioxins	organohalogen compounds	steroidal estrogens
Dithiocarbamates	organophosphates	strobilurin fungicides
endosulfans	organotins	sulfonamides
flame retardants	PAHs	synthetic musks
fluorobutane sulfonamides	parabens	terpene derivatives
Fluorooctane sulfonamides	PBBs	triazole fungicides
fluoroquinolones	PBDEs	trichothecene mycotoxins
fluorotelomer alcohols	PCBs	trihalomethanes
food additives	PCDDs	UV filters
food flavourings	PCDFs	veterinary drugs
furans	perfluorinated alkyl acids	veterinary pharmaceuticals
haloacetic acids	perfluorinated compounds	VOCs
haloacetonitriles	perfluoroalkylated substances	xylenes
haloaketones	pesticides	
halogenated hydrocarbons		

Appendix 5: World Health Organization List of Unique Endocrine Disrupting Chemicals

Source: World Health Organization. (2012). *State of the science of endocrine disrupting chemicals, 2012*. Inter-organization Programme for the Sound Management of Chemicals (IPCS). United Nations Environment Programme. http://www.who.int/ceh/publications/endocrine/en/

1,1,1-trichloropropanone	dieldrin	PCB114
1,1-dichloropropanone	dienestrol	PCB118
1,3-butadiene	diethylamine	PCB123
1,3-dichloropropanone	diethylhexyladipate	PCB126
1,4-dioxane	diethylhexylphthalate (DEHP)	PCB138
17α-ethynylestradiol	diethylhexylphthalate(DEHP)	PCB138+163
2,2'-azo-bis(isobutyronitrile)	diethylphthalate(DEP)	PCB149
2,4,6-tribromophenol (TBP)	dihydrodaidzein	PCB153
2,4-dichlorophenol	dihydrogenistein	PCB156
2-amino-1-methyl-6-phenylimidazo[4,5-b]pyridine(PhIP)	dihydrooxaphospha-phenanthrene (DOPO)	PCB157
		PCB163
	dihydroxyacetone	PCB167
2-amino-3-methylimidazo[4,5-f]quinoline (IQ)	diisodecylphthalate	PCB169
	diisononylphthalate(DiNP)	PCB170
2-phenylphenol	dimethoate	PCB171
3,5-dichlorophenol	dimethoxyphenylacetophenone	PCB18
3-OH-CB153	Dimethyldisulfide	PCB180
4-chloro-3-methylphenol	dimethylnitrosamine	PCB183
4-Chloroaniline	dimoxystrobin	PCB187
4-hydroxynonenal	dinitroaniline	PCB189
4MBC	DiNP	PCB19
4-nonylphenol	dioctylphthalate	PCB194
4-OH-CB107	diquat	PCB199
4-OH-CB146	diuron	PCB28
4-OH-CB187	divinylbenzene	PCB28/31
4-OH-PCB	DMSO	PCB47
4-OH-PCB107	endosulfan	PCB52
4'-o-methylequol	endosulfan I	PCB70
4-PnCDF	endosulfan II	PCB77
4tert-pentylphenol	enterodiol	PCB81
5-methylchrysene	enterolactone	PCB90/101
5-OH-equol	equol	PCB95
6-OH-BDE47	estradiol	PCBs
6-OH-daidzein	estriol	PCDDs
8-OH-daidzein	estrone	PCDFs
8-prenylnaringenin	ETBE (ethyl-t-butylether)	pentaBDE
Acetylsulfamethazine	ethanol	Pentabromobenzene
acrylamide	ethinylestradiol	Pentachloroanisole
aflatoxin	ethylbenzene	Pentachlorobenzene
alachlor	Ethylmethane sulfonate (EMS)	pentachlorophenol
Aldrin	ethylparaben	perfluorinated compounds
Alpha-pinene	F21388	perfluorocarboxylates C6 to C15
aluminium hydroxide	fenarimol	perfluorosulfonates C4, C6, C8 and
androstandiol	fenitrothion	C10
anthracene	finasteride	permethrin
antimony	fluoranthene	personal care products
Aroclor1254	fluoxetine	pestanal
arsenic	flutamide	pesticide metabolites
atenolol	flutriafol	pesticides
atrazine	fumonisin	PFAS
azoxystrobin	fumonisins	PFOA
BDE135	galaxolide	PFOS
BDE153	genistein	pharmaceuticals
BDE169	glycitein	phenanthrene
BDE176	gold	phenobarbital

BDE196	HBCD	phentoate
BDE197	HCB	phthalates
BDE202	HCH	picoxystrobin
BDE206	heptachlor	piperonyl butoxide
BDE207	Hexabromobenzene	plasticisers
BDE208	hexabromocyclodecane	platinum
BDE209	hexabromocyclododecane	PnCDD
BDE47	Hexachlorobenzene	p-Phenylenediamine
BDE53	hexachlorohexane	prochloraz
BDE99	Hexanal	procymidon
benz(a)anthracene	hexestrol	propioconazol
benzene	ibuprofen	propyl paraben
benzene hexachloride	imidacloprid	puerarin
benzo(b)fluoranthene	indeno[1,2,3-cd]pyrene	pyraclostrobin
benzo(g,h,i)perylene	indomethacin	pyrene
benzo(j)fluoranthene	iopromide	pyriproxyfen
benzo(k)fluoranthene	iron	quercetin
benzo[a]anthracene	iso-butylparaben	Resorcin
benzo[a]pyrene	Ivermectin	resorcinol
benzo[b]fluoranthene	kaempferol	resveratrol
benzo[ghi]perylene	kresoxim-methyl	rhodium
benzo[k]fluoranthene	LAS(Dodecylbenzenesulfonate)	roxithromycin
benzophenone-2	lead	saxitoxin
benzophenone-3	levonorgestrel	secoisolariciresinol
Benzotriazole	Limonene	selenium
benzotriazoles	lindane	silymarin
benzylbutylphthalate	linuron	soy extract
benzylparaben	malathion	sterigmatocystin
bezafibrate	malondialdehyde	sulfadiazine
bisphenol A	manganese	sulfadimethoxine
bisphenol A dimethacrylate	matairesinol	sulfamethazine
bisphenol F	medroxyprogesterone acetate	sulfamethizole
bromate	Melamine Polyphosphate	sulfamethoxazole
brominated flame retardants	melengestrol	sulfamethoxypyridazine
bromoacetic acid	mercury	sulfapyridine
bromoacetonitrile	mercury chloride	sulfasoxazole
bromodichloromethane	mestranol	sulfathiazole
bromoform	methoxychlor	taleranol
butylparaben	methylmercury	TBBPA
butylphenol	methylparaben	TBDE
cadmium	methyl-t-butylether (MTBE)	TCDD
carbamazepine	mirex	TCDF
carbaryl	monoacrylamide	tebuconazol
Carbendazim	mono-butylphthalate	tetrabromobisphenol A
carbofuran	mono-ethylhexylphthalate	tetrabromobisphenol A (TBBPA)
chlorate	mono-ethylhexyphthalate	thallium
chlorate hydrate	mono-ethylphthalate	thiachloprid
chlordane	mono-isononylphthalate	thiacloprid
chlorite	mono-methylphthalate	toluene
chloroacetic acid	monosodium glutamate (MSG)	tonalide
chloroacetonitrile	mycotoxins	trans-4-OH-equol
chlorobornanes	N,N-Dimethylaniline	trenbolone
chlorobromoacetic acid	N,N-dimethylurethane	triadimenol
chlorobromoacetonitrile	Napthalene	triazoles
chlorodibromomethane	NDMA (nitroso-dimethylamine)	tribromoacetic acid
chloroform	nickel	tribromoacetonitrile
chloropicrin	non-brominated flame retardants	tributyltin

chlorpyrifos	nonylphenol	trichloroacetic acid
chromium	norethynodrel	trichloroacetonitrile
chrysene	o,p'-DDD	triclosan
cis-4-OH-equol	o,p'-DDE	trifloxystrobin
copper	o,p'-DDT	trimethoprim
coumestrol	Ochratoxin A	triphenyltin
cyclohexamide	octaBDE 1,2,3	uranium
cyclopenta(c,d)pyrene	Octachlorostyrene	vinclozolin
cypermethrin	Octamethylcyclotetrasiloxane	VOCs
cyproconazol	octylphenol	zearalanol
daidzein	o-desmethylangolensin	zearalanone
DDE	OMC	zearalenone
DDT	ozone	zeranol
decaBDE	p,p'-DDE	zinc
dehydroequol	p,p'-DDT	α-HCH
deoxynivalenol	p,p-DDT	
DES	palladium	
dialkyl phthalate	parabens	
diazepam	paraquat	
diazinon	PBB153	
dibenz(a,h)anthracene	PBDE100	
dibenzo(a,e)pyrene	PBDE153	
dibenzo(a,h)-pyrene	PBDE154	
dibenzo(a,i)pyrene	PBDE183	
dibenzo(a,l)pyrene	PBDE203	
dibenzo[a,h]anthracene	PBDE206	
dibromoacetic acid	PBDE207	
dibromoacetonitrile	PBDE208	
dibromochloroacetic acid	PBDE209	
dibromochloroacetonitrile	PBDE28	
dibutylphthalate	PBDE47	
dichloroacetic acid	PBDE49	
dichloroacetonitrile	PBDE99	
dichloroaniline	PBDEs	
dichlorobromoacetic acid	PCB101	
dichlorobromoacetonitrile	PCB105	
dichloromethane	PCB110	
diclofenac		

Appendix 6: Total Production of Environmental Chemicals

Note: comprehensive databases on the annual and total production of anthropogenic ecotoxic substances are difficult to locate. Much information remains classified about US production of environmental chemicals of every description. Information about the quantities of methylmercury now in the environment are difficult to quantify and are not presently available.

Sources: ATSDR. (1994). *Toxicological profile for chlordane*. Agency for Toxic Substances and Disease Registry. Atlanta, GA. http://www.atsdr.cdc.gov/toxprofiles/tp31-c4.pdf

ATSDR. (1995b). *Toxicological profile for Mirex and Chlordecone*. Agency for Toxic Substances and Disease Registry. Atlanta, GA. http://www.atsdr.cdc.gov/toxprofiles/tp66-c4.pdf

ATSDR. (1996). *Toxicological profile for Endrin*. Agency for Toxic Substances and Disease Registry. Atlanta, GA. http://www.atsdr.cdc.gov/toxprofiles/tp.asp?id=617&tid=114

ATSDR. (2002c). *Toxicological profile for Aldrin/Dieldrin*. US Agency for Toxic Substances and Disease Registry. Atlanta, GA. http://www.atsdr.cdc.gov/toxprofiles/tp.asp?id=317&tid=56

ATSDR. (2007b). *Toxicological profile for heptachlor and heptachlor epoxide*. Department of Health and Human Services, Agency for Toxic Substances and Disease Registry. Atlanta, GA. http://www.atsdr.cdc.gov/toxprofiles/tp.asp?id=746&tid=135

ATSDR. (2013). *Toxicological profile for hexachlorobenzene*. U.S. Department of Health and Human Services, Agency for Toxic Substances and Disease Registry. Atlanta, GA. http://www.atsdr.cdc.gov/toxprofiles/tp.asp?id=627&tid=115

Bailey, R., van Wijk, D. and Thomas, P. (2009). Sources and prevalence of pentachlorobenzene in the environment. *Chemosphere*. 75(5). pg. 555-64. http://www.sciencedirect.com/science/article/pii/S0045653509000770

Besis, Athanasios and Samara, Constantini. (2012). Polybrominated diphenyl ethers (PBDEs) in the indoor and outdoor environments: A review on occurrence and human exposure. *Environmental Pollution*. 169. pg. 217-29. http://www.researchgate.net/profile/Constantini_Samara/publication/224948065_Polyb rominated_diphenyl_ethers_%28PBDEs%29_in_the_indoor_and_outdoor_environment s-- a_review_on_occurrence_and_human_exposure/links/53d8b4ad0cf2e38c63318c42.pdf

Breivik, K., Sweetman, A., Pacyna, J. and Jones, K. (2007). Towards a global historical emission inventory for selected PCB congeners — a mass balance approach: 2. Emissions. *Science of the Total Environment*. 290(1-3). pg. 199-224. http://www.sciencedirect.com/science/article/pii/S0048969701010750

Commission for Environmental Cooperation. (2006). *North American regional action plan for Lindane and other hexachlorocyclohexane (HCH) isomers.* http://www.cec.org/files/PDF/POLLUTANTS/LindaneNARAP-Nov06_en.pdf

Flint, S., Markle, T., Thompson, S. and Wallace, E. (2012). Bisphenol A exposure, effects and policy: A wildlife perspective. *Journal of Environmental Management*. 104. pg. 19-34. http://www.consbio.umn.edu/download/Flint_et_al_2012_BPA.pdf

Hasaneen, Mohammed Nagib. (2012). *Herbicides: Properties, synthesis and control of weeds*. Intech. http://cdn.intechweb.org/pdfs/25624.pdf

Li, Y. (2001). Toxaphene in the United States. *Journal of Geophysical Research*. 106(D16). pg. 17919-27. http://onlinelibrary.wiley.com/doi/10.1029/2000JD900824/pdf

Lowell Center for Sustainable Production. (2011). *Phthalates and their alternatives: Health and environmental concerns*. University of Massachusetts, Lowell, USA. http://www.sustainableproduction.org/downloads/PhthalateAlternatives-January2011.pdf

Lorz, P., Towae, F., Enke, W., et. al. (2007). *Ullman's encyclopedia of industrial industry: Phthalic acid and derivatives*. Wiley-VCH Verlag GmbH & Co., Weinheim, Germany.

National Toxicology Program. (2014). *Report on carcinogens, 13th edition*. US Department of Health and Human Services. http://ntp.niehs.nih.gov/pubhealth/roc/roc13/index.html

Paul, A., Jones, K. and Sweetman, A. (2009). A first global production, emission, and environmental inventory for perfluorooctane sulfonate. *Environmental Science and Technology*. 43(2). pg. 385-92. http://pubs.acs.org/doi/abs/10.1021/es802216n

UNEP. (2001). *Stockholm Convention on Persistent Organic Pollutants*. United Nations Environment Programme. http://www.chem.unep.ch/pops/POPs_Inc/dipcon/meetingdocs/25june2001/conf4_final act/en/FINALACT-English.PDF

US CDC. (2013). *Biomonitoring summary: Polybrominated diphenyl ethers and 2,2',4,4',5,5'-Hexabromodiphenyl(BB-153)*. National Biomonitoring Program. http://www.cdc.gov/biomonitoring/PBDEs_BiomonitoringSummary.html

Versar. (2010). *Review of exposure data and assessments for select dialkyl ortho-phthalates*. US Consumer Product Safety Commission, Bethesda, MD. http://www.cpsc.gov/PageFiles/126552/pthalexp.pdf

Vijgen, John. (2006). *The legacy of Lindane HCH isomer production*. International HCH & Pesticides Association. http://www.ihpa.info/docs/library/reports/Lindane%20Main%20Report%20DEF20JAN 06.pdf

Wang, Q., Zhao, L., Fang, X., et al. (2013). Gridded usage inventories of chlordane in China. *Frontiers of Environmental Science & Engineering*. 7(1). pg. 10-8. http://link.springer.com/article/10.1007/s11783-012-0458-z

Total Known Production of Stockholm Convention Chemicals:

Aldrin: production, import, export and stockpile amounts unknown (ATSDR 2002c)

Dieldrin: production, import, export and stockpile amounts unknown (ATSDR 2002c)

Chlordane: 100,000 to 1,000,000 pounds in US (ATSDR 1994); at least 5,490,000 pounds in China, still in production/use (Wang 2013)

Chlordecone: 3.6 million pounds produced in US from 1951-1975 (ATSDR 1995b)

Endrin: No data was collected on the production of Endrin (ATSDR 1996)

Heptachlor: No data was collected on the production of heptachlor (ATSDR 2007b) Hexabromobiphenyl (PBDE congener):

Hexabromocyclododecane (HBCD): 28,000 known tons annually in 2009-2010 (UNEP 2001)

Hexabromodiphenyl ether (PBDE congener):

Heptabromodiphenyl ether (PBDE congener):

Hexachlorobenzene (HCB): Little data available on total production, created as a byproduct of numerous industrial processes (ATSDR 2013)

Lindane (Gamma-HCH), Alpha hexachlorocyclohexane, Beta hexachlorocyclohexane: 4.8 Million tons of Gamma-HCH (Lindane) and byproduct HCH isomers from Lindane production, still in production in Romania (Commission for Environmental Cooperation 2006, Vijgen 2006)

Mirex: Over 4.8 million pounds produced in U.S., still in production (National Toxicology Program 2014)

Pentachlorobenzene: around 121,000 kilograms a year being released (Bailey 2009)

Technical endosulfan and its related isomers:

Tetrabromodiphenyl ether (PBDE congener):

Pentabromodiphenyl ether (PBDE congener):

Toxaphene: 720 kilotons from 1947 to 1986 (Li 2001)

DDT: 1.8 million tons since 1940 (US ATSDR 2002)

Perfluorooctane solfonyl fluoride (POSF): 122,500 tons (Paul 2009)

Perfluorooctane sulfonic acid (PFOS): 450-2,700 tons (from decay of POSF) (Paul 2009)

Hexachlorobenzene (HCB):

Polychlorinated biphenyls (PCB): 1.3 million tons (Breivik 2007)

Polychlorinated dibenzo-*p*-dioxins (PCDD):

Polychlorinated dibenzofurans (PCDF):

Total Known Production of Various Other Hazardous Chemicals

Glyphosate: 1.1 million tons/year in 2012, in production since 1950 (Hasaneen 2012)

Bisphenol A: 42 metric tons/year in 1942 in the US; 3.2 million metric tons/year in 2003 (Flint 2012)

PBBs (Polybrominated biphenyls): 13.3 million pounds in U.S. 1970 to 1976 (US ATSDR 2004b); 67,000 metric tons globally annually in 2000 (US CDC 2013)

PBDEs (Polybrominated diphenyl ethers): 310,000 tons annually in 2001 (Besis 2012)

All Phthalates: 11 billion pounds (Lowell Center for Sustainable Production 2011)

Phthalic anhydride: 3 million tons annually in 2000 (Lorz 2007)

Bis(2-ethylhexyl)phthalate (DEHP): over 3 billion kilograms annually in 2005 (Lorz 2007)

Dibutylphthalate (DBP): over 22 million pounds annually (Versar 2010)

Butylbenzylphthalate (BBP): below 40 million pounds annually in 2009 (Versar 2010)

Diisodecyl Phthalate (DIDP): 270 million pounds annually in 2002 (Versar 2010)

Diisoonyl Phthalate (DINP): 356 million pounds annually in 2002 (Versar 2010)

Appendix 7: Uses and Production Rates of Common Plastic Types (2014)

Source: Rossi, M. and Blake, A. (2014). *The plastics scorecard: Evaluating the chemical footprint of plastics.* Clean Production Action. http://www.ksat.com/content/dam/pns/ksat/news/defenders/2014/07/Plastics-Scorecard.pdf

Type	SPI#	Common Uses	Annual Production*
Polyethylene	2, 4	Plastic bags, packaging, bottles	78.07 million tons
Polypropylene	5	Textiles and fabric, packaging, automobiles	52.75 million tons
Polyvinyl Chloride (PVC)	3	Plumbing, house hardware, wiring, clothes	37.98 million tons
Polyethylene Terephthalate	1	Textiles and fabric, food packaging, bottles	18.99 million tons
Polystyrene	6	Packing peanuts, plastic tableware	10.55 million tons
Acrylonitrile Butadiene Styrene	7	Computer hardware, auto parts, injection molded toys like Lego bricks	8.44 million tons
Polycarbonate	7	CDs, glasses, lenses, windows	4.22 million tons
Miscellaneous	7	Various specialty plastics	77 million tons
Total:			288 million tons

* metric tons

Appendix 8: Hazard Ranked Plastic Polymers

Source: Lithner, D. (2011). *Environmental and health hazards of chemicals in plastic polymers and products*. University of Goethenburg, Department of Plant and Environmental Sciences. https://gupea.ub.gu.se/handle/2077/24978

Table 4. Hazard ranked plastic polymers that are composed of monomers with hazard classifications that belong to hazard levels *IV* or *V* (modified from paper V).

Rank	Hazard score	Polymer	Monomers (weight%)
1	13844	Polyurethane (PUR) [a], polyether based flexible foam, example	**Propylene oxide** (58); **Toluene-diisocyanate** (29); **Ethylene oxide** (7); HCF-134a (6)*
2	12379	Polyacrylonitrile (PAN) with comonomer, example acrylamide	**Acrylonitrile** (92); **Acrylamide** (8)
3	11521	Polyacrylonitrile (PAN)	**Acrylonitrile** (100)
4	10599	Polyacrylonitrile (PAN) with comonomer, example vinyl acetate	**Acrylonitrile** (92); Vinyl acetate (8)
5	10551	Polyvinyl chloride (PVC), plasticized, example with most toxic plasticiser	**Vinyl chloride** (50); **Plasticiser: Benzyl butyl phthalate (BBP)** (50)
6	10001	Polyvinyl chloride (PVC), rigid	**Vinyl chloride** (100)
7*	7384*	Polyurethane (PUR) [a], polyether based rigid foam, example	**Propylene oxide** (31); **4,4'-methylenediphenyl diisocyanate (MDI)** (52); Sorbitol (13)*; Cyclopentane (4)
8	7139	Epoxy resin DGEBPA [a], low mw (450), example with most toxic curing agent	**Bisphenol A** (45); **Epichlorohydrin** (37); **4,4'-methylenedianiline (MDA)** (18)
9	6957	Modacrylic, example with vinylidene chloride	**Acrylonitrile** (60); Vinylidene chloride (40)
10	6552	Acrylonitrile-butadiene-styrene (ABS) terpolymer	Styrene (58); **Acrylonitrile** (22); **1,3-butadiene** (20)
11*	5001*	Polyvinyl chloride (PVC), 50% non-classified plasticiser	**Vinyl chloride** (50); Diisodecyl phthalate DIDP (50)*
12	4515	Epoxy resin DGEBPA [a], low mw (450), example with least toxic curing agent	**Bisphenol A** (43); **Epichlorohydrin** (35); 4,4'-diamino diphenyl sulfone (DDS) (22)
13	4226	Epoxy resin DGEBPA [a], high mw (3750), example with non-classified curing agent	**Bisphenol A** (67); **Epichlorohydrin** (30); Dicyandiamide (3)
14	2788	Styrene-acrylonitrile (SAN) copolymer	Styrene (76); **Acrylonitrile** (24)
15	1628	High-impact polystyrene (HIPS)	Styrene (92); **1,3-butadiene** (8)
16	1500	Polyoxymethylene (POM), homopolymer	**Formaldehyde** (100)
16	1500	Phenol formaldehyde resins (PF) [a], example resol	**Phenol** (61); **Formaldehyde** (39)
17	1450	Phenol formaldehyde resins (PF) [a], example novolacs	**Phenol** (72); **Formaldehyde** (18); Hexamethylenetetramine (10)
18*	1414*	Unsaturated polyester [a] (UP), example with methyl metacrylate	**Phthalic anhydride** (31); **Methyl methacrylate** (30?); Propylene glycol (18)*; **Maleic anhydride** (21)
19**	1187**	Poly(m-phenyleneisophtalamide) (MPD-I) (Nomex®)	Isophthaloyl chloride (65)**; m-phenylenediamine (35)
19	1177	Polycarbonate (PC), example with phosgene	**Bisphenol A** (70?); **Phosgene** (30?)
20*	1117*	Unsaturated polyester [a] (UP), example with styrene	**Phthalic anhydride** (31); Styrene (30); **Maleic anhydride** (21); Propylene glycol (18)*
21**	1094**	Thermoplastic polyurethanes (TPU) polyester based rigid example	Adipic acid, Ethylene glycol & 1,4-Butanediol** (35); **4,4'-methylenediphenyl diisocyanate (MDI)** (49); 1,4-Butanediol (16)*
22	1021	Polymethyl methacrylate (PMMA)	**Methyl methacrylate** (100)
23	897	Polyphenylene sulfide (PPS)	**1,4-dichlorobenzene** (65); Sodium sulphide (35)
24*	882*	Melamine-formaldehyde resin (MF) [a]	**Formaldehyde** (59); Melamine (41)*
25	871	Polyoxymethylene (POM) copolymer, example with ethylene oxide	Trioxymethylene (96); **Ethylene oxide** (4)
26**	829**	Poly(p-phenyleneterephtalamide) (PPD-T) (Kevlar®)	Terephthaloyl chloride (66)**; p-phenylenediamine (34)
27*	750*	Urea-formaldehyde resin (UF) [a]	**Formaldehyde** (50); Urea (50)*
28**	610**	Polycarbonate (PC), example with diphenyl carbonate	**Bisphenol A** (50); Diphenyl carbonate (50)**
29**	556**	Thermoplastic polyurethanes (TPU) polyester based glycol soft example	Adipic acid, Ethylene glycol & 1,4-Butanediol (70); **4,4'-methylenediphenyl diisocyanate (MDI)** (24); 1,4-Butanediol (6)**

[a] Thermosetting plastics

In bold: monomers with level *IV* and/or level *V* hazard classifications. These are presented in Table 5.

*Contains ≥10 wt% non-classified substance, but with indication of low level of hazard according to SIDS initial assessment reports.

**Contains ≥10 wt% non-classified substance, for which ranking may be underestimated, due to elevated concern according to SIDS initial assessment reports, or lack of data.

410

Table 5. Monomers from Table 4 that have hazard classifications, which at least belong to hazard level *IV*. Classifications belonging to hazard levels *I-II* are not shown (modified from paper V).

Level IV and V monomers	Hazard classifications and category code (hazard levels III-V)	Hazard score
1,3-butadiene	**Carcinogenicity 1A (V), Mutagenicity 1B (V)**	20001
1,4-dichlorobenzene	**Aquatic chronic 1 (IV)**, Carcinogenicity 2 (III), Aquatic acute 1 (III)	1210
4,4'-methylenedianiline (MDA)	**Carcinogenicity 1B (V), Mutagenicity 2 (IV), Skin sensitization 1 (IV), Specific target organ toxicity -single exposure 1 (IV)**, Specific target organ toxicity - repeated exposure 2 (III), Aquatic chronic 2 (III)	13200
4,4'-methylenediphenyl diisocyanate (MDI)	**Respiratory sensitization 1 (IV), Skin sensitization 1 (IV)**, Carcinogenicity. 2 (III), Specific target organ toxicity - repeated exposure 2 (III)	2240
Acrylamide	**Carcinogenicity 1B (V), Mutagenicity 1B (V)**, Specific target organ toxicity - repeated exposure 1 (IV), Reproductive toxicity 2 ^{f} (III), Acute toxicity 3 ^{o} (III)	22240
Acrylonitrile	**Carcinogenicity 1B (V), Skin Sensitization 1 (IV)**, Acute toxicity 3 ^{o,d,i} (III), Serious eye damage 1 (III), Aquatic chronic 2 (III)	11521
Benzyl butyl phthalate (BBP) (Note: plasticiser)	**Reproductive toxicity 1B ^{FD} (V), Aquatic chronic 1 (IV)**, Aquatic acute 1 (III)	11100
Bisphenol A	**Skin sensitization 1 (IV)**, Reproductive toxicity 2 ^{f} (III), Serious eye damage 1 (III)	1210
Epichlorohydrin	**Carcinogenicity 1B (V), Skin sensitization 1 (IV)**, Skin corrosion 1B (III), Acute toxicity 3 ^{o,d,i} (III)	11400
Ethylene oxide	**Carcinogenicity 1B (V), Mutagenicity 1B (V)**, Acute toxicity 3 ^{i} (III)	20131
Formaldehyde	**Skin sensitization 1 (IV)**, Carcinogenicity 2 (III), Acute toxicity 3 ^{o,d,i} (III), Skin corrosion 1B (III)	1500
Hexamethylenetetramine	**Skin sensitization 1 (IV)**	1000
Maleic anhydride	**Respiratory sensitization 1 (IV), Skin sensitization 1 (IV)**, Skin corrosion 1B (III)	2110
Methyl methacrylate	**Skin sensitization 1 (IV)**	1021
m-phenylenediamine	**Mutagenicity 2 (IV), Skin sensitization 1 (IV), Aquatic chronic 1 (IV)**, Aquatic acute 1 (III), Acute toxicity 3 ^{o,d,i} (III)	3410
Phenol	**Mutagenicity 2 (IV)**, Acute toxicity 3 ^{o,d,i} (III), Specific target organ toxicity - repeated exposure 2 (III), Skin corrosion 1B (III)	1500
Phosgene	**Acute toxicity 2 ^{i} (IV)**, Skin corrosion 1B (III)	1100
p-phenylenediamine	**Aquatic chronic 1 (IV), Skin sensitization 1 (IV)**, Aquatic acute 1 (III), Acute toxicity 3 ^{o,d,i} (III)	2410
Propylene oxide	**Carcinogenicity 1B (V), Mutagenicity 1B (V)**	20061
Toluene-diisocyanate (TDI)	**Acute toxicity 2 ^{i} (IV), Respiratory sensitization 1 (IV)**, Carcinogenicity 2 (III)	3140
Vinyl chloride	**Carcinogenicity 1A (V)**	10001

^{o,d,i} toxic by oral, dermal and inhalation route
^{FD} may damage fertility and the unborn child (development)
^{f} suspected of damaging fertility
In bold: level *IV* and *V* classifications

• Substances causing acute toxicity to Daphnia magna leached from one third of all 83 tested plastic products and synthetic textiles even during the short term (1-3 d) leaching period in deionised water [I-III].
• The toxic leachates came mainly from products that were soft to semi-soft, i.e. plasticised PVC (11/13) and polyurethane (3/4), and from epoxy products (5/5), and from synthetic textiles made from various plastic fibres [I-III].
• Only one each of the 13 polyethylene, 10 polyester and 9 polypropylene leachates were acutely toxic [I-III].

• A considerable number of leachates from products intended for children (5/13) were toxic [I-III].
• None of the 12 leachates from articles for food or drinking water contact were acutely toxic [I, II].
• The toxic leachates from discarded electronic products came from the mixed material or the metal fraction, but none came from the pure plastic fraction [IV].
• Toxicity Identification Evaluation, performed on some leachates, indicated that the major toxicants were hydrophobic organics for the plastic product [I, II] and synthetic textile [III] leachates, and metals for the electronic product leachates [IV].
• Many other toxic responses than acute toxicity are highly relevant in plastic leachates [V], and leaching is suspected to be more likely to occur at low concentrations during a long period of time.
• The plastic polymers ranked as most hazardous are made of monomers classified as either carcinogenic or both carcinogenic and mutagenic (category 1A or 1B) [V].
• These belong to the polymer families of polyurethanes, polyacrylonitriles, PVC, epoxy resins, and styrenic copolymers (ABS, SAN and HIPS), and have a large global production (1-37 million tons/year). PVC accounts for 17% of the global production of plastics [V].
• The polymers that ranked as least hazardous, i.e. polypropylene and polyethylene, account for 54% of the global production of plastics [V]. A considerable number of polymers (31 out of 55) are made of monomers that belong to the two highest of the ranking model's hazard levels, i.e. levels IV and V [V].
• Polymers which are composed of level IV monomers, and also have a large global production (1-5 million tons/year), are phenol formaldehyde resins, unsaturated polyesters, polycarbonate, polymethyl methacrylate, and urea-formaldehyde resins [V].
• The most toxic plastic product leachates [I, II, III] were dominated by the polymers which had been ranked as the most hazardous [V], however, in many cases additives were suspected to be a more likely cause of toxicity than monomers.
• For several of the identified hazardous substances used in polymer production the risks ought to be evaluated for decisions on need for risk reduction measures, substitution, or even phase out [V].
• The hazard ranking model is a useful tool for comparing substances, mixtures or articles which can be used in hazard and risk assessment.
• There is a need to assess the risks from exposure in a wider context, including plastic pollution in the environment, degradation products, hazardous additives and mixture toxicity.

Appendix 9: European Plastics Demand by Resin Type (2012)

Source: Plastics Europe. (2013). *Plastics: The facts, 2013*. Association of Plastic Manufacturers. http://www.plasticseurope.org/documents/document/20131014095824-final_plastics_the_facts_2013_published_october2013.pdf

Type	SPI#	Uses	Market percent
Polyethylene Terephthalate (PET)	1	Bottles, food packaging	6.5%
High Density Polyethylene (HDPE)	2	Jugs and bottles, caps, food packaging	12%
Polyvinyl Chloride (PVC)	3	PVC windows, pipes, boots	10.7%
Low Density Polyethylene (LDPE)	4	Shopping bags, wire cables, agricultural mulch plastic	17.5%
Polypropylene	5	Car parts, flower pots, binders	18.8%
Polystyrene	6	Yogurt containers, glasses frames	7.4%
Miscellaneous	7	Sponges, toys, Teflon, etc.	19.8%

Appendix 10: Examples of Toxic and Potentially Toxic Leachates from Plastics

Sources: Chanda, Manas. (2012). *Plastics technology handbook, fourth edition (plastics engineering)*. CRC Press, FL.

Groß, R., Bunke, D., Gensch, C., et al. (2008). *Study on hazardous substances in electrical and electronic equipment, not regulated by the RoHS directive*. Öko-Institut for Applied Ecology. http://ec.europa.eu/environment/waste/weee/pdf/hazardous_substances_report.pdf

Kusy, R. and Whitley, J. (2005). Degradation of plastic polyoxymethylene brackets and the subsequent release of toxic formaldehyde. *American Journal of Orthodontic Dentofacial Orthopedics*. 127(4). pg. 420-7. http://www.ncbi.nlm.nih.gov/pubmed/15821686

Lithner, D., Damberg, J., Dave, G. and Larsson, Å. (2009). Leachates from plastic consumer products: Screening for toxicity with Daphnia magna. *Chemosphere*. 74(9). pg. 1195-200. http://www.ncbi.nlm.nih.gov/pubmed/19108869

Lithner, D. (2011a). *Environmental and health hazards of chemicals in plastic polymers and products*. University of Goethenburg, Department of Plant and Environmental Sciences. https://gupea.ub.gu.se/handle/2077/24978

Lithner, D., Nordensvan, I. and Dave, G. (2012). Comparative acute toxicity of leachates from plastic products made of polypropylene, polyethylene, PVC, acrylonitrile-butadiene-styrene, and epoxy to Daphnia magna. *Environmental Science Pollution Research International*. 19(5). pg. 1763-72. http://www.ncbi.nlm.nih.gov/pubmed/22183785

Loyo-Rosales, J., Rosales-Rivera, G., Lynch, A., et al. (2004). Migration of nonylphenol from plastic containers to water and a milk surrogate. *Journal of Agriculture Food Chemistry*. 52(7). pg. 2016-20. http://www.ncbi.nlm.nih.gov/pubmed/15053545

Lyon, F. (2008). *IARC Volume 97: Monographs on the evaluation of carcinogenic risks to humans*. International Agency on Research. http://monographs.iarc.fr/ENG/Monographs/vol97/mono97.pdf

Nakashima, E., Isobe, A., Kako, S., et al. (2012). Toxic metals derived from plastic litter on a beach. In: *Interdisciplinary Studies on Environmental Chemistry— Environmental Pollution and Toxicology*. Kawaguchi, M., Misaki, H., Sato, T., et al. eds. pg. 321-8. http://www.terrapub.co.jp/onlineproceedings/ec/06/pdf/PR639.pdf

Plastics Europe. (2013). *Plastics: The facts, 2013*. Association of Plastic Manufacturers. http://www.plasticseurope.org/documents/document/20131014095824-final_plastics_the_facts_2013_published_october2013.pdf

Rutkowski, Joseph and Levin, Barbara. (1986). Acrylonitrile-Butadiene-Styrene copolymers (ABS): Pyrolysis and combustion products and their toxicity: A review of the literature. *Fire and Materials*. 10. pg. 93-105. http://fire.nist.gov/bfrlpubs/fire86/PDF/f86017.pdf

Smiley, Robert. (2002). *Ullmann's encyclopedia of industrial chemistry: Phenylene- and toluenediamines*. Wiley-VCH, Weinheim.

US Environmental Protection Agency. (2013b). *1,3-Butadiene*. US EPA, Washington, DC. http://www.epa.gov/airtoxics/hlthef/butadien.html

US Environmental Protection Agency. (2013c). *Phthalic anhydride*. US EPA, Washington, DC. http://www.epa.gov/airtoxics/hlthef/phthalic.html

World Public Union. (2013). *Plastic numbers to avoid: BPA numbers*. World Public Union. http://www.worldpublicunion.org/2013-06-01-NEWS-plastic-numbers-to-avoid-bpa-numbers.html

Leachate:	Found in:
Metals:	
Chromium	Polyethylene, Polypropylene, PVC[1]
Cadmium	Polyethylene, Polypropylene, PVC, Polystyrene[1]
Tin	Polyethylene, Polypropylene, PVC, Polystyrene[1]
Antimony	Polyethylene, Polypropylene, PVC, Polystyrene[1]
Antimony Trioxide	Polyethylene[6]
Lead	Polyethylene, Polyethylene, Polypropylene, PVC, Polystyrene[1]
Silver	Polycarbonate (compact discs)[4]
Monomers used in Production:	
Bisphenol A	PVC, Polystyrene, Polycarbonate[2], Reactive epoxy, ABS[5]
DEHP	PVC[2]
Styrene	PVC[2]

Methyl methacrylate	Poly(methyl methacrylate)[3], Unsaturated Polyester[7]
Tetrafluoroethylene	PTFE (Teflon)[3]
Formaldehyde	Bakelite, Novolacs, Resol and other phenol-formaldehyde and melamine-formaldehyde resins[3,7], Polyoxymethylene[13]
Hexabromocyclododecane	Polystyrene[5]
Medium-chained chlorinated paraffins	PVC[5]
Short-chained chlorinated paraffins	Flame retardant in numerous rubbers, paints, adhesives, sealants and plastics[5]
Butylbenzylphthalate	PVC[5]
Dibutylphthalate	Most plastics, printer ink, paints[5]
Nonylphenol and nonylphenol ethoxylates	HDPE, PVC[6]
Diethylhexyl adipate	PVC, Polyethylene[6]
Propylene oxide	Polyether foam, Polyurethane[7]
Acrylonitrile, Acrylamide	ABS, Modacrylic, Polyacrylonitriles[7]
Vinyl chloride	PVC[7]
4,4'-methylenediphenyl diisocyanate (MDI)	Thermoplastic polyurethanes, Polyurethane[7]
4,4'-methylenedianiline	Epoxy resin, polyurethane[7]
1,3-butadiene	ABS[7], Rubbers, acrylics[8]
Epichlorohydrin (decays into 3-MCPD)	Epoxy resin DGEBPA (used on teabags)[8]
Phthalic anhydride	Unsaturated polyester[7], PVC, rubber, dyes, lacquer[9]
Maleic anhydride	Unsaturated polyester, Polybutylene terephthalate (PBT)[7]
Isophthaloyl chloride	Poly(m-phenyleneisophtalamide) or "Nomex"[7]
m-phenylenediamine	Nomex, epoxy resins[7]

Phosgene	Polycarbonate[7]
Sodium sulphide	Polyphenylene sulfide (PPS)[7]
Trioxymethylene	Polyoxymethylene plastic[7]
Ethylene oxide	Polyoxymethylene plastic[7], Polyester, Polyethylene terephthalate[10]
p-phenylenediamine	Kevlar (p-phenyleneterephtalamide)[7], urethanes, hair dye[11]
Diphenyl carbonate	Polycarbonate[7]
Adipic acid	Nylon, Thermoplastic polyurethanes[7]
O-acetyl tributyl citrate	PVC, rubber[5]
Di(2-ethylhexyl) phosphate	Cellulose ester plastics[5]
Tri(2-ethylhexyl) phosphate	PVC, nonylcarbonate[5, 12]
Tri-2-ethylhexyl trimellitate	PVC[5]
Polymer degradation products:	
Hydrogen Cyanide	Nylon, Polyacrylonitrile, Polyurethane[3], ABS[14]
Hydrogen Chlorine	PVC[3]
Hydrogen Fluoride	Polyvinylidene fluoride, PTFE[3]

([1]Nakashima 2012; [2]World Public Union 2013; [3]Lithner 2012; [4]Lithner 2009, [5]Groß 2008, [6]Loyo-Rosales 2004, [7]Lithner 2011a, [8]US EPA 2013b, [9]US EPA 2013c, [10]Lyon 2008, [11]Smiley 2002, [12]Chanda 2012, [13]Kusy 2005, [14]Rutkowski 1986)

Production of Plastic by Country (2012)

China: 23.9%
Europe: 20.4%
NAFTA: 19.9%
Asia (other than China): 15.8%
Middle East and Africa: 7.2%
Latin America: 4.9%
Japan: 4.9%
Commonwealth of Independent States (former USSR): 3% (Plastics Europe 2013)

Appendix 11: World Coal Production by Country

Source: US EIA. (2015). *International energy statistics database*. US Energy Information Administration.
http://www.eia.gov/cfapps/ipdbproject/IEDIndex3.cfm?tid=1&pid=7&aid=1

(in thousands of tons)

Country:	2008	2009	2010	2011	2012
China:	3,099,061	3,301,803	3,560,635	3,878,012	4,017,920
United States:	1,171,809	1,074,923	1,084,368	1,095,628	1,016,458
India:	570,010	614,918	619,843	633,774,	649,644
Europe:	796,478	763,584	739,872	779,441	773,613
Australia:	432,383	449,631	467,823	443,390	463,783
Russia:	336,163	304,228	354,615	354,869	390,152
World:	7,470,959	7,601,609	7,999,455	8,443,803	8,687,297

Appendix 12: US Field production of crude oil

Source: US EIA. (2015). *U.S. field production of crude oil*. US Energy Information Administration.
http://www.eia.gov/dnav/pet/hist/LeafHandler.ashx?n=PET&s=MCRFPUS1&f=M

U.S. Field Production of Crude Oil

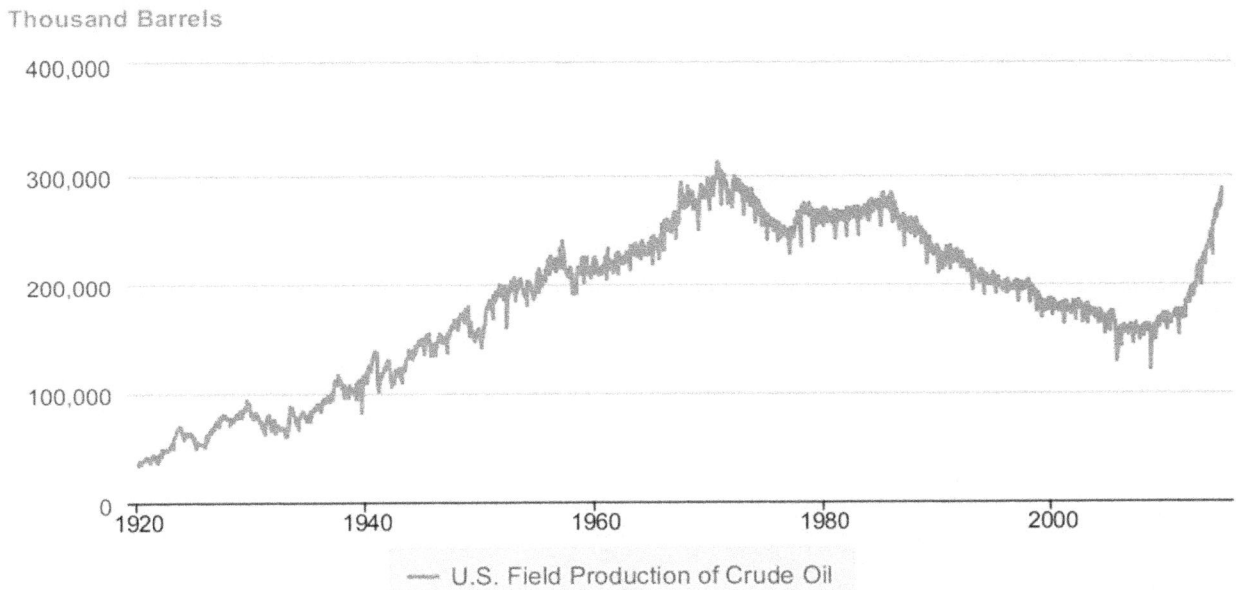

Thousand Barrels

Appendix 13: World Crude Oil Production

Source: US EIA. (2015). *International energy statistics database*. US Energy
Information Administration.
http://www.eia.gov/cfapps/ipdbproject/IEDIndex3.cfm?tid=1&pid=7&aid=1

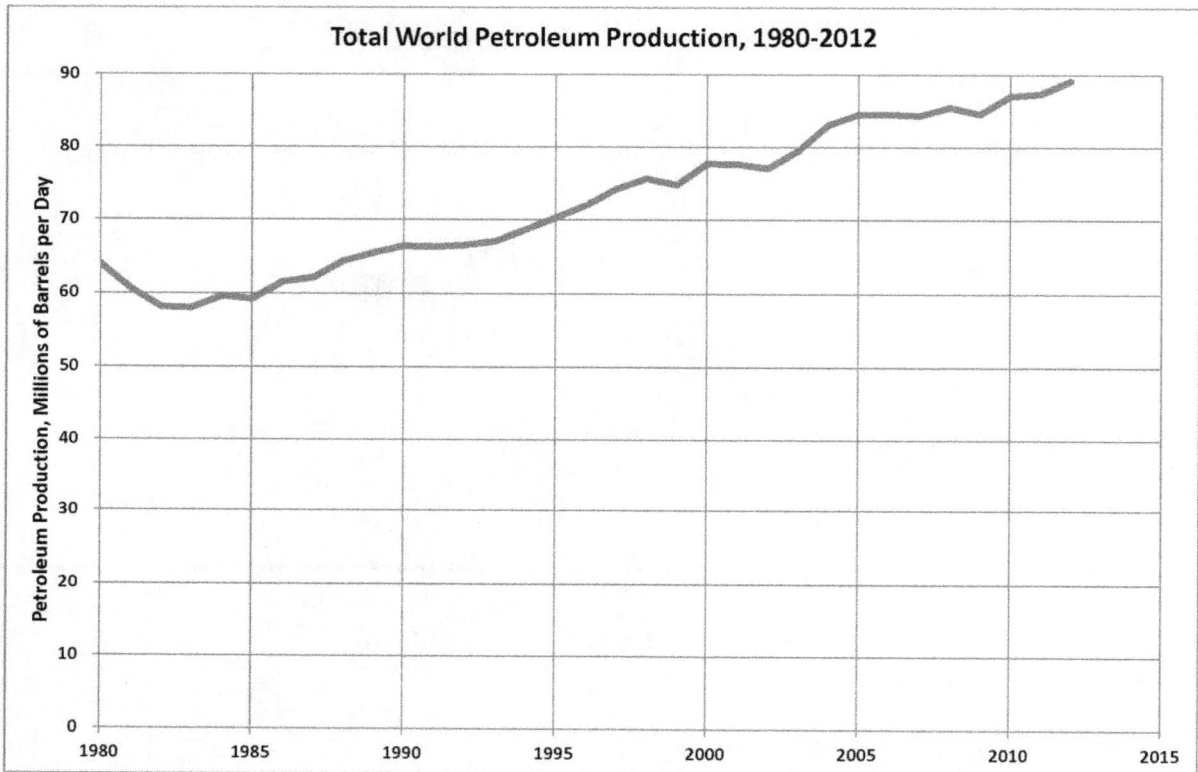

Total World Petroleum Production, 1980-2012

Appendix 14: Concentrations of organic chemicals reported in sewage sludges

Harrison, E., Oakes, S., Hysell, M. and Hay, A. (2006). Organic chemicals in sewage sludges. *Science of the Total Environment*. 367. pg. 481-97. http://cwmi.css.cornell.edu/Sludge/organicchemicals.pdf

Table 1
Concentrations of organic chemicals reported in sewage sludges and sources of those data

	Range mg/kg dry wgt	Data sources[a]
Aliphatics—short chained and chlorinated		
Acrylonitrile	0.0363–82.3	[1]
Butadiene (hexachloro-1,3-)[SSL]	ND–8	[1–4]
Butane (1,2,3,4-diepoxy)	ND–73.9	[5]
Butanol (iso)	ND–0.165	[5]
Butanone (2-)	ND–1540	[5]
Carbon disulfide[SSL]	ND–23.5	[5]
Crotonaldehyde	ND–0.358	[5]
Cyclopentadiene (hexachloro)[SSL]	<0.005	[2]
Ethane (hexachloro)[SSL]	0.00036–61.5	[3]
Ethane (monochloro)	ND–24	[3]
Ethane (pentachloro)	0.0003–9.2 g	[3]
Ethane (tetrachloro)	<0.1–5.0	[6]
Ethane (trichloro) isomers[SSL]	ND–33	[7]
Ethylene (dichloro)[SSL]	<0.01–865	[3,8]
Ethylene (monochloro)	<0.025–110	[2,3]
Ethylene (tetrachloro)[SSL]	ND–50	[1–3,5,7,8]
Ethylene (trichloro)[SSL]	ND–125	[2,3,5,7]
Hexanoic acid	ND–1960	[5]
Hexanone (2-)	ND–12.7	[5]
Methane (dichloro)[SSL]	ND–262	[3,5,8,9]
Methane (monochloro)	ND–30	[5]
Methane (tetrachloro)[SSL]	ND–60	[2,3,5–7]
Methane (trichloro)[SSL]	ND–60	[2,5–7]
Methane (trichlorofluoro)	ND–3.97	[5]
N-alkanes (polychlorinated)	1.8–93.1	[10]
N-alkanes	ND–758	[5]
Organic halides absorbable (AOX) and extractable (EOX)	1–7600	[7,11–13]
Pentanone (methyl)	ND–0.567	[5]
Polyorganosiloxanes	8.31–5155	[14–18]
Propane (dichloro) isomers[SSL]	ND–1230	[1,3,5]
Propane (trichloro)	0.00459–19.5	[1,3]
Propanenitrile (ethyl cyanide)	ND–64.7	[5]
Propanone (2-)	ND–2430	[5]
Propen-1-ol (2-)	ND–0.0312	[5]
Propene (trichloro)	<0.0010–167	[1]
Propene chlorinated isomers[SSL]	0.002–1230	[3,5]
Propenenitrile (methyl)	ND–218	[5]
Squalene	ND–16.7	[5]
Sulfone (dimethyl)	ND–0.784	[5]
Chlorobenzenes		
Benzene (dichloro) isomers[SSL]	ND–1650	[2,3,5,8,19,20]
Benzene (hexachloro)[SSL]	ND–65	[1,2,4,7,11,20–22]
Benzene (monochloro)[SSL]	ND–846	[3,5,19]

	Range mg/kg dry wgt	Data sources[a]
Chlorobenzenes		
Benzene (pentachloro)	<0.005–<0.01	[2,20]
Benzene (tetrachloro)	<0.001–0.22	[2,20]
Benzene (trichloro) isomers[SSL]	ND–184	[2,3,5,19,20]
Flame retardants		
Brominated diphenyl ether congeners (BDEs)	<0.008–4.89	[23–30]
Cyclododecane (hexabromo) isomers	<0.0006–9.120	[31]
Tetrabromobisphenol A	<0.0024–3322	[32]
Tetrabromobisphenol A (dimethyl)	<0.0019	[32]
Monocyclic hydrocarbons and heterocycles		
Acetophenone	ND–6.92	[5]
Aniline (2,4,5-trimethyl)	ND–0.220	[5]
Benzene[SSL]	ND–11.3	[3,5,33]
Benzene (1,4-dinitro)	ND–4.4	[5]
Benzene (ethyl)[SSL]	ND–65.5	[3,5]
Benzene (mononitro)[SSL]	ND–1.55	[2,5]
Benzene (trinitro)	12	[34]
Benzenethiazole (2-methylthio)	ND–64.4	[5]
Benzenethiol	ND–3.25	[5]
Benzoic acid[SSL]	ND–835	[5]
Benzyl alcohol	ND–156	[5]
Analine (chloro) (P-)[SSL]	ND–40.2	[5]
Cymene (P-)	ND–84.3	[5]
Dioxane (1,4-)	ND–35.3	[5]
Picoline (2-)	ND–365	[5]
Styrene[SSL]	ND–5850	[3,5]
Terpeniol (alpha)	ND–2.56	[5]
Thioxanthe-9-one	ND–19.6	[5]
Toluene[SSL]	ND–1180	[3,5,6,8,9,34,35]
Toluene (chloro)	1.13–324	[5]
Toluene (2,4-dinitro)[SSL]	ND–10	[2,5,34]
Toluene (para nitro)	100	[34]
Toluene (trinitro)	12	[34]
Xylene isomers[SSL]	ND–6.91	[5,8,33,35–37]
Nitrosamines		
N-nitrosdiphenylamine[SSL]	ND-19.7	[5]
N-nitrosodiethylamine	ND–0.0038	[38]
N-nitrosodimethylamine	0.0006–0.053	[38]
N-nitrosodi-*n*-butylamine	ND	[38]
N-nitrosomorpholine	ND–0.0092	[38]
N-nitrosopiperdine	ND–trace	[38]
N-nitrosopyrrolidine	ND–0.0042	[38]
Organotins		
Butylitin (di)	0.41–8.557	[39–44]
Butyltin (mono)	0.016–43.564	[39–44]

421

Table 1 (*continued*)

	Range	Data
	mg/kg dry wgt	sources [a]
Organotins		
Butyltin (tri)	0.005–237.923	[9,39–44]
Phenyltin (di)	0.1–0.4	[42,43]
Phenyltin (mono)	0.1	[42,43]
Phenyltin (tri)	0.3–3.4	[42,43]
Personal care products and pharmaceuticals		
Acetaminophen	0.0000006–4.535	[45]
Gemfibrozil	ND–1.192	[45]
Ibuprofen	0.000006–3.988	[45]
Naproxen	0.000001–1.022	[45]
Salicylic acid	0.000002–13.743	[45]
Antibiotics		
Ciprofloxacin	0.05–4.8	[46,47]
Doxycycline	<1.2–1.5	[47]
Norfloxacin	0.01–4.2	[46,47]
Ofloxacin	<0.01–2	[47]
Triclosan (4-chloro-2-(2,4-dichloro-phenoxy)-phenol and related compounds	ND–15.6	[25,48–50]
Fluorescent whitening agents		
BLS (4,4'-bis(4-chloro-3-sulfostyryl)-biphenyl)	5.4–5.5	[51]
DAS 1 (4,4'-bis[(4-anilino-6-morpholino-1,3,5-triazin-2-yl)-amino] stilbene-2,2'-disulfonate)	86–112	[51]
DSBP (4,4'-bis (2-sulfostyryl)biphenyl)	31–50	[51]
Fragrance material		
Acetyl Cedrene	9.0–31.1	[52]
Amino Musk Ketone	ND–0.362	[37]
Amino Musk Xylene (AMX)	ND–0.0315	[37]
Cashmeran (DPMI) (6,7-dihydro-1,1,2,3,3-pentamethyl-4(5H)-indanone)	ND–0.332	[34,37]
Celestolide (1-[6-(1,1-Dimethylethyl)-2,3-dihydro-1,1-methyl-1H-inden-4-yl]-ethanone)	0.010–1.1	[34,37,53,54]
Diphenyl Ether	ND–99.6	[5,52]
Galaxolide (HHCB) (1,3,4,6,7,8-Hexahydro-4,6,6,7,8,8-hexamethylcyclopenta[g]-benzopyran)	ND–81	[25,34,37, 52–56]
Galaxolide lactone (1,3,4,6,7,8-Hexahydro-4,6,6,7,8,8-hexamethylcyclopenta[g]-2-benzopyran-1-one)	0.6–3.5	[54]
Hexyl salicylate	Trace–1.5	[52]

Table 1 (*continued*)

	Range	Data
	mg/kg dry wgt	sources [a]
Fragrance material		
Hexylcinnamic Aldehyde (Alpha)	4.1	[52]
Methyl ionone (gamma)	1.1–3.8	[52]
Musk Ketone (MK) (4-tertbutyl-3,5-dinitro-2,6-dimethylacetophenone)	ND–1.3	[37,52,57]
Musk Xylene (1-*tert*-butyl-3,5-dimethyl-2,4,6-trinitrobenzene)	ND–0.0325	[57]
OTNE (1-(1,2,3,4,5,6,7,8-octahydro-2,3,8,8-tetramethyl-2-naphthalenyl))	7.3–30.7	[52]
Phantolide (1-[2,3-Dihydro-1,1,2,3,3,6-hexamethyl-1H-inden-5-yl]-ethanone)	0.032–1.8	[34,37, 53,54]
Tonalide (1-[5,6,7,8-Tetrahydro-3,5,5,6,8,8-hexamethyl-2-naphthalenyl]-ethanone)	ND–51	[25,37, 52–55]
Traseolide (ATII) (1-[2,3-Dihydro-1,1,2,6-tetramethyl-3-(1-methyl-ethyl)-1H-inden-5-yl] ethanone	0.044–1.1	[53,54]
Pesticides		
Aldrin[SSL]	ND–16.2	[1–5,21,22, 33,58,59]
Azinphos Methyl	ND–0.279	[5]
Benzene (pentachloronitro)	ND–8.83	[5]
Captan	ND–0.968	[5]
Chlordane[SSL]	ND–16.04	[1,3,5]
Chlorobenzilate	ND–0.104	[2,5]
Chloropyrifos	ND–0.529	[5]
Ciodrin	ND–0.093	[5]
Cyclohexane isomers (lindane and others[SSL]**)**	ND–70	[1–7,9,11,21, 22,59–62]
DDT and related congeners[SSL]	ND–564	[1–5,7,9, 11,21,22,33, 58,60–62]
Diallate	ND–0.394	[2,5]
Diazinon	ND–0.151	[5]
Dicrotophos (Bidrin)	ND–0.550	[5]
Dieldrin[SSL]	ND–64.7	[1–7,21,22, 33,60,61]
Dimethoate	ND–0.340	[2,5]
Disulfotone	<0.0050	[2]
Endosulfans	ND–0.280	[2,4,5,21]
Endrin[SSL]	ND–1.17	[1,2,4,5,21, 22,59]
Famphur	<0.0050–0.400	[2]

(*continued on next page*)

Table 1 (*continued*)

	Range	Data
	mg/kg dry wgt	sources[a]

Pesticides

	Range	Data sources[a]
Heptachlor epoxides[SSL]	ND–0.780	[1,2,5,21]
Heptachlor[SSL]	ND–16	[2,3,5,21,22]
Isobenzan	ND–0.130	[4]
Isodrin	ND	[4]
Isophorone[SSL]	<0.0050–0.08294	[2]
Leptophos	ND–0.319	[5]
Methoxychlor[SSL]	<0.015–0.330	[2]
Mevinphos (phosdrin)	ND–0.148	[5]
Naled (Dibrom)	ND–0.484	[5]
Naphthoquinone (1,4-)	<0.0050	[2]
Nitrofen	ND–0.195	[5]
Parathion (ethyl)	<0.0050–0.380	[2]
Parathion (methyl)	<0.0050–0.070	[2]
Permethrin isomers	<0.15–163	[20,63]
Phenoxy herbicides[SSL]	ND–7.34	[1,2,5]
Phenoxypropanoic acid (trichloro)	ND–0.121	[5]
Phorate (*O,O*-diethyl *S*-[(ethylthio) methyl] phosphorodithioate)	<0.0050–0.200	[2]
Phosphamidon	ND–0.232	[5]
Pronamide (dichloro (3,5-)-*N*-(1,1-dimethylpropynyl) benzamide)	<0.0050–0.008	[2]
Pyrophosphate (tetraethyl)	ND–20	[5]
Quintozene	ND–0.100	[4]
Safrol (iso)	<0.0050–0.750	[2]
Safrole (EPN)	ND–0.545	[2]
Toxaphene[SSL]	51	[3]
Trichlorofon	ND–2.53	[5]
Trifluralin (Treflan)	ND–0.235	[5]
Phenol[SSL]	ND–920	[2,3,5,7,8,36,66]
Phenol chloro congeners[SSL]	<0.003–8490	[1–3,5–9,33,35,49,61,66–68]
Phenol chloro methyl congeners	ND–136	[2,3,5,8,9,61,64]
Phenol methyl congeners[SSL]	ND–1160	[2,3,5,7–9,34,66]
Phenol nitro methyl congeners	0.2–187	[5]
Phenols nitro congeners[SSL]	<0.003–500	[2,3,8]

Table 1 (*continued*)

	Range	Data
	mg/kg dry wgt	sources[a]

Phthalate acid esters/plasticizers

	Range	Data sources[a]
Bis(2-chloroethyl) ether[SSL]	<0.020–0.130	[2]
Bis(2-chloroisopropyl) ether	<0.150–5.700	[2]
Bis(2-cloroethoxy) methane	<0.020–0.240	[2]
Di(2-ethylhexyl) adipate	<0.100–0.450	[2]
Phthalates[SSL]	ND–58,300	[2,3,5–9,28,33,36,58,69–73]

Polychlorinated biphenyls, naphthalenes, dioxins and furans

	Range	Data sources[a]
Aroclor 1016	0.2–75	[6,74]
Aroclor 1248	ND–5.2	[5,6,33,58]
Aroclor 1254	0.0667–1960	[1,5]
Aroclor 1260	ND–433	[1,5,6,58,60]
Biphenyl (decachloro)	0.11–2.9	[1]
Biphenyls (polybrominated)	431	[3]
Dibenzofuran	ND–59.3	[5]
Dioxins and furans (polychlorinated dibenzo)	ND–1.7	[5,8,72,75–81]
PCB congeners	ND–765	[2–5,7,11,13,21,22,28,35,53,59,61,71,72,79,81–87]
Phenylether (chloro)	<0.020	[2]
Terphenyls and naphthalenes (polychlorinated)	ND–11.1	[2,3,5,9,28,53]
		74,88,89]
Benzidine	12.7	[3]
Benzo(*a*)anthracene[SSL]	ND–99	[2,3,5,8,21,53,82,88–90]
Benzo[*ghi*]perylene	ND–12.9	[1,2,5–8,21,22,28,53,88–91]
Benzofluoranthene congeners[SSL]	0.006–34.2	[3,89]
Benzofluorene congeners	ND–8.1	[62,89]
Benzopyrene congeners[SSL]	ND–24.7	[1–3,5–8,11,21,22,28,33,53,62,82,88–91]

Table 1 (*continued*)

	Range mg/kg dry wgt	Data sources [a]
Polynuclear aromatic hydrocarbons		
Biphenyl	ND–15,300	[3,5,53]
Chrysene[SSL]	ND–32.4	[3,5,8,21,53, 82,88,90]
Chrysene+triphenylene	0.01–14.7	[2,89]
Dibenzoanthracene congeners[SSL]	ND–13	[2,3,8,21,53, 88,89,91]
Dibenzothiophene	ND–1.47	[5]
Diphenyl amine	ND–32.6	[5]
Fluoranthene[SSL]	ND–60	[1–3,5–8,21, 22,28,33,53,62, 82,88–90]
Fluorene[SSL]	<0.01–8.1	[2,8,21,53, 82,88]
Fluorene (nitro)	0.941	[28]
Indeno(1,2,3-c,d) pyrene[SSL]	ND–9.5	[2,7,8,21,22, 28,53,88–91]
Naphthalene[SSL]	ND–6610	[2,3,5,6,8,21, 36,53,62,88]
Naphthalene methyl isomers	ND–136	[2,5,28,53]
Napthalene methyl congeners		
Napthalene nitro congeners	ND–0.0798	[28]
Perylene	ND–69.3	[3,5,53,89,91]
Phenanthrene	<0.01–44	[2,3,5,6, 8,21,28,53, 62,82,88–90]
Phenanthrene methyl isomers	ND–37.4	[5,53]
Pyrene[SSL]	0.01–37.1	[2,3,5,6, 8,21,53, 82,88–90]
Pyrene (phenyl)	0.06–6.86	[1]
Retene (7-isopropyl-1-methylphenanthrene)	0.260	[28]
Total PAH	ND–199	[9,11,28, 72,86]
Triphenylene	ND–15.4	[5]
Sterols, stanols and estrogens		
Campestanol (5a+5b)	3.0–14	[55]
Campesterol	6.3	[55]
Cholestanol (5a-)	22.7	[49,87]
Cholesterol	57.4	[55]
Coprostanol	216.9	[55]
Estradiol (17b)	0.0049–0.049	[92,93]
Estrone	0.016–0/0278	[92,93]
Ethinylestradiol (17a)	<0.0015–0.017	[92,93]
Sitostanol (5a-b+5b-b-)	14.1–93.9	[55]
Sitosterol (b-)	29.6–31.1	[55]
Stigmastanol (5a-+5b)	1.9–12.9	[55]
Stigmasterol	6.7	[55]

Table 1 (*continued*)

	Range mg/kg dry wgt	Data sources [a]
Surfactants		
Alcohol ethoxylates	ND–141	[70,94,95]
Alkylbenzene sulfonates	<1–30,200	[6,7,9, 70–72,74, 85,94,96–98]
Alkylphenolcarboxylates	10–14	[92]
Alkylphenolethoxylates	ND–7214	[2,7,25,28, 49,69,71,72, 85,90,92, 94,99–101]
Alkyphenols (nonyl and octylphenol)	ND–559,300	[2,6,9,18,25, 28,36,49,64, 69,74,92, 95,99–107],
Coconut diethanol amides	0.3–10.5	[70]
Poly(ethylene glycol)s	1.7–17.6	[70]
Triaryl/alkyl phosphate esters		
Cresyldiphenyl phosphate	0.61–179	[3]
Tricresyl phosphate	0.069–1650	[3]
Tricresyl phosphate	<0.020–12.000	[2]
Tri-*n*-butylphosphate	<0.020–2.400	[2]
Triphenylphosphate	<0.020–1.900	[2]
Trixylyl phosphate	0.027–2420	[3]

Appendix 15: Number of chemicals reported in sewage sludges of various classes

Source: Harrison, E., Oakes, S., Hysell, M. and Hay, A. (2006). Organic chemicals in sewage sludges. *Science of the Total Environment*. 367. pg. 481-97. http://cwmi.css.cornell.edu/Sludge/organicchemicals.pdf

Number of chemicals reported in sludges in each class, number of studies from which data were obtained, number that are priority pollutants, target compounds or for which there are SSLs, and number for which maximum reported concentrations in sludges exceed an SSL

	# chem	# of studies	# PP chem	# TC chem	# chem with SSLs	# chem that exceed an SSL
Aliphatics	58	19	16	17	16	15
Chlorobenzenes	11	13	6	7	5	5
Flame retardants	29	11	0	0	0	
Monocyclic HC	34	12	7	12	11	10
Nitrosamines	7	1	2	1	1	1
Organotins	6	7	0	0	0	
PCPs	36	17	0	0	0	
Pesticides	71	20	18	19	18	15
Phenols	40	20	10	14	9	8
Phthalate	19	16	9	8	6	6
PCBs	108	38	5	6	0	
PAHs	52	25	18	18	13	8
Sterols and stanols	16	3	0	0	0	
Surfactants	23	33	0	0	0	
Triaryl/alkyl phosphate.esters	6	2	0	0	0	
Total	516	113[a]	91	102	79	68

[a] Note: # of studies is not a sum of the list above because some studies include data for more than one class.

Appendix 16: Compounds found in combustion gases

Source: Andersson, Berit. (2003). *Combustion products from fires: Influence from ventilation conditions*. Lund University, Sweden, Department of Fire Safety Engineering.
http://lup.lub.lu.se/luur/download?func=downloadFile&recordOId=642023&fileOId=642045

Table 23. Compounds found with GC-MC analysis, in the combustion gases.

Compound / Sample description	Chlorobenzene 005	019	022	023	025	TMTM 007	008	009	010	014	036	037	038	Dimethoate 011	012	013	015	CNBA 017	018	Nylon 032	034
benzene	x	x	x	x	x			x			x	x	x	x	x	x	x	x	x		x
toluene	x			x		x	x	x	x	x	x	x	x	x		x			x	x	x
tetracloroethene	x			x																	
clorobenzene	x	x	x	x	x													x	x		
ethynylbenzene	x														x	x			x		
etenylbenzene	x			x																	
phenol	x	x	x						x	x		x							x		
clorophenol	x		x															x	x		
1,3-diclorobenzene	x																	x	x		
1,4-diclorobenzene	x																	x	x		
1H-purine 2,6-	x																				
1,2-dichlorobenzene	x			x																	
1-chloro-4-etynyl-benzene	x			x																	
1-chloro-3-etynyl-benzene	x																		x		
2-cloro-benzonitrile																		x	x		
naphtalene	x		x	x		x	x	x	x		x	x	x	x	x	x	x	x	x		x
1,2,4-trichlorobenzene	x																				
dodecane										x											x
4-chlorophenol	x	x																	x		
1-methyl naphtalene	x					x		x													
1-methyl naphtalene	x		x	x																	
1-ethylidene 1-H indene	x																				
2,4,6-trimethyloctane	x	x				x			x	x		x									
1-chloronaphtalene	x	x																			
2-chloronaphtalene																		x	x		
1-ethylnaphtalene	x																				
biphenylene	x	x	x													x					
3-phenyl-2-propenylchloride	x																				
1,8-dimethylnaphtalene	x																				
acenaphthylene	x			x																	
3-methyl-tridecane	x			x																	
2,6,11-trimethyl-dodecane																			x		
2-meth.-,1-(1,1-dimeth.eth.)propanoic acid	x																				
dimethyl-heptadecane																			x		
allyl ethyl ester phthalic acid	x																				
3,6-dimethylundecane	x																				
2,5,6-trimethyldecane	x																				
heptadecane	x	x	x				x	x	x	x		x							x	x	
octanoic acid	x																				
eicosane	x		x							x									x		
3-(ethenyloxy)methyl-heptane	x		x																		
1-hexadecanol	x	x	x																		
buth.oct.ester1,2-benzendicarb.acid	x	x																			
dimethylphenyl ester benzoic acid	x																				x	
buthyl 2-methylpropyl ester -	x																				
1,2 benzene dicarboxylic acid	x	x								x											
diisooct.ester 1,2 benzenedicarb.acid	x	x	x																		
1,3,5,7-cyclooctatetraene						x	x	x		x		x			x	x					
benzonitrile							x	x	x									x	x		

426

Table 23. (continued)

Compound / Sample description	Chlorobenzene					TMTM								Dimethoate				CNBA		Nylon	
	005	019	022	023	025	007	008	009	010	014	036	037	038	011	012	013	015	017	018	032	034
2,6-(1,1-isobut.)-4-methylcarbamatphenol											X										
1,2-propendienylbenzen											X										
bis(2-ethylhexyl)phtalat											X									X	
2-methyldodecane							X	X	X		X		X								
2-methyl-nonadecane																					X
6-ethyl-2-methyl-octane																					X
2,8-dimethyl-undecane										X											X
2-meth-,1-(1,1-dimeth.eth)-2-meth-1,3-prop.acid	X	X	X	X																	
diethyl phtalat	X	X	X	X																	
dioctyl ester	X		X	X																	
6-methyl-octadecane															X		X				
hexane											X										
methyl ester thiocyanic acid											X										
nonanal						X				X	X	X	X								
ethyl-benzaldehyde											X										
tetramethyl-thiourea						X	X	X	X	X	X	X	X	X		X					
tetradecane											X										
thiram						X	X	X	X	X	X	X	X								
trimethyl-thiourea						X					X	X									
4-(2,2,3,3-tetramethylbutyl)-phenol											X										
nonyl-phenol											X										
docosane											X										
dibenzofuran																		X	X		
xylene														X	X	X	X	X	X		
cyclohexanone														X	X	X	X				
1-ethyl-3-methyl-benzene																X					
benzaldehyde																X					
1H-indene														X		X					
1-methyl-1H-indene																X					
1-ethylidiene-1H-indene																X					
thiophene						X	X	X					X								
methyl ester, thiocyanic acid								X	X		X	X									
(dimethylamino)-acetonitrile								X													
dimethyl-cyanamide								X													
N,N-dimethyl-methanethioamide						X	X	X	X		X	X	X								
N,N-dimethyl-benzenamide								X													
benzothiophene							X	X		X			X								

427

Appendix 17: Temporal Trends in Obesity

Sources: OECD. (2014). *Obesity update: June 2014*. Organization for Economic Cooperation and Development.
http://www.medicosypacientes.com/gestor/plantillas/articulos/archivos/imagenes/www.oecd.org_els_health-systems_Obesity-Update-2014.pdf

US CDC. (2010). *Prevalence of overweight among children and adolescents: United States, 2003-2004*. Centers for Disease Control.
http://www.cdc.gov/nchs/data/hestat/overweight/overweight_child_03.htm

Weight-control Information Network. (2014). *Overweight and obesity statistics*. US Department of Health and Human Services. Washington, DC.
http://win.niddk.nih.gov/publications/PDFs/stat904z.pdf

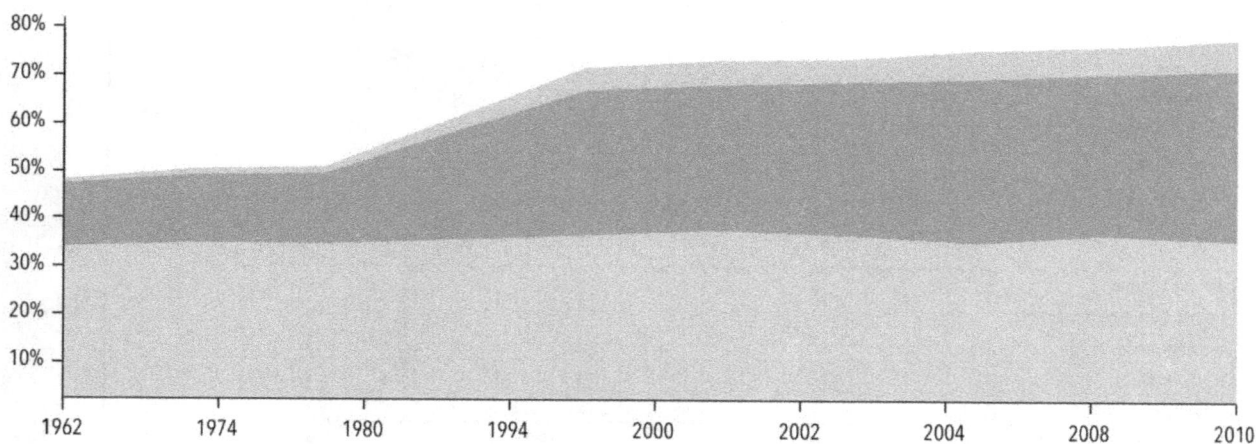

Trends in Overweight and Obesity among Adults, United States, 1962-2010[†]   ▓ Overweight ■ Obesity ▒ Extreme Obesity

(Weight-control Information Network, 2014)

428

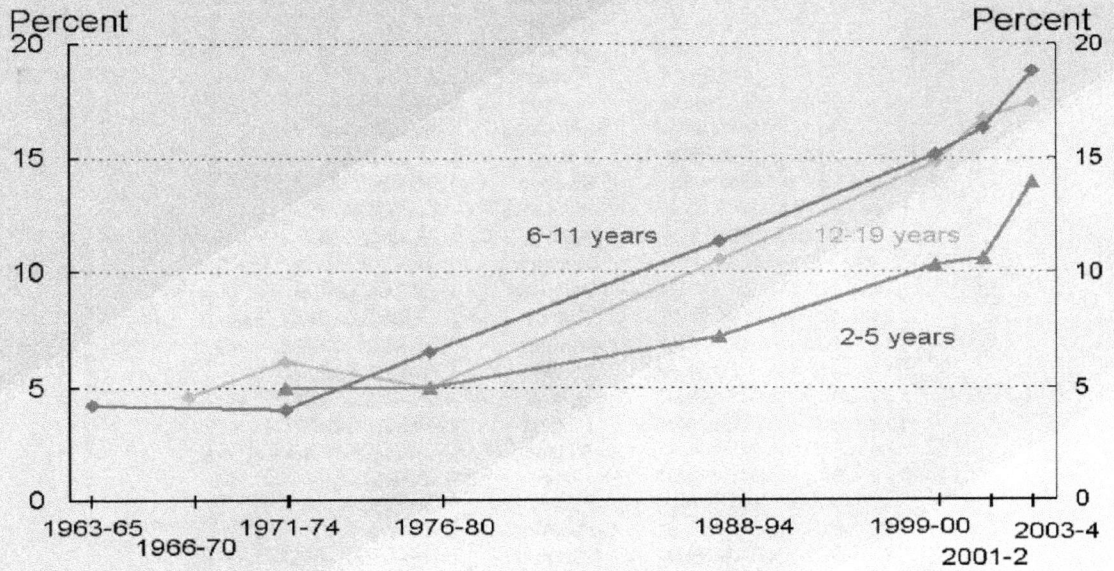

Trends in Child and Adolescent Overweight

Note: Overweight is defined as BMI >= gender- and weight-specific 95th percentile from the 2000 CDC Growth Charts.
Source: National Health Examination Surveys II (ages 6-11) and III (ages 12-17), National Health and Nutrition Examination Surveys I, II, III and 1999-2004, NCHS, CDC.

(US CDC 2010)

Obesity Among Adults, 2012 or Nearest Year

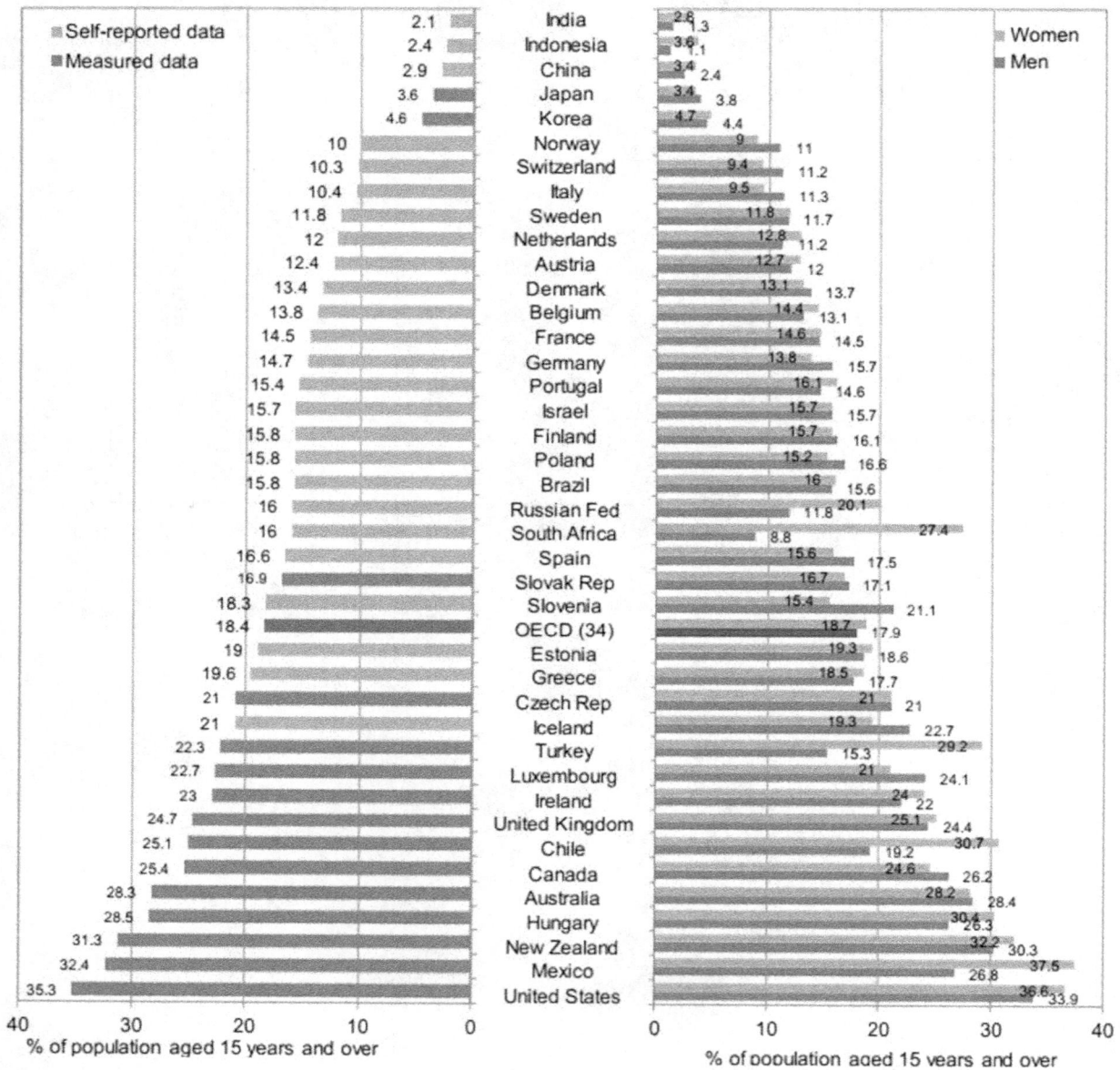

Left chart legend: Self-reported data; Measured data

Right chart legend: Women; Men

Country	Total	Women	Men
India	2.1	2.8	1.3
Indonesia	2.4	3.6	1.1
China	2.9	3.4	2.4
Japan	3.6	3.4	3.8
Korea	4.6	4.7	4.4
Norway	10	9	11
Switzerland	10.3	9.4	11.2
Italy	10.4	9.5	11.3
Sweden	11.8	11.8	11.7
Netherlands	12	12.8	11.2
Austria	12.4	12.7	12
Denmark	13.4	13.1	13.7
Belgium	13.8	14.4	13.1
France	14.5	14.6	14.5
Germany	14.7	13.8	15.7
Portugal	15.4	16.1	14.6
Israel	15.7	15.7	15.7
Finland	15.8	15.7	16.1
Poland	15.8	15.2	16.6
Brazil	15.8	16	15.6
Russian Fed	16	20.1	11.8
South Africa	16	27.4	8.8
Spain	16.6	15.6	17.5
Slovak Rep	16.9	16.7	17.1
Slovenia	18.3	15.4	21.1
OECD (34)	18.4	18.7	17.9
Estonia	19	19.3	18.6
Greece	19.6	18.5	17.7
Czech Rep	21	21	21
Iceland	21	19.3	22.7
Turkey	22.3	29.2	15.3
Luxembourg	22.7	21	24.1
Ireland	23	24	22
United Kingdom	24.7	25.1	24.4
Chile	25.1	30.7	19.2
Canada	25.4	24.6	26.2
Australia	28.3	28.2	28.4
Hungary	28.5	30.4	26.3
New Zealand	31.3	32.2	30.3
Mexico	32.4	37.5	26.8
United States	35.3	36.6	33.9

% of population aged 15 years and over

(OECD 2014)

World Obesity Trends over Time
(Top: Adults, Bottom: Children)

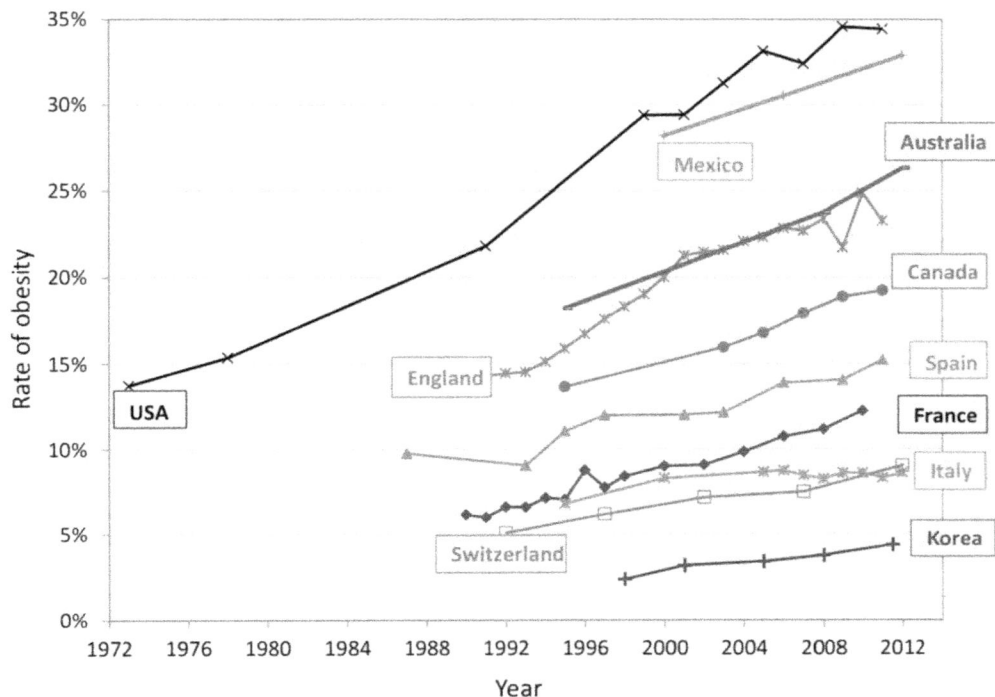

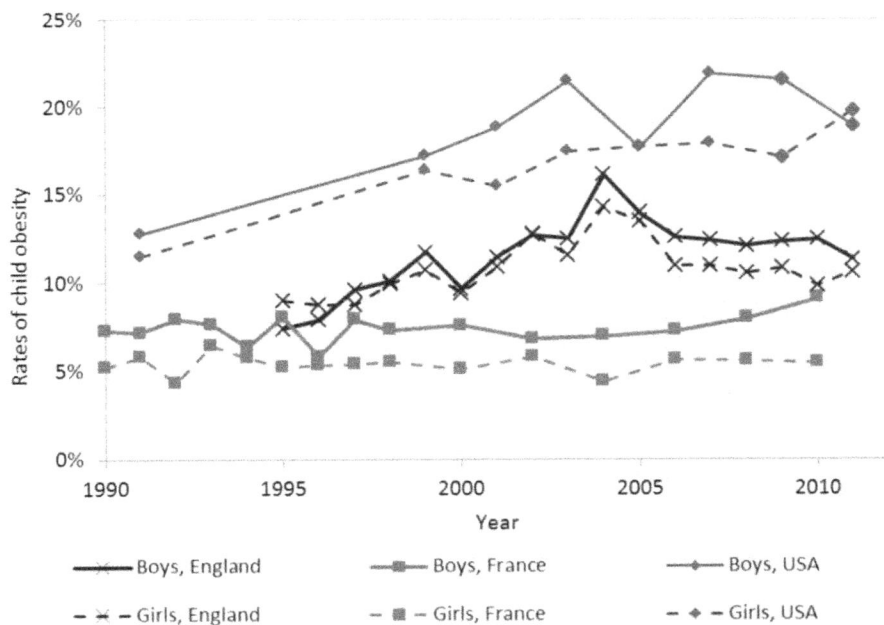

(OECD 2014)

Appendix 18: Temporal Trends in Diabetes

Source: US CDC. (2011). *Diabetes: Successes and opportunities for population-based prevention and control at a glance, 2011*. Centers for Disease Control. http://www.cdc.gov/chronicdisease/resources/publications/AAG/ddt.htm

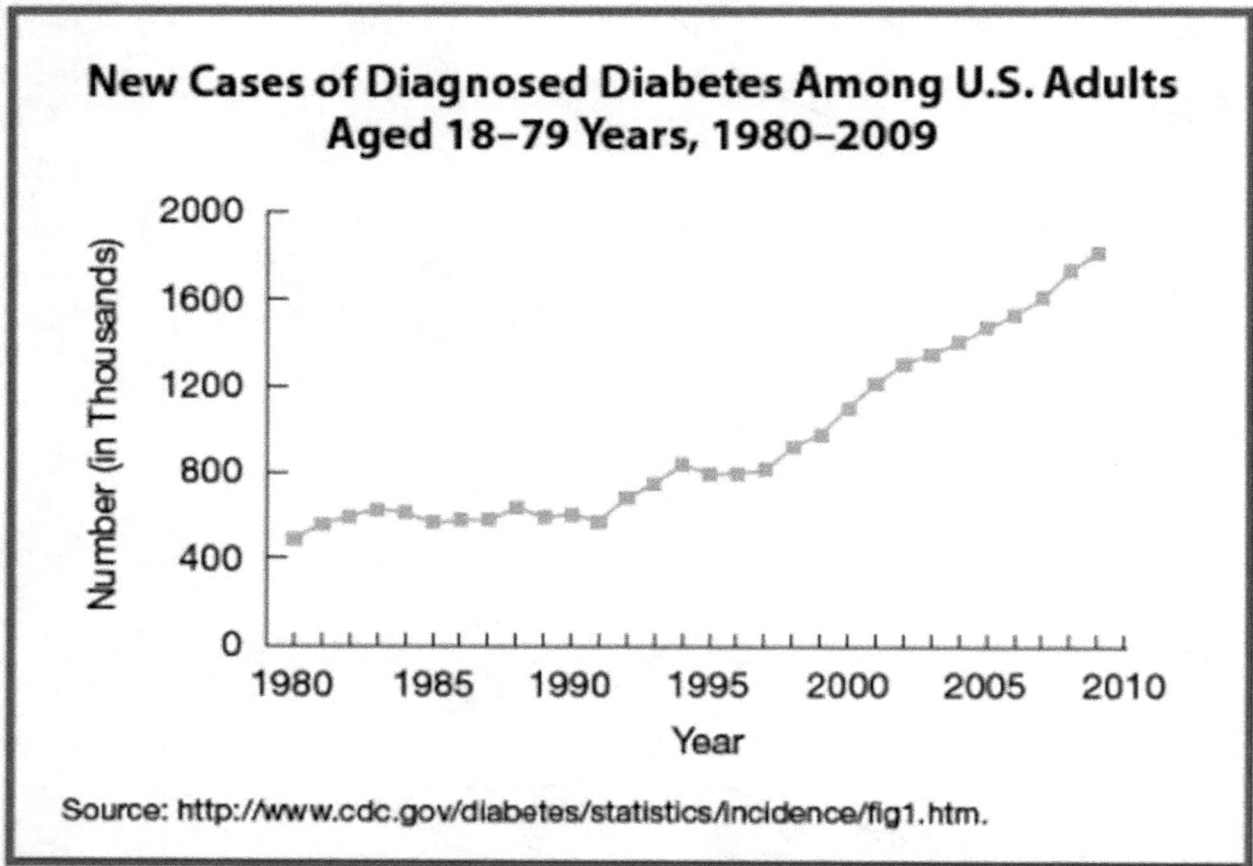

New Cases of Diagnosed Diabetes Among U.S. Adults Aged 18–79 Years, 1980–2009

Source: http://www.cdc.gov/diabetes/statistics/incidence/fig1.htm.

Appendix 19: Temporal Trends in Autism

Source: Blaxill, Mark F. (2004). What's going on? The question of time trends in autism. *Public Health Reports.* 119.
http://www.publichealthreports.org/issueopen.cfm?articleID=1413

Figure 1. Reported prevalence of autism and autistic spectrum disorders (ASDs), by midpoint year of birth, United Kingdom and United States, 1954–1994

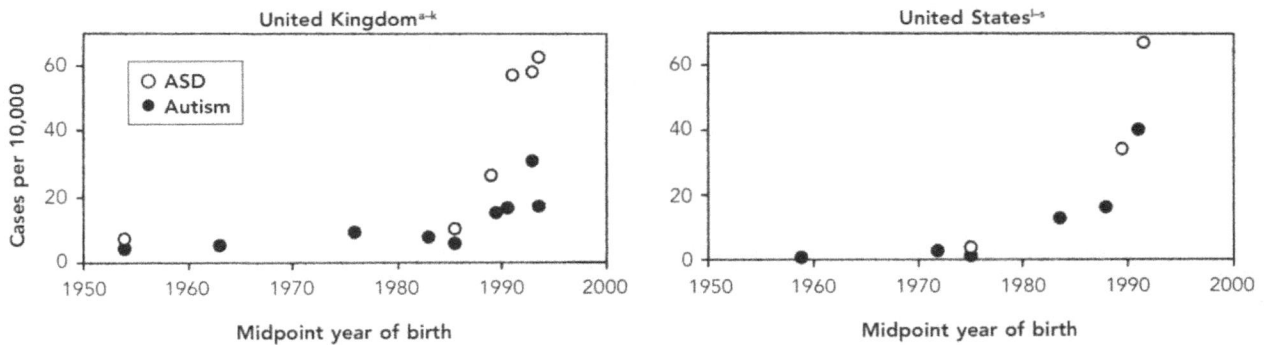

Appendix 20: Conceptual Model of the Sea-Surface Microlayer Ecosystem

Source: GESAMP. (1995). *The sea-surface microlayer and its role in global change*. Report No. 59. GESAMP, The Joint Group of Experts on the Scientific Aspects of Marine Environmental Protection. London. http://www.gesamp.org/data/gesamp/files/media/Publications/Reports_and_studies_59/gallery_1358/object_1388_large.pdf

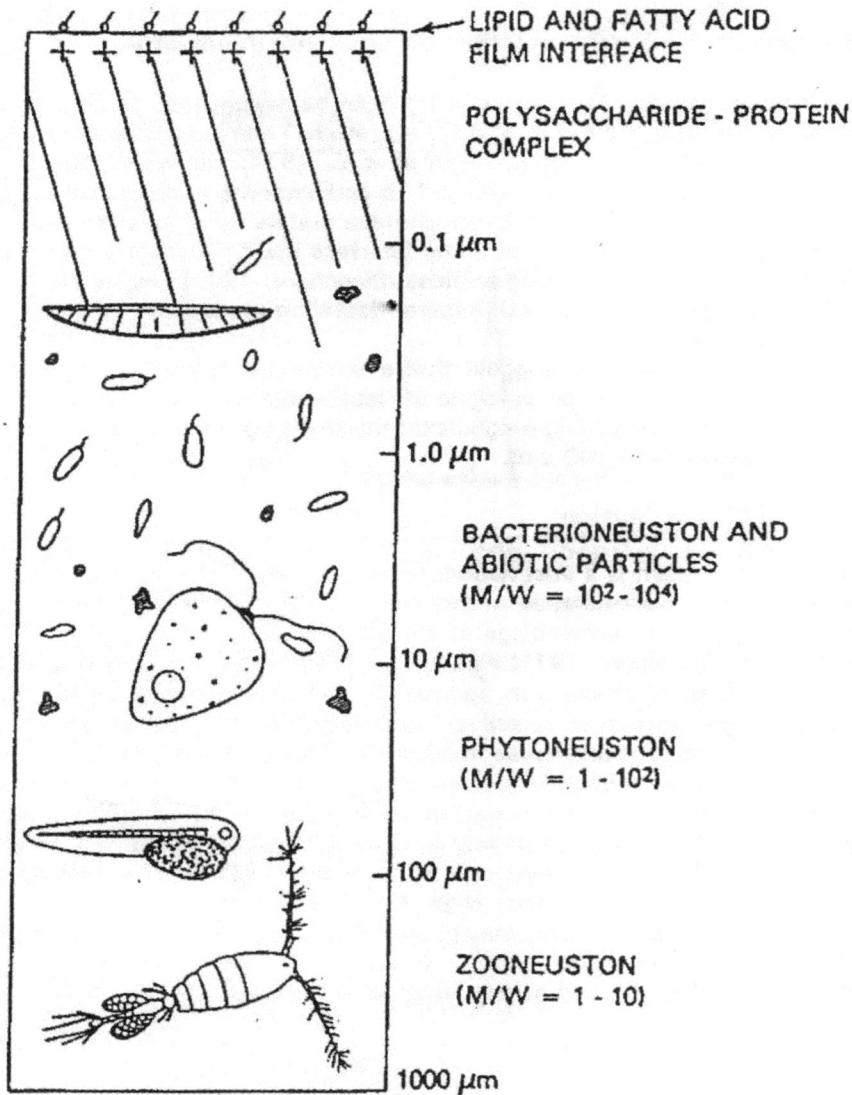

M/W = typical microlayer to water concentration ratios based on a number of studies.

Appendix 21: Diagrams from *State of the science of endocrine disrupting chemicals, 2012* on methylmercury and EDCs

Source: World Health Organization. (2012). *State of the science of endocrine disrupting chemicals, 2012*. Inter-organization Programme for the Sound Management of Chemicals (IPCS). United Nations Environment Programme. http://www.who.int/ceh/publications/endocrine/en/

Watras, C., Bloom, N., Hudson, R., et. al. (1994). Sources and fates of mercury and methylmercury in Wisconsin lakes. Lewis Publishers, Boca Raton, FL.

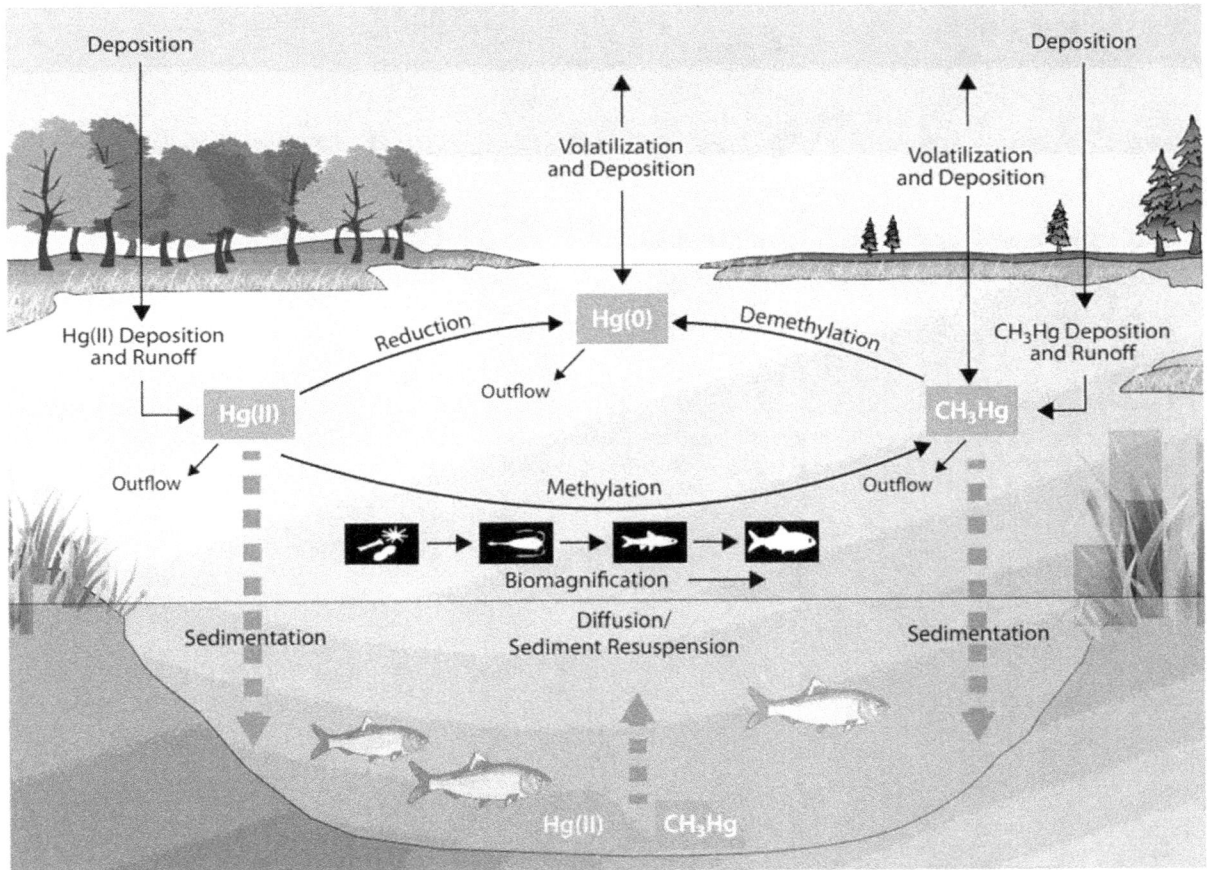

The complex cycling of mercury and methylmercury in the aquatic environment (modified from Watras & Huckabee, 1994).

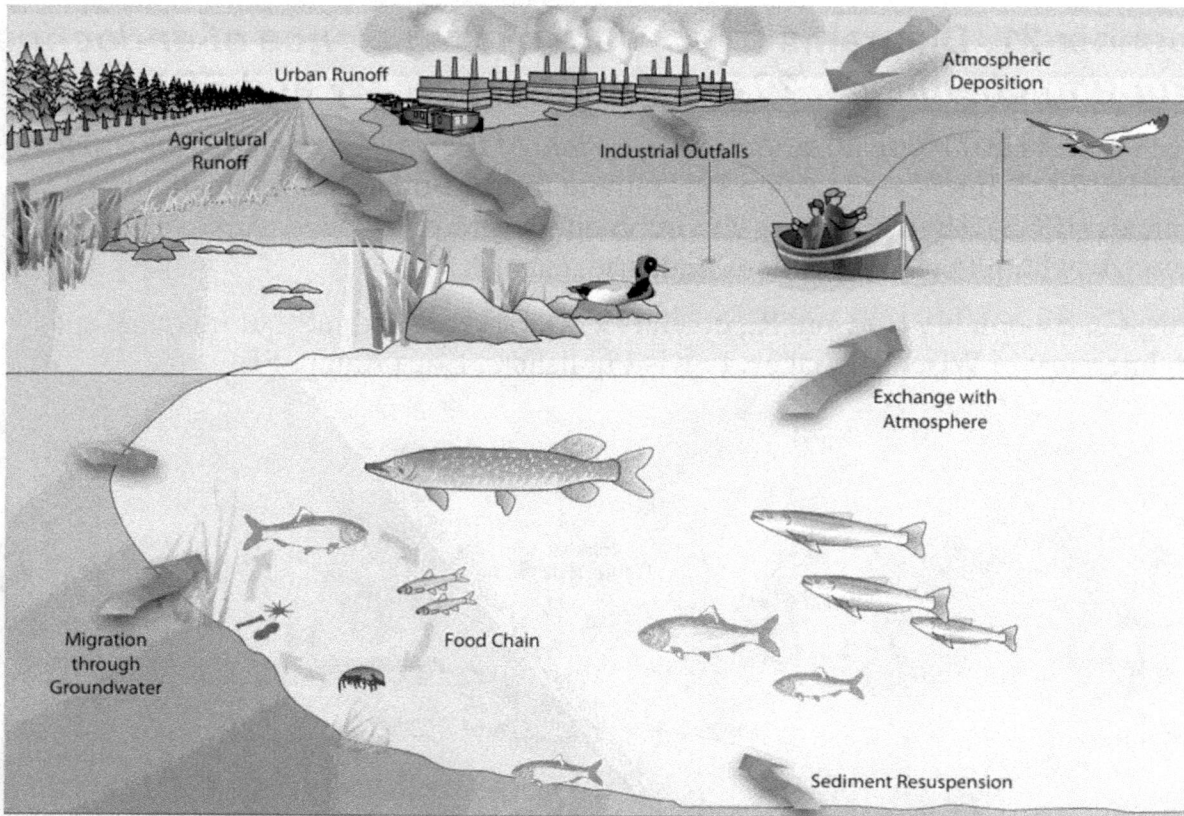

Exposure of fish and wildlife in urban regions due to continuous release of EDCs in effluents and to the atmosphere (Redrawn based on a figure from Chapter 4 of The Great Lakes: An Environmental Atlas and Resource Book, www.epa.gov/greatlakes/atlas/).

www.ingramcontent.com/pod-product-compliance
Lightning Source LLC
Chambersburg PA
CBHW051202200326
41519CB00025B/6982